Animals and Society

ANIMALS AND SOCIETY

An Introduction to Human-Animal Studies

Margo DeMello

COLUMBIA UNIVERSITY PRESS NEW YORK

Columbia University Press
Publishers Since 1893
New York   Chichester, West Sussex
cup.columbia.edu
Copyright © 2012 Columbia University Press
All rights reserved

COVER IMAGES: © istockphoto / Photos by Jason Ruis (top) and Jeff Chevrier (bottom)
COVER DESIGN: Martin Hinze

Library of Congress Cataloging-in-Publication Data

DeMello, Margo.
    Animals and society: an introduction to human-animal studies / Margo DeMello.
        p. cm.
    Includes bibliographical references and index.
    ISBN 978-0-231-15294-5 (cloth : alk. paper)—ISBN 978-0-231-15295-2 (pbk. : alk. paper)—
    ISBN 978-0-231-52676-0 (ebook)
1. Human-animal relationships. 2. Human-animal relationships–History. 3. Animals and
civilization–History. I. Title.
    QL85.D48 2012
    599.15–dc23

                                                                    2012013347

References to Internet Web sites (URLs) were accurate at the time of writing. Neither the author
nor Columbia University Press is responsible for URLs that may have expired or changed since
the manuscript was prepared.

Contents

Preface

While I have wanted to write a human-animal studies (HAS) textbook for years—since the first time I taught an HAS class in 2003—the drive to complete this book came from my work with the Animals and Society Institute (ASI), and my role on the Human-Animal Studies Executive Committee. In October 2004, a group of fifteen scholars working in the field of HAS first came together in conjunction with the International Compassionate Living Festival to discuss this rapidly growing field, and our roles in it. At that first meeting, we strategized about the further development of HAS and how we could help to enhance the presence of this newly developing discipline in academic institutions across the country.

Since that meeting, our committee, now made up of myself, Ken Shapiro (ASI), Carrie Rohman (Lafayette College), Cheryl Joseph (Notre Dame de Namur University), Christina Risley-Curtiss (Arizona State University), Kathie Jenni (University of Redlands), Paul Waldau (Religion and Animals Institute), Georgina Montgomery (Michigan State University), and Robert Mitchell (Eastern Kentucky University), has gone on to create the Human-Animal Studies Fellowship program, an annual fellowship that brings together scholars in HAS for a six-week intensive period each summer, now held in conjunction with Wesleyan University. We also wrote an edited collection, *Teaching the Animal: Human-Animal Studies Across the Disciplines* (Lantern 2010), that includes concrete information on teaching HAS in a variety of natural science, social science, and humanities courses.

In the few years since we have been meeting, we have seen the field of human-animal studies grow by leaps and bounds. There are now more courses offered at more colleges and universities than ever before, more conferences devoted to HAS, more college programs, institutes, journals, list serves, veterinary programs, legal centers, and organizations. Clearly, interest in HAS is exploding. But one thing the discipline still needed was a textbook for college students.

This book is intended to fill the gap in the field by giving professors a comprehensive overview of the field, and by giving students an easy-to-read text covering many of the issues related to the question of animals in human society. I also hope that the existence of this text will encourage more professors to develop new courses that focus on the human-animal relationship. But I especially hope that readers—both students and instructors—will recognize that there is an important place within the college curriculum for looking at animals, their relationship with humans, and the very real implications of those relationships.

Acknowledgments

Cultural studies critic Cary Wolfe recently wrote that "trying to give an overview of the burgeoning area known as animal studies is, if you'll permit me the expression, a bit like herding cats" (2009: 564). Herding cats, indeed!

Without the aid of numerous people and organizations, this textbook could not have been written. I owe thanks to more people than I can possibly mention, but will here thank those who most directly helped me to get this book off the ground. First, I have to acknowledge the Human-Animal Studies Committee—without its establishment, I would not even have gotten involved in the field. The Human-Animal Studies Committee would not have itself gotten established without the support of the Animals and Society Institute (ASI), a research and educational organization that advances the status of animals in public policy and promotes the study of human-animal relationships. I owe my deepest thanks to my fellow committee members for inspiring me with their own work, but especially to Ken Shapiro, ASI's executive director, for his support for this project. All proceeds raised from the sale of this textbook will, in fact, be donated to ASI to support the organization's important work. In fact, if it weren't for meeting Ken at an animal rights conference in 2002, I would not have known about the existence of human-animal studies, and most likely would not have returned to teaching after having left the field a few years earlier. If it weren't for Ken, my life would be very different today.

I am also thankful to Wendy Lochner at Columbia University Press for expressing an interest in this book, and for her work expanding Columbia's Animal Studies series. I first began shopping this book around in 2005, and was told by each publisher I approached that the market was too small to support such a text. Wendy recognized both the importance of, and growth of, the field of HAS and knew that with the growth of HAS there would indeed be an audience for this book. Through Wendy, six anonymous readers provided feedback on this manuscript and I am grateful to all of them. I am especially grateful to philosopher Ralph Acampora, whose comments on chapter 18 were invaluable in improving that chapter.

I am also grateful to the following colleagues and friends for providing essays on their own work that are included in these chapters. Thanks to Susan McHugh, Annie Potts, Walter Putnam, Molly Mullin, Garry Marvin, Cheryl Joseph, Cynthia Kay Chandler, Clinton Sanders, Ken Shapiro, David Nibert, Carol Gigliotti, Laura Hobgood-Oster, Philip Armstrong, Robert W. Mitchell, and Kathie Jenni for generously contributing in this way.

I want to thank as well everyone who provided photos for this book. That includes Anita Carswell, Annette Evangelista, Carol Adams, Christine Morrissey of Harvest Home Animal Sanctuary, Criss Starr of New Mexico House Rabbit Society, Dr. Carolynn Harvey, Drew Trujillo, Ed Turlington, Ed Urbanski and Yvonne Boudreaux of Prairie Dog Pals, Elizabeth Terrien, Great Ape Trust, Jonnie Russell, Karen Diane Knowles, Kate Turlington, Kerrie Bushway, Lynley Shimat Lys, Mark Dion through the Tanya Bonakdar Gallery, Mary Cotter, Mercy for Animals, People for the Ethical Treatment of Animals, Robin and Christopher Montgomery, Suzi Hibbard, The Jane Goodall Institute, Thomas Cole, Tracy Martin, Vicki DeMello, and Yvette Watt. I am also grateful to the many photographers who contributed their work for free to Wikimedia Commons; many of the images in this book came from that site. I especially want to extend my appreciation to Dan Piraro, creator of the syndicated comic strip *Bizarro*, for generously granting me the use of his award-winning cartoons for this book. I would also like to thank Linda Walden, Tina Otis, Rich Sievers, Alison Giese, and especially Jenni Bearden for their help.

Finally, I am personally grateful for the support, love, and encouragement of my parents, Robin and Bill, who have always encouraged my efforts no matter how unusual, and my husband, Tom Young. And last but certainly not least, I owe perhaps the greatest thanks to the nonhuman animals—rabbits, cats, dogs, and birds—who have shared my life. I cannot imagine where I would be personally or professionally if it were not for their presence in my life.

ANIMALS AND SOCIETY

CONSTRUCTING
ANIMALS
Animal Categories

I

Human-Animal Studies

IF YOU WERE SURFING the news and popular culture sites on the Internet during the first two weeks of November 2010, you would have heard about the aftermath of the November 2 midterm elections, President Obama's trip to the G20 financial summit, a devastating volcano in Indonesia, and the spread of cholera in Haiti. But you would have also heard about how comic actor Dick Van Dyke fell asleep on a surfboard in the water off of a Virginia beach a few years back and found himself stranded in the ocean but was saved when a group of porpoises pushed his surfboard back to shore. Another story that November was of a Texas auto mechanic who was killed by a drunk driver back in June, but whose dog, Spot, loyally awaited his master's return on the country road that he used to drive home. There was a story about a demand made by People for the Ethical Treatment of Animals that Amazon stop selling books on dog fighting on the popular e-commerce website. After Halloween there were a number of stories about zoo animals eating pumpkins as a holiday treat. There was an article about how pet trusts—trusts that allow people to leave money for the care of their companion animals after they die—are becoming increasingly popular with pet owners who want to ensure that once they die, their animals will be treated with dignity and respect. There were articles about animals harming humans—a bull killing a man and injuring his wife, and one of a pet monkey biting a woman's face and being consequently euthanized. And, as there are every week, there were multiple stories about animal cruelty cases—in one, thirty-five starving horses and donkeys were removed from a property where officials also found dead animals, and others involved suspects knifing and burning dogs to death. And, related to the G20, you might have read about how the Korean government, the host of the second 2010 G20 summit, placed goldfish into the water supply to ensure that the water was safe for the international conference, prompting protests from animal rights organizations. These stories are just a brief indication of how important, and prevalent, animals are in our lives today.

What Is Human-Animal Studies?

We are surrounded by animals. Not only are we ourselves animals, but our lives, as humans, are intimately connected with the lives of nonhuman animals. Animals share our homes as companions whom often we treat as members of the family; we may even buy clothing for them, celebrate their birthdays, and take them with us when we go on vacation. We can view animals on the Animal Planet network or television shows such as *Cats 101* or *Meerkat Manor* and subscribe to magazines such as *Dog Fancy* or *Rabbits USA*. We eat animals, or their products, for almost every meal, and much of our clothing and most of our shoes are made up of animal skins, fur, hair, or wool. We wash our hair with products that have been tested on animals and use drugs that were created using animal models. We visit zoos, marine mammal parks, and rodeos in order to be entertained by performing animals, and we share our yards—often unwillingly—with wild animals whose habitats are being eroded by our presence. We refer to animals when we speak of someone being "blind as a bat" or call someone a "bitch." We include them in our religious practices and feature them in our art, poetry, and literature. Political protest ignites because of disagreements over the status and treatment of animals. In these and myriad other ways, the human and nonhuman worlds are inexorably linked.

For thousands of years, animals of all kinds have figured prominently in the material foundations and the ideological underpinnings of human societies. **Human-animal studies** (HAS)—sometimes known as **anthrozoology** or **animal studies**—is an interdisciplinary field that explores the spaces that animals occupy in human social and cultural worlds and the interactions humans have with them. Central to this field is an exploration of the ways in which animal lives intersect with human societies.

Human-animal studies is not the study of animals—except insofar as the focus of our study is both nonhuman and human animals. But unlike ethology, comparative psychology, zoology, primatology, or the various animal behavior disciplines, HAS is not about studying animals per se. Rather, we study the interactions between humans and other animals, wherever and whenever we find them. On the other hand, our work is informed by those disciplines that do take animals as the object of study. Their work on the behavior of animals, animal learning, cognition, communication, emotions, and culture has been hugely influential in recent years, both within HAS and outside of our field. By understanding more about the behaviors and mental and emotional processes of animals, we can better understand human interactions with them.

Much of human society is structured through interactions with nonhuman animals or through interactions with other humans about animals. Indeed, much of human society is based upon the exploitation of animals to serve human needs. Yet until very recently, academia has largely ignored these types of interactions. This invisibility—in scholarly inquiry—was perhaps as great as the presence of animals in our daily lives.

This presence, however, becomes difficult to ignore when we consider the magnitude of animal representations, symbols, stories, and their actual physical presence in human societies and cultures. Animals have long served

---

BOX 1.1

DEFINITIONS

Animal rights: A philosophical position as well as a social movement that advocates for providing nonhuman animals with moral status and, thereby, basic rights.

Animal studies: Generally used, at least in the natural sciences, to refer to the scientific study of, or medical use of, nonhuman animals, as in medical research. In the humanities, it is the preferred term for what the social sciences calls HAS.

Anthrozoology: The scientific study of human-animal interaction, and the human-animal bond.

Critical animal studies (CAS): An academic field of study dedicated to the abolition of animal exploitation, oppression, and domination.

Ethology: The scientific study of animal behavior.

Human-animal studies: The study of the interactions and relationships between human and nonhuman animals.

Figure 1.1. Annette Evangelista poses with Ms. Bunny Penny. (Photograph courtesy of Drew Trujillo.)

as objects of study—in biology, zoology, medical science, anthropology, and the like—but were rarely considered to be more than that, and were even more rarely considered to be "subjects of a life" rather than objects of study. One possible reason has to do with the human use of nonhuman animals: When we grant that animals have subjectivity, including their own interests, wants, and desires, it becomes more difficult to justify many of the practices that humans engage in with animals, such as meat consumption or medical experimentation.

This book is designed to bring into the realm of scholarly inquiry the relationships that exist between humans and other animals. A major focus of this volume is the social construction of animals in American culture and the way in which these social meanings are used to perpetuate hierarchical human-human relationships such as racism, sexism, and class privilege. Another major focus of this book is animal-human interactions in several major social institutions: the family, the legal system, the political system, the religious system, and the educational system. We will also examine how different human groups construct a range of identities for themselves and for others through animals. Finally, this text examines several of the major philosophical positions about human social policy regarding the future of human-animal relations. What are the ethical, ecological, and societal consequences of continuing our current patterns of interaction into the twenty-first century?

History of HAS

Human-animal studies is one of the newest scholarly disciplines, emerging only in the last twenty years in academia. But unlike fields like political science, anthropology, English, and geology, HAS is multidisciplinary and interdisciplinary. That is, it is a field of study that crosses disciplinary boundaries and is itself composed of several disciplines. In other words, HAS scholars are drawn from a wide variety of distinct disciplines (interdisciplinary), and HAS research uses data, theories, and scholarship from a variety of disciplines (multidisciplinary).

Human-animal studies and the related field of **critical animal studies** are the only scholarly disciplines to take seriously and place prominently the relationships between human and nonhuman animals, whether real or virtual. Like feminist scholars in the 1970s did with the categories of "woman," "female," and "feminine," HAS and CAS scholars have been inserting "the animal" into the humanities, social sciences, and natural sciences. As humans' dependence on nonhuman animals increases and as our relationship with them changes in the twenty-first century, *not* examining this relationship within the context of academia seems bizarre—especially given the increased presence of animal advocacy in the world around us.

Like women's studies and African-American studies, which rose alongside feminism and the civil rights movement, respectively, HAS has risen parallel to the animal protection movement, and indeed borrows heavily from that movement. (CAS is much more explicitly connected to the movement, however.) The publication of two major philosophical works on animals—Peter Singer's *Animal Liberation* (1975), followed by Tom Regan's *The Case for Animal Rights* (1983)—led to an explosion of interest in animals among academics, animal advocates, and the general public. We can see that the rise of HAS in academia, especially over the last decade, is related directly to the philosophical debate regarding animals as worthy of ethical inquiry.

Outside of philosophy, a number of scholars began writing about animals in the decades to follow, including historians, anthropologists, sociologists, psychologists, geographers, and feminist scholars. The 1980s, for example, saw a number of books released by historians that focused on the history of various practices or attitudes toward animals. One of the earliest approaches was historian Keith Thomas's 1983 book *Man and the Natural World: A History of the Modern Sensibility*, which explores the origin of the concept of "nature" in Western thought. Robert Darnton published *The Great Cat Massacre and Other Episodes in French Cultural History* in 1984, in which Darnton considers the torture of cats by a group of eighteenth-century Parisian working-class men who used cats as a substitute for their feelings of hatred toward their boss and his cat-loving wife. Coral Lansbury's *The Old Brown Dog: Women, Workers, and Vivisection in Edwardian England* (1985) looks at the intersection between class and gender in the early anti-vivisection movement in Victorian England. Harriet Ritvo and J. M. Mackenzie followed in 1987 and 1988 with *The Animal Estate* and *The Empire of Nature: Hunting, Conservation and British Imperialism*, respectively, both of which look at the history of the British relationship with animals. Also published in this decade was James Serpell's *In the Company of Animals* (1986), which focuses on pet keeping in a cross-cultural context. Two major works that had a major impact in the field were Yi-Fu Tuan's 1984 classic *Dominance and Affection: The Making of Pets*, which examines the power relations inherent in the human-pet relationship, and Donna Haraway's 1989 work *Primate Visions: Gender, Race, and Nature in the World of Modern Science*, which examines how the prevailing narratives about human origins—based on the work of primatologists—reflected and maintained ideologies of class, nationality, gender, and race. Also, and fundamentally, the 1980s saw the launch of the first journal devoted to HAS—*Anthrozoös* (1987), published by the International Society for Anthrozoology.

Beginning in the 1990s, the field began to grow, with major new works published throughout the decade. Aubrey Manning and James Serpell's edited collection *Animals and Human Society: Changing Perspectives* (1994) looks at the role animals play in human societies, with chapters ranging from images of animals in medieval times to nineteenth-century attitudes toward animals. That same year, feminist scholars Carol Adams and Josephine Donovan's edited collection *Animals and Women: Feminist Theoretical Explorations* became the first major work to consider the issue of feminism and animals. And Arnold Arluke and Clinton Sanders's *Regarding Animals*, published two years later in 1996, is now considered a classic in the field of human-animal studies. This sociological study of the human-animal relationship explores, among other things, the ways that people who work with animals (such as lab workers, animal shelter workers, or dog trainers) cope with their work and the complex relationships that form in that context. In 1997, Jennifer Ham and Matthew Senior published *Animal Acts: Configuring the Human in Western History*, a landmark publication that addresses animality in literary theory. The following year, geographers Jennifer Wolch and Jody Emel's *Animal Geographies: Place, Politics, and Identity in the Nature-Culture Borderlands* was the first book to approach human-animal relationships from a cultural geography perspective and looks at those geographies where animal and humans meet (and conflict). Also in 1998, the first major text on the link between animal cruelty and human violence—Randall Lockwood and Frank Ascione's edited collection *Cruelty to Animals and Interpersonal Violence*—was published. This book remains the classic text on this important subject and was followed in 1999 by *Child Abuse, Domestic Violence and Animal Abuse*, edited by Frank Ascione and Phil Arkow, which further addresses this link. Clinton Sanders's *Understanding Dogs: Living and Working with Canine Companions*, also published in 1999, focuses on how people perceive dogs and invest them with meaning.

The first decade of the twenty-first century has seen dozens more books released in HAS, in virtually every discipline, demonstrating the enormous growth of the field. In 2000, Anthony Podberscek, Elizabeth Paul, and James Serpell's book *Companion Animals and Us: Exploring the Relationships between People and Pets* focuses exclusively on the people-pet relationship. This first decade has also been marked by the rise of a number of key texts in critical animal studies, such as Steve Best and Anthony Nocella's *Terrorists or Freedom Fighters: Reflections on the Liberation of Animals* (2004), Lisa Kemmerer's *In Search of Consistency: Ethics and Animals* (2006), and Carol Gigliotti's *Leonardo's Choice: Genetic Technologies and Animals* (2009).

Human-Animal Studies as a Way of Seeing

Sociologists are well-known for a way of seeing the social world that is known as the **sociological imagination**. Coined by C. Wright Mills (1959), the sociological imagination provides us with a way of seeing our lives in social context; it allows us to see the ways in which social forces shape our lives.

Similarly, HAS is also a field of study, like sociology, and a way of seeing. HAS is defined by its subject matter—human-animal relationships and interactions—but also in part by the various ways in which we understand animals themselves. Although HAS is not about understanding animal behavior (although we do, as mentioned, draw on the findings of ethology), we do want to understand animals in the context of human society and culture. We explore the literary and artistic usage of animals in works of literature or art, the relationship between companion animals and their human families, the use of animals as symbols in religion and language, the use of animals in agriculture or biomedical research, and people who work with animals. Our focus then is to look at animals wherever they exist within the human world.

Like anthropology, which is known as a **holistic** science because it studies the whole of the human condition (biology, culture, society, and language) in the past, present, and future, human-animal studies, too, is holistic. For example, in *Stories Rabbits Tell: A Natural and Cultural History of a Misunderstood Creature* (2003), Susan Davis and I cover the rabbit's evolution as a species; the domestication of the rabbit; the rabbit's role in the folklore, myths, and religions of people around the world; rabbits in art, literature, kitsch, and toys; the use of rabbits in the meat, fur, pet, and vivisection industries; and, finally, the rabbit as beloved house pet. We saw our task as trying to expose the "real" rabbit, as well as all of the various constructions, uses, and interpretations of the rabbit. Our book, then, focuses on the broadest possible construction of the human-rabbit relationship.

As in the earlier example, HAS scholars try to understand how animals are **socially constructed**. On one level, animals surely exist in nature. However, once they are incorporated into human social worlds they are assigned to human categories, often based on their use to humans, and it is these categories (lab animal, pet, and livestock) that shape not only how the animals are seen but also how they are used and treated. To take it one step further, we may ask: What is an animal *outside* of culture? As sociologist Keith Tester wrote, "A fish is only a fish if you classify it as one" (1991:46). Moreover, these classifications are not neutral—they are politically charged in that they serve to benefit some (humans, some animals) at the expense of others (other animals).

> [In] a certain Chinese encyclopedia entitled the *Celestial Emporium of Benevolent Knowledge* . . . it is written that animals are divided into (a) those that belong to the Emperor, (b) embalmed ones, (c) those that are trained, (d) suckling pigs, (e) mermaids, (f) fabulous ones, (g) stray dogs, (h) those that are included in this classification, (i) those that tremble as if they were mad, (j) innumerable ones, (k) those drawn with a very fine camel's hair brush, (l) others, (m) those that have just broken a flower vase, (n) those that resemble flies at a distance.
> —JORGE LUIS BORGES, *OTHER INQUISITIONS* (AUSTIN: UNIVERSITY OF TEXAS PRESS, 1964)

What does this (fictional) classification scheme tell you? This system, like many of the systems of classification that we use today, tells us that being an animal in human society has little to do with biology and almost everything to do with human culture. Animals' physical identity is less important to their status and treatment than their symbolic identification and their social meaning. In addition, how we classify animals shapes how we see animals, and how we see them shapes how we classify them. By closely examining the social categories that we have constructed around animals, we can come closer to understanding our relationship to those animals.

In part, human-animal studies is about getting to the core of our representations of animals and understanding what it means when we invest animals with meanings. What do animals mean to us then? They mean a great many things. Perhaps it would help to begin with a brief look at all of the places in which animals exist. The spatial distribution of nonhuman animals is one way to understand how animals have been incorporated into human societies and to get at the implications of that incorporation.

Where Are Animals?

Animals exist in our homes primarily as pets. Pet keeping, as a cultural practice, involves the incorporation of animals into human families and human domestic space. What is expressed in these relationships and what does it tell us about human societies and cultures? The presence of companion animals, in particular, shapes and constrains relationships—where we live, what we do, where we vacation, who our friends are, etc. Here, animals often act as surrogate humans (to socialize children and to teach them empathy), as substitute children for childless and older adults, and as companions for the elderly and disabled. Anthropomorphism (the attribution of supposedly human qualities to nonhuman animals) is perhaps expressed in its most complex form in humans' relationships with their pets.

Animals exist on farms and, increasingly, on **factory farms**, where they are turned into meat. One of the key relationships between animals (wild and domesticated) and humans is that humans kill animals in order to consume them. In fact, for many people the most common interaction that they have with other animals is when they eat them. All societies, in different ways, express concern about this relationship. Not all animals are regarded as appropriate sources of food. For example, some are regarded as too close to humans to be acceptable as food, and others are regarded as too disgusting to be eaten. Why should an animal be "tasty" in one society and tabooed in another? What exactly does it mean to turn a living creature into "meat"? What taboos and moral concerns do human societies express about killing and eating other animals? And what can we say about the fact that over ten billion land animals per year are raised exclusively to be slaughtered for food?

Animals exist on fur farms. The use of animal skin for clothing and shelter is, like the use of animals for meat, one of the most ancient of all animal uses and predates the domestication of animals by many thousands of years.

Like farms where animals are raised to become food, fur farms are becoming larger, profits are greater, and the methods used to raise animals for fur have become more industrial.

Animals exist in scientific laboratories. Animals have been used in science since antiquity, when ancient scholars dissected animals to understand how human bodies work. Today, approximately twenty million animals per year are used in scientific and medical research. HAS scholars who focus on this issue look at the relationship between researcher and animal, the social construction of "laboratory animals," the social and cultural contexts of animal experimentation in the West, the moral status of the ethics of animal experimentation, and scientific justifications for experimentation.

Animals exist "at work." Since the **Neolithic Revolution**, humans have worked with animals, and today many people still do. Indeed, the first animal to have been domesticated was the dog, which was domesticated as a hunting partner to humans. A few thousand years later, human civilization emerged in part thanks to the domestication of large ruminants—animals that were valuable to humans as sources of meat, milk, and, crucially, labor. Without large domesticated animals such as cattle and horses that could be harnessed to a plow, ridden, or attached to a cart, it is hard to imagine how societies could have developed as they did. Even though much of animal labor has been replaced by technology in our postindustrial world, animals still fulfill important roles for humans. One focus of HAS research, for instance, is the use of animals in a therapeutic context—as companions and aids to the disabled and as therapy animals in schools, hospitals, or even prisons.

Animals exist in zoos, marine mammal parks, and other venues in which they perform for human entertainment. Animals are made to race and fight against each other, some are ridden in a variety of performances and sports, some are made to do tricks in circuses, some are challenged by humans in events such as bullfights and rodeos, and some are judged with respect to other animals in events such as herding trials and hunting. What meanings are expressed in such performances? What can we understand from examining humans watching animals, and participating with animals, in these contexts? Animals do not represent themselves in human societies, but human societies certainly make representations of them in a variety of ways and give them cultural meaning. HAS scholars who look at performing animals often focus on the cultural representation and exhibition of animals, particularly wild animals, which can be interpreted as a story that humans tell about themselves through the medium of animals.

Animals increasingly exist in virtual worlds—on television, on YouTube, and in movies. In 1996, Discovery Communications created a new cable television network called Animal Planet, which is watched by millions of people in more than seventy countries. Animal Planet's programming includes classic nature documentaries, as well as dozens of "reality TV" shows that focus on, for example, emergency veterinarians, animal cruelty officers, pet psychics, dog trainers, zookeepers, and the animals with whom they work. In addition, since the launch of YouTube in 2006, animal videos have become some of the most popular features on the Internet. Sites such as Cute Overload and I Can Has Cheezburger? feature nothing but cute photos of animals with funny captions. There is even a blog called Cute Things Falling Asleep that has as its stated aim capturing cute animals as they are falling asleep. From their website: "This blog is called Cute Things FALLING Asleep, not Cute Things Already Sleeping. The whole idea is to watch cute things fighting, and losing, the battle to stay awake." For many people around the world, watching virtual animals is one of our most enjoyable pastimes.

Animals exist "in the wild," or outside of human society, although those arenas in which animals can truly exist in the wild are becoming fewer and farther between. Even when they are wild, animals constantly touch our lives—during our own outdoor activities such as hiking or picnicking, through hunting, when we are birding or whale watching, when we find that deer or rabbits have eaten some of our garden, or even when we encounter dead animals on the side of the road. What is the impact of these wild encounters on both human societies and wildlife?

Sometimes animals exist "out of place," as geographers Chris Philo and Chris Wilbert (2000) term it when an animal is found outside of society's prescribed place for them. For instance, in the United Kingdom, all dogs must be owned by people. Stray dogs, then, are essentially illegal according to English law and must be rounded up. Other cultures have no such requirement. In many cultures, feral cats also live in a sort of liminal existence—they are neither domesticated nor wild. In many countries, they must be rounded up or killed. Pigeons are another animal that, once highly prized as messengers, have now been reinterpreted as urban pests—not truly wild, but not domesticated either.

Animals exist in the myths, legends, and folktales of people around the world. The religions of all societies incorporate animals (negatively and positively) into their cosmologies, beliefs, practices, and symbolism. Animals are worshipped, made the object of taboos, sacrificed, and associated with gods, spirits, and other supernatural beings. HAS scholars explore how religious thought and practice make sense of the animal world and use it to comment on the human condition.

Animals exist as cultural symbols and as linguistic metaphors, similes, and slurs. Every language has animal expressions that are used to refer to people, practices, and beliefs. Why are human languages so rich with animal references, and what does that say—about humans and animals? Linguists, anthropologists, and feminist theorists have all discussed the meaning of such practices, and in particular how women and ethnic minorities are often targeted through animal slurs. In addition, HAS scholars are interested in how certain terms—such as slaughter, for example—are used in reference to humans.

Animals exist as mirrors for human thought; they allow us to think about, talk about, and classify ourselves and others. For that reason, the study of race, class, and gender has been enhanced by examining the role of animals in human society. Animals, and our use of them, play a vital part in racialization and the construction of gender. Likewise, class, in numerous ways, is tied to how we relate to the animal world.

In addition, animals affect our court system and our laws. Human-animal conflicts and controversies are often dealt with in the courts, and federal, state, and municipal codes include a range of laws that deal with such issues as how one can legally treat animals, how many animals one can live with, dog bites, animal control policies, hunting laws, and more. These laws may protect humans from animals (as in animal nuisance laws), or they may protect animals from humans (as in animal cruelty laws). We also have laws that define crimes against whole species like the **Endangered Species Act** or the **Convention on International Trade in Endangered Species**. Animal issues often become law enforcement issues, such as when an animal attacks a human, when a cockfighting ring is uncovered, when dogs and cats are abandoned, or when a lion is found loose in a city. Scholars who look at issues surrounding animals and the law note that animals are generally considered to be property under the law and thus have no rights of their own; even wild animals are considered public property.

Finally, one of the results of living lives so connected to—and dependent on—animals is that we need a variety of ways to deal with the conflicts inherent in this complicated relationship. The animal rights and welfare movements are a result of this. In recent decades, the issue of animal rights has engaged the attention, emotions, and thoughts of a wide public. In many Western societies, animals have come to be regarded as an oppressed minority, and various organizations have been fighting for a change in that status. Many HAS scholars devote their attention to understanding these relatively new social justice movements.

Why does it matter where animals live? Where animals are found is where, almost without exception, humans are found. And where they are found

Figure 1.2. Rocky, a rooster with crooked toes, lives at Harvest Home Animal Sanctuary. (Photograph courtesy of Christine Morrissey.)

influences how they are categorized—farm animal, zoo animal, laboratory animal, wild animal, or pet—which in turn determines how those animals will be treated. Those animals whose lives are hidden to us—because they are raised, live, or die in hidden spaces such as factory farms or biomedical labs—are subject to a very different type of treatment than animals whose lives are lived where the public can see them. One reason why meat eating is still largely unexamined in our society is because how animals that are raised for food live and die is relatively invisible to the general public.

In many ways, what we will see in this book is that human societies are largely dependent on animals—for food, clothing, security, labor, and pleasure. How did we end up this way? Does it matter? Can this be reversed? And do we want it to be?

Defining the Animal

If HAS is defined by its subject matter—the interaction between humans and animals—one problem is definitional. How do we define "human"? How do we define "animal"? These questions are not as simple as they look and, in fact, point to one of the primary issues of the field: What makes an animal an animal?

Animals are defined through human linguistic categories—pet, livestock, and working animal—and those categories themselves are related to how the animal is used by humans. In addition, these categories are often related to where animals are spatially located: in the house, on the farm, in the lab, on television, in the "wild," etc. To complicate matters even further, "human" is generally defined as what is *not* animal—even though, biologically speaking, humans *are* animals. As psychologist Ken Shapiro has pointed out (2008:7), the name of the field itself—human-animal studies—is "as incoherent as saying 'carrots and vegetables.'"

This linguistic conundrum—how we define animal and how, even, we define this field—has implications that go beyond semantics. Calling animals "animals" when in fact they should, at the very least, be defined as "other animals" may be part of what keeps nonhuman animals subjugated. The artificial boundary between "us" and "them" is certainly part of what allows humans to use other animals for human benefit. If we were to grant a continuity among the various species, then the use of some animals for meat and others for labor (and still others as family members) would be, perhaps, harder to justify. Philosopher Lisa Kemmerer has suggested the term "anymal" to refer to nonhuman animals (Kemmerer 2006), but this does not really get us any closer to answering the question: What is an animal? Folklorist Boria Sax (2001), on the other hand, suggests that we define animals as a *tradition*. Here he means using the artistic, mythic, legendary, and literary aspects of the animal in our definition, which would link animals to the ideas, practices, and events that make up human culture.

We have already discussed (and will address at greater length later in this text) the social construction of animals. One task of many HAS scholars is to *deconstruct* those constructions: to unpack the various layers of meaning that we have imposed onto animal bodies and to try to see the animal within. What is a chicken when we no longer think of the chicken as a food item—a breast, a leg, a nugget? Who, indeed, is the chicken? Why it is so difficult to even conceive of, let alone answer, that question goes back to the power of the social category: When an animal is known, as is the chicken, in only a limited, commodity form, it is inconceivable to think of the animal in any other way. The British artist Banksy played with this issue when he created an installation called "The Village Pet Store and Charcoal Grill" in New York City in 2008. In the windows of his shop, Banksy created animatronic displays for the public to view, including one that featured a chicken and her "babies": a set of chicken nuggets walking around, being hatched, and drinking from containers of dipping sauce. Another display featured a female rabbit applying lipstick to her lips in front of a mirror. Both displays made visible some of the invisibilities of society's use of animals today: how we transform live chickens into "McNuggets" and how rabbits are killed in the process of testing lipstick and other cosmetics.

The same is true for dogs. As Americans, we so cherish the dog that it would be almost impossible for most of us to conceive of the dog as a food item—bred only to be slaughtered and consumed in a variety of foods. Yet when dogs are perceived differently—such as pit bulls, which are seen in American society as a uniquely dangerous type of beast—they take on a very different character, and thus a whole new way of interacting with them (or avoiding them) emerges.

Understanding Animals and Their Uses

Although human-animal studies was born out of the interest in animals in society, which also led to the rise of the animal protection movement, it is not about animal advocacy. HAS theorists look at all of the ways that animals play a role in human society and culture—good and bad. We try to make visible what was once invisible or what is so taken for granted that we never even consider it. HAS exposes the often ugly side of the human-animal relationship and then allows you—the student—to use that information as you will in your own life.

On the other hand, many HAS scholars were drawn to the field precisely because of their interest in, or passion for, animals, so it should not be surprising to learn that many HAS scholars are indeed animal advocates. Just as scholars of women's studies, ethnic studies, or gay and lesbian studies often are advocates in those fields, many HAS scholars do indeed care about animals and use the knowledge gained from HAS and related fields such as ethology to try to advocate for a better life for them.

Does this make human-animal studies less objective than other scholarly fields? The reality is that no scholarship is truly objective, and HAS is no exception. Academics—whether in the humanities, social sciences, or natural and physical sciences—bring with them into their work their own values, biases, and agendas. Academic scholarship is simply not value-free as we once thought. Further, HAS scholars cannot avoid the sometimes ugly reality of the treatment of animals in human societies, and like sociologists, many do take a stand at times. Like sociologists who, thanks to their understanding of social problems, often play a role in crafting policies intended to ameliorate human suffering, HAS scholars may also play such a role. At the very least, the knowledge obtained from certain types of HAS research can be used in this way.

A closely related field to human-animal studies is CAS. CAS is an academic field of study dedicated to the abolition of animal exploitation, oppression, and domination. Unlike HAS, it is not only an academic discipline but one that has an explicit political agenda: to eliminate the oppression of nonhuman animals in all social contexts. The Institute for Critical Animal Studies, for example, publishes an academic journal, the *Journal for Critical Animal Studies*, and hosts conferences, but it also engages in university-based activist activities such as bringing vegan food onto college campuses or working to end academic animal research.

On the other hand, there is nothing in the field of HAS that demands that researchers, instructors, or students take an advocacy or political position of

Figure 1.3. A racehorse being "schooled" between races at Del Mar to acclimate it to the noise and the crowd. (Photograph courtesy of Robin Montgomery.)

any kind. In addition, whether HAS scholars are drawn from geography, sociology, anthropology, psychology, religious studies, or English literature, the work that we do adheres to the requirements of evidence-based scholarship. HAS, regardless of the interests of the researcher, is about rigorous academic scholarship, which goes a long way toward blunting the criticism that HAS is really animal rights in disguise.

Methodological Problems

Unlike other disciplines that are often defined by their methodology, HAS is, as we said earlier, defined by its subject matter: the human-animal relationship. In fact, because of the interdisciplinary nature of HAS, it is difficult to locate any theoretical paradigms or methodologies that are shared by scholars across the disciplines. Instead, it makes more sense to look at those theoretical approaches that are common *within* specific disciplines. Also, in terms of methodology, because of the diversity of fields represented in HAS, there is no one methodology used within all of the disciplines. How then, do we know what we know?

This question becomes even more complicated when we go back to the subject matter at hand: the human-animal relationship. Understanding the human side of that relationship is one thing—sociologists and anthropologists, for example, can draw on classic methods such as **participant observation** or surveys in order to understand human attitudes toward animals, and literary critics or art historians can analyze the role played by animals within literature or art. But how can we ever understand the feelings, attitudes, and perceptions of the animals themselves? Relying on the work of ethology and comparative psychology—fields that study the behavior of nonhuman animals—is one approach, but one not without its own set of problems.

Anthropologist Talal Asad, in a discussion about the problematics of cultural translation within anthropology, discusses how the translation of other cultures can be highly subjective and problematic due in part to the "inequality of languages" (1986:156). The ethnographer is the translator and the author of that which is being translated because it is he or she who has final authority in determining the meaning of the behavior being studied. Cultural translation, thus, is inevitably enmeshed in conditions of power, with the anthropologist inevitably holding the power in the relationship.

This same problem exists, arguably to a much greater extent, when trying to understand—and put into human words—the minds of nonhuman animals. Animal behavior studies, for example, still rely primarily on objective accounts by scientists who attempt to suppress their own subjectivity. The suppression of the author from the text (through the use of technical, reductionist language and passive voice) contributes to the objectification of the animal and, in some ways, makes it even harder to see the animal within the account. Still, these studies, especially those conducted in recent years that move away from the starkly reductionist studies of the past, can be extremely valuable in terms of the insight they can give us into animal behavior, and, hopefully, animal emotions and consciousness.

That we do not share a common language with nonhuman animals—although we can certainly communicate with them—makes it even harder to access their minds. The philosopher Ludwig Wittgenstein once wrote, "If a lion could talk, we could not understand him" (1994:213). And as we said earlier in this chapter, the fact that our perception of animals is heavily colored by the social construction of those same animals makes it even harder to try to understand them in and of themselves.

What does the worldview of a dog look like? Can we imagine what it is like to see through a dog's eyes? Or to smell through a dog's nose? Whether or not we can truly answer these questions (we certainly cannot use surveys or other traditional methods to answer them), HAS scholars need to remind themselves that other ways of knowing and being in the world exist. For instance, Ken Shapiro (2008) writes of what he calls "**kinesthetic empathy**," in which a person attempts to empathize with an animal by understanding their bodily experiences. According to Shapiro, taking the role of the nonhuman animal in this manner allows a "general access to the intended world of the other" (2008:191). This also involves an understanding of the animal's personal history—his or her biography—that adds to our understanding and, thus, to our empathy. From this empathy Shapiro begins to gain an understanding of his dog as an individual and tries to understand his dog's relationship to him (rather than his relationship to his dog). A method like this allows us to avoid the charge of anthropomorphism, or projecting human characteristics onto animals.

Theoretical Starting Points

Some disciplines have theoretical and methodological approaches that are more amenable to the study of human-animal relations than others, and, not surprisingly, those disciplines—found primarily in the **social sciences** and **humanities**—have a greater foothold with HAS than other approaches.

One of those fields is sociology. Sociology's key theoretical paradigms—**symbolic interactionism**, **functionalism**, and **conflict theory**—are all applicable to HAS and are all seen in some classic HAS research in the field. As sociologist Cheryl Joseph writes:

> Given sociology's premise that human beings are social *animals* whose behaviors are shaped by the individuals, groups, social structures, and environments of which we are part, it seemed both logical and timely to enjoin the discipline with the study of other animals in the context of human society. (2010:299)

For example, functionalism, which focuses on social stability and the function of the social institutions, can be used to analyze the roles that animals play in human society and the basis for human attitudes toward those animals. Conflict theory, another macro approach widely used in sociology, derives from the work of Karl Marx and focuses on conflict and power struggles within society. It too can be used within HAS and is especially valuable when looking at the exploitation of animals for human economic gain.

Because this is such a central element of how humans and animals interact, conflict theory is widely used by sociologists interested in animals today.

Another important approach in the sociological study of humans and animals is symbolic interactionism. Symbolic interactionism is a microlevel theory—that is, it focuses on person-to-person interactions rather than on large social forces—and is a perfect approach to use to study the interactions and relationships between humans and nonhuman animals. In addition, symbolic interactionism looks at how humans construct the social world and create meaning within it via interaction and the use of symbols. This approach allows sociologists to not only study the interaction between humans and animals but also to analyze the meanings given to those interactions. In addition, sociology's use of participant observation, combined with interviews, surveys, and other tools, provides a way for us to observe and analyze humans' interactions with, and attitudes about, animals, as in the classic book *Regarding Animals* by Arluke and Sanders (1996). In addition, in another classic study, Alger and Alger (1999) use participant observation to study cats, and find that these animals can communicate with humans via the same symbols used by humans.

Anthropology, or the study of humankind, is a four-field discipline that encompasses cultural anthropology, biological anthropology, linguistic anthropology, and archaeology. As such, it has no overriding theoretical paradigms or methodologies. However, cultural anthropology—the study of contemporary human cultures—shares with sociology not only some of its methods—most notably, participant observation—but also a focus on humans as social animals. Unlike in sociology, however, animals have historically played a large role in anthropological studies, albeit in a secondary fashion. For example, nonhuman primates have been used by biological anthropologists as a lens to understand human behavior and evolution. Cultural anthropologists have looked at animals as resources within human social and economic systems, as symbolic stand-ins for nature and savagery, as totems and symbols, and as mirrors for creating cultural and personal identity. As Donna Haraway once wrote, "We polish an animal mirror to look for ourselves" (1991:21). Classic examples of this type of work include Clifford Geertz's analysis of the Balinese cockfight (1994), Emiko Ohnuki-Tierney's analysis of the monkey in Japanese society (1990), or Elizabeth Lawrence's analysis of the rodeo (1982). Anthropology, then, is **anthropocentric**: In spite of the field's interest in animals, animals themselves have rarely been seen as an object of inquiry in and of themselves.

Another problem is that cultural anthropology traditionally excludes animals from the realm of anthropological study because **culture**—socially transmitted knowledge and behavior that are shared by a group of people—is thought by most anthropologists to be the one item separating humans from other animals. Although humans are biological *and* cultural beings, animals are thought to be biological creatures that lack culture. On the other hand, at least anthropology explicitly acknowledges that humans *are* animals—and that is a start. (We will address alternative perspectives on the question of animals and culture in chapter 17.)

Even though cultural anthropologists do not yet grant nonhuman animals culture, cultural anthropology still has many qualities that make it a particularly appropriate field from which to engage in human-animal studies. It is intersubjective, relying on a give-and-take relationship between researcher and subject rather than an objective, objectifying approach that separates researcher from subject and subject from object. Anthropology's holistic approach, participant observation methodology, and its rejection of reductionism allow for a representation of the "other" that is historically and socially contingent. This makes anthropology an ideal discipline from which to understand the role of animals in human society. In fact, work by scholars such as Rebecca Cassidy and Molly Mullin (2007), who look at the continuing importance of animal domestication to people and animals, or Patricia Anderson, who looks at companion parrots (2003), does focus on animals as more than just windows to understanding human identity construction. Another good example is John Knight's edited volume *Animals in Person* (2005), which is a cross-cultural look at the human-animal relationship in a variety of settings and forms. In that volume, for example, Peter Dwyer and Monica Minnegal discuss the importance of pigs in Papua New Guinea, not just as an economic resource but also as animals capable of instilling grief in those who raise and eat them.

In recent years, geographers have contributed to human-animal studies through their focus on the spaces in which animals live and how those spaces help determine the nature of the human-animal relationship. Like the recent work in anthropology, geographers take into account animal agency in these discussions—because animals encounter humans as much as humans encounter animals. Two major collections that deal with this issue are Jody Emel and Jennifer Wolch's *Animal Geographies: Place, Politics, and Identity in the Nature-Culture Borderlands* (1998) and Chris Philo and Chris Wilbert's *Animal Spaces, Beastly Places* (2000), both of which move from a look at animals as they exist in "wild spaces" and focus on the human locations in which they now exist—whether rural, suburban, or urban—from zoos to laboratories to farms to homes. In addition, these texts address how it is that animals have moved into the places in which they now exist and what kinds of power relationships keep them there. As Jody Emel and Julie Urbanik write, animals' "'places' were determined not only by their own wants and desires, but also through practices of imperialism, masculinity and femininity, class, racialization, livelihood strategies, economies of scale, and so forth" (2010:203). Animal geographers also address other issues, such as when and where animals are "out of place"—and how that is decided. Here, geographers are also interested in what it is that animals want and where they want to go. Questions such as this one are rarely addressed in academic accounts of animals and demonstrate the extent to which human-animal studies have evolved in just a short amount of time.

Like anthropology, psychology is another social science that has historically included animals as part of its focus. However, for much of its history, animals have been used as objects from which to understand humans, even though human-animal interaction—and animals themselves—have largely been absent. For example, experimental psychologists conducted scientific research on animals (that were seen as stand-ins for humans) in order to understand such phenomena as attachment, perception, and learning. For example, an infamous psychological series of studies was the work done by psychologist Harry Harlow in the 1950s, now known as the "mother love" studies, in which he separated infant rhesus monkeys from their mothers in order to study the effects of maternal deprivation on the infants' development. (Harlow's work has been heavily criticized for its cruelty and for proving what most people already knew: that raising babies without any sort of parental care or contact results in psychological disturbed individuals—a fact that Harlow himself acknowledged.) Another classic psychological area that utilizes animals is **behaviorism**, in which animals are trained to engage in certain activities in order to understand human learning (and sometimes to improve animal lives). Today, psychologists still conduct research on animals, but other psychologists (for example, comparative psychologists) address issues of animal intelligence, emotion, and cognition through much less invasive techniques such as observation in the field. Even though all of these approaches use animals, they are not generally considered human-animal studies because they do not look at the human-animal relationship.

Today, there is a broad list of topics that psychologists do cover in HAS. For example, many psychologists focus their studies on the human-animal relationship and address the ways in which animals serve as substitute family members, working partners, and **social lubricants**—enabling social interaction with other humans (Netting, New, and Wilson 1987). Related to this approach are studies that focus on human-animal attachment, in which, for example, researchers study human grief over animal death or the physical and emotional benefits accrued (to humans) from the **human-animal bond**. How animals are used in the human socialization process is another area of research, and psychologists have focused on the role that animals play in child development. Others look at interspecies communication and the ways in which humans and animals use voice, body, and symbols to communicate with each other and deepen their bond. Another focus of study in psychology is the attitudes that people have toward animals, how those attitudes are formed, what characteristics those attitudes are correlated with, and how those attitudes can change. For instance, what is the relationship between living with animals and attitudes such as empathy or compassion?

An important area of psychological work (as well as the related field of social work) today focuses on animal abuse and the link between animal abuse and human violence. Scholars such as Frank Ascione and Randall Lockwood have demonstrated the connection between these two forms of violence, and their work has led to policy changes and legislation around the country (Lockwood and Ascione 1998; Ascione and Arkow 1999; Ascione 2005; Ascione 2008). Known as "**the link**," this is one of the most talked-about areas of HAS and has a wide variety of social work and law enforcement implications.

Within the humanities are a number of fields in which human-animal studies plays a major role, such as cultural studies. Because of its own interdisciplinary and multidisciplinary nature, cultural studies is well suited to the "out-of-the-box" nature of the field. Cultural studies scholars use the findings from a variety of disciplines (such as biology, ethology, and environmental science) and combine them with analyses of representations of animals in culture (such as on television and in film) in order to understand the cultural construction of animals and how those constructions are historically and culturally contingent. Like sociologists, cultural studies scholars seek to understand how the human-animal relationship has been shaped by social forces, attitudes, structures, and institutions. Donna Haraway's *Primate Visions* (1989), Steve Baker's *Picturing the Beast* (1993), and Jonathan Burt's *Animals in Film* (2002) are all good examples of this approach.

History is another field with a long tradition of human-animal studies. Historian Georgina Montgomery and sociologist Linda Kalof (2010) position the interest in animal histories to E. P. Thompson's (1968) call for historians to engage in "history from below," or history based on the experiences of regular people rather than generals and kings. Today that call has been answered by historians who focus on the histories of animals.

Here again, the question of animal agency comes into play because traditionally animals were not thought to even *have* a history, much less play a role in creating it. But historians working in HAS treat animals as historical subjects. Addressing historical periods from the perspective of animals, as well as humans, involves a very different type of perspective. Of course, animals do not leave written records, so reconstructing the history of animals necessarily involves relying on human records, making their history somewhat derivative in nature. Related to history is **zoöarchaeology**, which, like archaeology, relies on material remains (rather than written accounts) to reconstruct the past but specifically focuses on the remains of animals. But how to get at the motivations driving the behavior of animals in the past is still easier said than done.

Harriet Ritvo's *Animal Estate* (1987), which analyzes the roles played by animals in Victorian England, and Coral Lansbury's *The Old Brown Dog* (1985) were some of the first major HAS texts published by historians and illustrated the complex ways in which animals figured in the lives of humans in the past. One interesting historical study is Katherine Grier's 2006 book *Pets in America: A History*, which looks at the rise of pets in the United States and the role that they played in transmitting values to American children during the Victorian era. According to Grier's work, parents and moralists saw having a relationship with a companion animal as a way to instill positive virtues in a child, which she calls the "domestic ethic of kindness." The idea that children can learn positive values by relating to animals is still with us today, as humane organizations and pet industry promoters encourage parents to bring home an animal as a way to teach responsibility, kindness, and nurturing behavior to children.

Gender studies and women's studies, like cultural studies, are interdisciplinary fields and, as such, are especially appropriate for human-animal studies. In addition, both highlight difference and focus on how difference is constructed and represented. Feminist scholars have addressed the ways that sexism and **speciesism** parallel each other and shape one another and are based on the assumption that there are essential, meaningful differences between say, men and women or humans and animals. HAS studies grounded in feminism also address the process of **othering**, which allows for the assignment of different characteristics to different groups. These differences are then used to justify the domination of certain categories of people or animals based on their supposed essential natures.

Feminist and animal rights activist Carol Adams (1990, 1994) has written extensively on the parallels between the control of women's bodies and the control of animal—and primarily female—bodies, and how again, the two are inextricably linked. Adams calls this the "sex-species system," which objectifies and sexualizes female and animal bodies, especially through meat consumption and pornography, both of which reduce the female-animal body into meat that is consumed literally or metaphorically. Ultimately, Adams links extreme violence toward animals with violence toward women. Biologist Lynda Birke (1994) also looks at the construction of difference in women and animals and the implications of othering within science.

Philosophy is perhaps the one discipline whose approaches and theories are widely used in other HAS disciplines. For instance, HAS courses, regardless of whether they are taught in sociology, anthropology, psychology, or geography, often have a section devoted to the ethical issues related to animals. As such, those theorists—ancient and modern—who have written about the moral status of animals are used and read in a wide variety of HAS courses. Philosophers going as far back as Pythagoras have examined the human-animal relationship and questioned whether society's treatment of animals is ethical. In more recent years, the publication of Peter Singer's *Animal Liberation* (1975) and Tom Regan's *The Case for Animal Rights* (1983) have spurred not only the modern animal rights movement but modern philosophical writings as well, all of which we will cover in chapter 18.

Real-World Implications of Human-Animal Studies

HAS is more than an academic field. Like women's studies or ethnic studies, HAS research has real-world policy implications. Some HAS studies provide specific data that directly inform a particular policy area. Examples include recent research on the use of elephants in zoos (Bradshaw 2007), disaster planning that includes companion animals (Irvine 2007), and "dangerous dog" legislation (Bradley 2006), all of which can and should be used in crafting public policy on those issues.

One area where human-animal studies has directly been used to craft social policy has been in the link between animal abuse and human-human violence. Policy implications of this important area of research have led to the development of programs for battered women where they can bring their companion animals (so they do not have to leave them at home when they flee their batterer) and programs that train veterinarians, social workers, and law enforcement professionals to detect the signs of abuse in a home—whether that abuse is aimed at animals or humans. Other scholars have focused on **animal hoarding** and have tried to understand the risk factors for this condition.

Animal-caused violence is another area that has been studied by HAS scholars. Dog bites and dog attacks are two of the most heavily researched areas. Scholars have looked at the risk factors for dogs that bite and have ana-lyzed the legislation and social policies that have been enacted in commu-nities as a result of so-called "dangerous dogs." This research (for example, Bradley 2007) has found that dogs that are chained, dogs that are bullied, and dogs that are neglected are more likely to bite than dogs that are well treated, regardless of the breed. This research suggests that **dangerous dog legislation**—legislation banning breeds of dogs such as pit bulls—cannot truly solve the problem of dog bites.

Another real-world application of the work that has come out of HAS is in the area of animal-assisted therapy. Like the work on the link regarding "dangerous dogs," animal-assisted therapy is human-centered in that animals are used to help humans in a variety of situations—those who are hospi-talized, disabled, elderly, or alone. Swim-with-dolphin programs, therapeu-tic horseback-riding programs, and animals in hospitals, hospices, nursing homes, and prisons are just a few examples of the types of programs that have resulted from this line of research.

Another important application of HAS research has to do with ani-mals in shelters. At least four million animals per year are euthanized at animal shelters nationwide in the United States. (The Humane Society of the United States puts that estimate at six to eight million animals per year.) This is an important issue in terms of the deaths to animals and the trauma to workers, and it is an important economic issue as well. It costs millions of dollars each year for animal care workers to catch, care for, and ultimately euthanize all of those animals. Research that looks at all of the factors associated with the breeding, abandonment, and adoption of companion animals can be an important factor in helping to alleviate this enormous problem. Some scholars have looked at things such as tem-perament testing for domestic cats, what factors lead people to abandon animals at shelters (see Kass et al. 2001), and whether things such as dog training classes or other shelter support programs can lead to permanent placement of animals in homes. Other scholars have focused on animal shelter workers (see Arluke and Sanders 1996), who have the difficult job of caring for animals thrown away by the general public. HAS scholars use the term **compassion fatigue**, which refers to the fatigue felt by shelter workers, animal welfare volunteers, veterinarians, and others who work in the caring professions. Other studies have looked at how movies and television shows can lead to fads in pet ownership, such as the rise in Dal-matian purchases after the movie *101 Dalmatians* or the rise in Chihuahua purchases after *Beverly Hills Chihuahua* or the Taco Bell television com-mercials featuring a talking Chihuahua.

Other disciplines that are affected by the work of human-animal studies scholars include animal welfare science, animal law, and humane education, all of which use scholarship on human-animal relationships in a real-world context—to help better the lives of farm animals, to craft law and social policy, and to educate children.

Ultimately, HAS can be a powerful tool not only to better understand the human-animal relationship but also to actively affect the policy decisions and legislation that shape the ways in which we interact with animals.

Suggested Additional Readings

Adams, Carol and Josephine Donovan, eds. 1994. *Animals and Women: Feminist Theoretical Explorations*. Durham, NC: Duke University Press.

Arluke, Arnold and Clinton Sanders. 1996. *Regarding Animals*. Philadelphia: Temple University Press.

Davis, Susan and Margo DeMello. 2003. *Stories Rabbits Tell: A Natural and Cultural History of a Misunderstood Creature*. New York: Lantern Books.

Ingold, Tim, ed. 1988. *What Is an Animal?* London: Routledge.

Kalof, Linda and Brigitte Resl, eds. 2007. *A Cultural History of Animals*, vols. 1–6. Oxford: Berg.

Manning, Aubrey and James Serpell, eds. 1994. *Animals and Human Society: Changing Perspectives*. London: Routledge.

Regan, Tom. 2001. *Defending Animal Rights*. Urbana: University of Illinois Press.

Serpell, James. 1996. *In the Company of Animals: A Study of Human–Animal Relationships*. Cambridge: Cambridge University Press.

Singer, Peter. 2002. *Animal Liberation*. Rev. ed. New York: Harper Perennial.

Wolch, Jennifer and J. Emel, eds. 1998. *Animal Geographies: Place, Politics and Identity in the Nature–Culture Borderlands*. New York: Verso.

Suggested Films

A Natural History of the Chicken. DVD. Directed by Mark Lewis. Washington, DC: Devillier Donegan Enterprises, 2000.

Websites

Banksy's Pet Store: http://thevillagepetstoreandcharcoalgrill.com
Cute Overload: http://www.cuteoverload.com
Cute Things Falling Asleep: http://www.cutethingsfallingasleep.org
LolCats: http://www.icanhascheezburger.com

Coming to Animal Studies

SUSAN MCHUGH
University of New England

Literary animal studies started for me when the question of animal agency arose in a survey-course discussion of a short, forgettable William Wordsworth poem entitled "Nutting." A shy undergraduate, I hesitantly volunteered an interpretation of the text as reflecting the squirrel's thoughts on the subject of seasonal change.

"That's insane," said the truly venerable professor, as the class fell silent. "Animals don't think, and they certainly don't write poetry."

Twenty years or so later, this moment of candor remains stunning, only for different reasons. My reading certainly failed to take into account the poem's original context and attests instead to sensibilities peculiar to late twentieth-century America, where (unlike in Wordsworth's England) squirrels abounded. But this rebuke says so much more.

I tell this story a lot because it has a bit of a happy ending: I'm now a literary and cultural theorist, one of a handful whose research focuses on animal stories. And, long after this scene unfolded, I discovered that my professor and I were both right, but in different ways. Creatures such as the poet's rodent locate a peculiar paradox in the history of literary criticism: Although animals abound in literature across all ages and cultures, only in rarified ways have they been the focal point of systematic literary study. Serving at once as a metaphor for the poetic imagination and voicing the limits of human experience, a figure such as Wordsworth's squirrel gains literary value as dissembling the human, as at best metaphorically speaking of and for the human. Challenging this tradition of reading animals in literature only ever as humans-in-animal-suits (as I did, however inadvertently) also necessarily involves coming to terms with a discipline that appears organized by the studied avoidance of just such questioning.

Among literary scholars today, I am far from alone in taking a historical approach in order to grapple with this problem of animals in literary history. We ask: What changed so that people became able to read animals as having their own stories, as having history, in the broadest sense? Looking at the ways in which stories of companion species mutate across time, my own research tracks how these mutations intersect with other social changes, and in so doing I make the case that narrative and species forms emerge only ever through complex historical and cultural interrelationships. Such work often requires digging through archives, even sometimes rethinking archives, especially when it comes to challenging our most casual assumptions about the nature of particular cross-species relationships.

To take just one example, when the Oxygen network chose in its 2008 Olympic programming to give comprehensive coverage to equestrian events—a first in television history for the only Olympic sports in which men and women compete against each other on equal terms—it seemed to many a no-brainer. Girls and horses have a natural affinity, right? This assumption has long troubled me, because my research has uncovered precious few antecedents to Enid Bagnold's 1935 bestselling novel National Velvet, *known to most today through the classic 1944 film featuring Elizabeth Taylor in the title role as an unlikely (and then still illegal) winner of the Grand National Steeplechase. If the girl-horse connection is so obvious, then why were women riders barred from these competitive international arenas, for instance, by members of the British Jockey Club until 1978? And why are there so few images and stories of girls and horses before the twentieth century? The search for answers to these questions set me on a course of painstaking archival research.*

Funded by the John H. Daniels Fellowship for Research in Residence at the National Sporting Library, I spent a month combing their collections to find an answer to this literary and cultural conundrum. Tracing the history of jumping sports back to foxhunting stories, I found an intriguing clue in the 1865 novel Mr. Facey Romford's Hounds *by R. S. Surtees (whose novels, despite influencing Charles Dickens, Virginia Woolf, and many other famous novelists, have fallen out of print), in which accomplished horsewoman Lucy Glitters is accused of being a "pretty horsebreaker." Writing in a more prudish era, Surtees's few literary critics made glancing references to this character's similarities to a notorious courtesan of the day, in whose biographies I at last discovered that "pretty horsebreaker" was a euphemism for the most common form of paid work then available to female riders.*

Like car-show models in the next century, the job of pretty horsebreakers was to attract attention to themselves (and away from their mounts' defects) while riding through city parks and thoroughfares clothed in tight-fitted riding habits, the kind of public, professional display of the body that in Victorian England was tantamount to prostitution. By the end of the nineteenth century, more sympathetic depictions of women as stunt-show or "circus riders" such as Finch Mason's 1880 story "The Queen of the Arena" emphasized instead the fictional (and often factual) associations of such workers with the daring and athleticism required of cross-country riding to hounds. But the damage was done already by the tainted sexuality coloring the earlier representations, charting a precarious way to the greater involvement of girls and women in riding sports by the twentieth century.

Relating this historic struggle for equality to representational patterns in human-animal companionship and to the increasingly sexualized depictions of violence toward girls and horses together in more recent narratives such as

International Velvet *(1978) and* The Horse Whisperer *(novel 1996; film 1999), my research uncovered how this literary and cultural history informs the novelty of* National Velvet, *if not the declining popularity of horseracing, in a chapter of my book* Animal Stories *(2011). Admittedly, the histories such as these that I piece together uncover these and other ugly realities of domestic life for humans and animals alike. But I continue to find inspiration where these stories loft hopes for better futures that are precisely made possible through these shared lives and perhaps most poignantly where they offer important lessons about endangered and foregone cross-species relations.*

In this light, it might seem strange that my research has focused largely on one particularly ubiquitous kind of animal, whose stories stereotypically always end in death. At the risk of sounding downright foolish, I admit that it wasn't until after I published my book Dog *(2004) and several essays on* Canis familiaris *in literature as well as fine art and genetic science that it dawned on me that having been mauled by a dog as a small child might explain my profound sympathy for and curiosity about stories of this particular species. Yet, because I never thought about this work as being about me (never mind my facial scarring from that incident), it was easy to collaborate with and learn from other scholars, especially those working in other disciplines, to figure out the complex social scenes built into all representational histories of literatures and cultures, particularly when they involve members of other species.*

Some scholars are concerned that this kind of intellectual crossbreeding will weaken the role of literature in university culture. Whether such efforts work overall to reinvigorate the humanities from within the discipline of literary studies or this kind of engagement with literary animals leads to a more comprehensive dismantling of all structures of knowledge remains to be seen. My own experiences indicate that the scale and immediacy of the problems of representing animals require a variety of different archives of texts, methodological tools for interpreting them, and theories to establish their broader relevance in order to change ways of reading and writing about as well as living with animals for the better. For my own work, it won't matter so much if literary studies continues as a distinct discipline of the humanities or becomes utterly reconfigured through postdisciplinary frameworks so much as that it remain open to different ways of joining in these broader struggles of human-animal research.

2

Animal-Human Borders

Animals and Humans: The Great Divide?

One of my favorite websites is http://www.icanhascheezburger.com, a website featuring **lolcats**—pictures of cats (and sometimes other animals) with funny captions supposedly said by the cats and written in a form of grammar called "lolspeak." People not familiar with lolcats may find the idea of attributing human words to cats to be somewhat ridiculous, but the practice of giving nonhuman animals human characteristics is thousands of years old and can be found in the myths, folktales, symbolism, and artwork of peoples around the world (as we will discuss in chapters 14 to 16). On the other hand, much of human culture—especially recent culture—is built on the assumption that humans are *not* like animals; we may attribute human thoughts, words, or behaviors to nonhuman animals, but humans are unique and set apart from the animal world.

The question is why? On what basis does the human-animal division rest?

From a strict biological perspective, humans *are* animals. In other words, we—like dogs, cats, or insects—are multicellular, eukaryotic creatures that use carbon for growth, move independently, sexually reproduce, and must eat other organisms to survive. As members of the animal kingdom, we are further classified into a number of animal groupings such as vertebrates, mammals, and primates. But this is not the system of categorization that most of us use in our daily lives. Most of us just call ourselves human, and

Figure 2.1. Lolcats are photographs of cats with funny captions in odd grammar superimposed on them. This one is a lolcamel. (Photograph courtesy of Lynley Shimat Lys.)

use the separate term "animal" to refer to all other animals on the planet. How did we come to understand that humans are in some way unique in the animal world (a belief known as **human exceptionalism**) and if so, in what ways? Is being an animal a domain, which includes humans, or a condition, which excludes humans? And how is it that we know?

Interestingly, the divide between humans and all other animal species is neither universally found nor universally agreed upon. It is neither an exclusively behavioral nor biologically determined distinction but has, at times, included biology, behavior, religious status, and kinship. Ultimately, we will see that this divide is a social construction. It is culturally and historically contingent; that is, depending on time and place this border not only moves but the reasons for assigning animals and humans to each side of the border change as well.

Non-Western Understandings

It is easy to believe that the conceptual boundary that we are discussing between humans and animals is universally found in human societies.

However, anthropological studies and historical research have shown that there is quite a bit of cultural and historical diversity in this regard. In many non-Western societies, nature and animals are not necessarily categories that are easily to the opposite of culture or humans. In fact, many cultures see (or saw) animals as potential clan members, ancestors, separate nations, or intermediaries between the sacred and profane worlds. Many of those cultures share a belief in **animism**, a worldview that finds that humans, animals, plants, and inanimate objects all may be endowed with spirit.

For instance, there are societies that do not recognize "animals" as a distinct category of beings. Hunter-gatherers in particular tend to see humans and animals as related rather than distinct, and humans are as much a part of nature as are animals (Ingold 1994). In these cultures, there is no rigid division between society and nature or even between persons and things. Rather than attempting to control nature, hunters seek to maintain proper relations with animals. Of course they have to hunt, but they often seek the animals' permission or pardon when they do so; some cultures even feel that animals present themselves willingly to respectful hunters. Many Australian aboriginal tribes believe that nature, in which humans and animals coexist, needs human intervention in the form of hunting and rituals for balance (Flannery 1994).

Even in non-Western societies that do share our human-animal categories, the borders between human and animal are often fluid, with animals (and often humans) being seen as having the power to reincarnate into other species. For instance, the nomadic Turkic people of Central Asia believe that they descended from wolves; a she-wolf named Asena was said to have nursed a baby human, leading her to give birth to a race of half-wolf/half-human babies who became the Turkic people. Similarly, there are many societies that have animals as creator figures; in Hinduism, myths feature a divine cow mother created by Brahma, and Prithu, a god who, disguised as a cow, created the earth's plants. Among the Tlingit, the raven is the creator god. In other cultures, people can be turned into animals by gods or by magic, such as the Greek belief that Artemis transformed a nymph into a bear (that later became the constellation Ursa Major).

Among ancient Egyptians, the border between human and animal was present but not absolute. Egyptians thought that animals, like humans, worshipped Ptah, the god who created humans and animals. In addition, cats were considered to be deities, and the sun god Ra could manifest himself as a cat. Because of the importance of cats in ancient Egypt, the penalty for killing a cat was death; Egyptians shaved off their eyebrows in mourning when a beloved family cat died. Beginning in about the fourteenth century BCE, cats were mummified alongside humans. The mummified bodies were treated similarly to humans, with their faces painted onto the bandages and, in some cases, burial masks made of bronze placed over their faces.

A variety of cultures have animal gods or spirits that can manifest themselves in either human or animal form. The Ainu people of Japan are another culture with animals as gods; for example, the bear is the head of all of the gods, and when the gods visit the earth, they take on the appearance of animals. A number of cultures had horse gods, including Koda Pen, worshiped by the Gonds of India, and Epona, worshipped by the Gauls. Fish gods were worshipped by the Canaanites, the Greeks, and the Japanese. The Hawaiians worshipped a hawk god; the Algonquins worshipped a hare god named Michabo. Other cultures had gods that were part human and part animal such as Pan, the Greek god who was half man/half goat.

Animals also play a role in the kinship systems of people around the world. For example, many Native American and Australian groups recognize animals as **totems**—important genealogical figures to whom members of a clan trace their ancestry, and who provide protection. Some Native American cultures also believe that animals and humans share the same culture even though their bodies are different. Animal totems are found in Africa as well; for example, the Kadimu of Kenya believe that they are descended from pythons. Other cultures believe that the souls or spirits of the dead are incarnated in animals. For instance, the Thai believe that white elephants may contain the souls of the dead, and among the Zulu, Kafirs, Masai, and Nandi of Africa, the snake is seen as the incarnation of dead ancestors.

Finally, **transmigration**, in which a person transforms into an animal, is a common tenet in shamanistic cultures and those cultures with a belief in witchcraft. **Shamans** and witches are thought to possess this ability; some Amazonian tribes see their shamans as jaguars in (temporary) human form. Witches cannot only transform themselves into animals but often have **animal familiars** that serve their interests. One interesting example of this belief is the milk hare, a witch found in Scandinavian and Northern European folklore. Milk hares are spirits that witches use to steal milk from their neighbors' cows; if the spirit, or hare, is caught and killed, she will revert back to being a woman. Similar beliefs are found throughout Europe, but instead of hares, cats are the most common spirit animal.

The belief in transmigration goes back thousands of years. Pythagoras, a Greek philosopher who lived in the fifth century BCE, believed that human souls transmigrated into nonhuman animals after death, which formed the core of his argument against eating meat. In addition, Hinduism holds a belief in reincarnation that extends beyond the borders of species. The Hindu belief in **pantheism**—that the natural and human worlds are one

Figure 2.2. A calcite statue of Amenhotep III and the crocodile god Sobek, Luxor Museum. (Photograph courtesy of Jerzy Strzelecki, Wikimedia Commons.)

Figure 2.3. This Musqueam Totem pole is at the University of British Columbia campus in Vancouver, Canada, within the traditional territory of the Musqueam Indian band. (Photograph courtesy of Leoboudv, Wikimedia Commons.)

and the same—is reflected not only in the notion of reincarnation and the interconnectivity of all species, but also in their belief in **ahimsa** (nonviolence). However, Hindus and Buddhists, who also share a belief in nonviolence, still see nonhuman animals as inferior to humans.

Why then does the Western system that we subscribe to remove humans from the realm of animals?

Speciesism and the Rise of the Human-Animal Border

The absolute divide between human and animal as well as the differential valuing of human and animal most likely arose with the domestication of animals. Anthropologists recognize two generalized modes of production that refer to how people make a living: collection and production. Collection refers to the gathering of wild plant foods and the hunting of wild animals and entails an intimate *interaction* with nature. On the other hand, production, which involves the domestication of and control over plants and animals, entails *intervention* with nature. Even **pastoral societies**, which herd animals, tend to relate differently to their animals than do agricultural societies. Among herders in pastoral societies, there is a mutual dependence between human and animal. For example, the Nuer, an East African pastoral tribe, believe that they should not kill an animal solely for food; if they do so, the cow may curse them (Evans-Pritchard 1940). With the rise of agriculture, especially in Europe, the Middle East, and Asia, a new concept of animals and humans emerged, with humans transcending and controlling animals and nature. With this change, animals no longer exist in the same world as humans—they exist in nature, which humans have now transcended.

In the West, the separation of human and animal is reinforced in classical Greek thought and biblical accounts of creation. In *Politics*, the Greek philosopher Aristotle (1943) distinguished animal from human because of the human ability to speak, which forms the basis for humans' ethical existence. Thus animals, because they lack this ability, were created to serve human needs. In addition, Aristotle saw the soul as having three parts: a nutritive part that plants have, a sensitive part that animals have, and a rational part that only humans have.

Early Christians (and before them, Jews) borrowed Greek thought about animals—especially theories such as Aristotle's that devalued the body in favor of a higher consciousness lacking in animals —as well as the concept of the soul, which was borrowed from Egyptians. They then created a theology that reifies human difference from and superiority over animals. In the Book of Genesis, the distinction is clear: "And God said, Let us make man in our image, after our likeness: and let them have dominion over the fish of the sea, and over the fowl of the air, and over the cattle, and over all the earth, and over every creeping thing that creepeth upon the earth" (Genesis 1:26). St. Augustine, writing in the fourth and fifth centuries, also felt that humans are valued precisely because of their link to the divine (Augustinus 2000), a link that is not shared by animals (or women, for that matter). The medieval notion of the **great chain of being** (or *scala naturae*), borrowed from Aristotle, in which God created all of life according to a hierarchy of higher and lower beings—with man just beneath God, and animals below humanity—further reinforces this view.

This division between human and animal was further solidified through the writings of St. Thomas Aquinas, a thirteenth-century theologian who maintained that the world is divided into persons who have reason and thus immortal souls, and nonpersons that are essentially things that can be used in any way to serve the interests of people. Persons are persons because they are rational, and thus have intrinsic value and ought to be respected; animals, being irrational, have only instrumental value and can be used in any way humans see fit (Thomas 1906).

The Roman Catholic Catechism teaches that "nonhuman animals, like plants and inanimate things, are by nature destined for the common good of the past, present, and future community" (*Catechism of the Catholic Church* 1994, paragraph 2415), and this view has been mirrored in a number of statements and encyclicals put out by the Church in the twentieth century. For instance, Pope John Paul II said in 1984 that "it is certain that animals are intended for man's use," and in 2001, the Vatican's Pontifical Academy for Life stated, "humans enjoy a unique and superior dignity, and God has placed nonhuman creatures at the service of people [so] the sacrifice of animals is justified as long as there will be a 'relevant benefit for humans'" (Preece and Fraser 2000). In addition, according to Catholic teachings, even though animals (as well as plants) have souls of a sort, they are not eternal souls and cease to exist when the animal dies. The belief that only humans have eternal souls is linked to Aristotle's teachings because according to Christian teaching the human soul is the seat of reason and intelligence; animal souls, however, lack all reason. Finally, we see in the development of thought in European Christendom the idea that humans are good because of their closeness to God; animals, on the other hand, lack that basic goodness. Further, humans who do bad things are considered to be animal-like or beastly, and the lower side of human nature is thought to be the animal side.

This distinction between human and animal became universal throughout the West and was strengthened through social practice and philosophical thought. In Europe in particular, animals were thought to have been created expressly for human exploitation. Nature was considered a force to be subdued, and Christian clergy, going back to Aquinas and the great chain of being, were especially inclined to emphasize that humans were radically different from and superior to all other creatures. With **animality** posited as something inferior to humankind and as something to be conquered and exploited, early modern Europeans made concerted efforts to maintain distinct boundaries between themselves and animals; upper-class English families did not allow their babies to crawl, for example, because it was seen as animal-like. **Bestiality** was thus an extremely serious crime, often considered a capital offense. This concern with the human-animal boundary has also been used to explain medieval Europeans' fear of werewolves, and their preoccupation with monsters and mythical beasts—especially half-human, half-animal creatures.

Still, most humans during this time period had ample contact with animals, and animals' and humans' lives were intimately connected. In fact, during the Middle Ages animals were routinely charged with crimes and were tried in the same courts as humans. They could also be excommunicated from the Church. Sometimes those animals convicted of crimes were executed, as in the case of the homicidal pig who was hanged for his crimes in 1567 (Beirne 1994); at other times, they were sentenced to prison. Animals also were made to pay the price for the collective guilt of European communities—public rituals in which cats were burned to death or thrown from rooftops illustrate the animal's role as public scapegoat. And finally, even when people were convicted of crimes, animals often were punished as well. When a person was convicted of a particularly heinous crime, he could be hanged alongside two dogs that were also hanged for his sins. In all of these examples, the human-animal boundary was a serious cause for social concern.

As European philosophy evolved through the Renaissance and early modern period, the division between animals and humans widened and the justifications for this border became more sophisticated, but also hearkened back to Aristotle's formulation. As we saw earlier, Aristotle found that it was rationality and the capacity for speech that separated humans from nonhuman animals (and slaves from nonslaves), and that it was the purpose of the animal or slave to serve the more rational creature. René Descartes, the French seventeenth-century philosopher, claimed that mentality and the ability to speak were the primary characteristics separating humans from animals (Descartes 1991). Because animals are incapable of using language, Descartes considered them to be essentially machines—mindless automata which operate without higher thought or consciousness. A century later, Immanuel Kant, the eighteenth-century German philosopher, wrote that rationality was the key characteristic that separated humans from other animals; because animals lack the capacity for rational moral choice, they are not moral agents and thus have no moral standing (Kant 1785). Even in the twentieth century, we see that these arguments are still popular among philosophers. For example, the German philosopher Martin Heidegger (1971) not only saw the capacity for language as the key separation between humans and animals, he also considered language to be central to the capacity to know, understand, and rationally interact with the world.

Today, the issue of morality has come to be one of the key characteristics by which we distinguish between humans and animals. Humans and

BOX 2.1

HISTORY OF BESTIALITY

Images of bestiality have existed since the Stone Age, when carvings and drawings depicting humans having sex with animals were depicted. Bestiality was practiced in societies around the world and many peoples, such as the Hittites, had rules specifying which animals could be used for sex and which were disallowed. The Babylonians, Greeks, Egyptians, and Romans all practiced it to some extent. The ancient Hebrews were the first to forbid sex with all animals and this prohibition extended into Christendom.

In the Middle Ages, bestiality was a crime that could be punished by death—both the human and the animal would often be killed; in Denmark, perpetrators were strangled and burned to death. Even with laws such as these throughout Europe, bestiality was evidently quite common. For example, it was so common for boys to have sex with cows and sheep that the Catholic Church tried to have boys excluded from being shepherds in the seventeenth century. For centuries, going back to the Old Testament, laws prohibiting bestiality and laws prohibiting homosexuality were the same because the acts were considered to be equally offensive and were often lumped together under the label "sodomy."

Bestiality was extremely common in Asia and the Middle East; in some cultures it was socially acceptable, in others it was not. There is an old Arab belief that sex with animals increases a man's virility, cures diseases, and increases the size of his penis. It was also common in the New World; it was so common among the Incas, for instance, that bachelors were forbidden to keep female alpacas. Other cultures apparently saw bestiality as normal. Such was once the case with a number of Native American tribes of the Plains states and the Southwest.

Today, different countries have different laws, many of which are grounded in questions of animal cruelty rather than biblical notions of morality. For instance, bestiality—defined as penetration of any animal by the penis of a human, or any human by the penis of an animal—is illegal in the United Kingdom, whereas in Sweden it is perfectly legal (although perhaps not socially acceptable). In the United States, bestiality is not addressed by federal legislation; instead, 30 states have criminalized it—the most recent being Florida in 2011—and the rest have not.

animals may both be sentient creatures but only humans possess the capacity for moral or ethical behavior. Today, the second key characteristic that divides animal from human goes back, once again, to Aristotle.

Humans alone are thought to possess intelligence, language, self-awareness, and agency. According to Aristotle, those characteristics give humans, and humans alone, the basic rights of life and freedom from persecution and pain. In addition, these characteristics give humans the right to control those who lack them (i.e., other animals). Perhaps humans, then, can be defined best by their relationship to animals and, in particular, to their superiority over them.

Evolution and the Continuity Between the Species

In the mid-nineteenth century, naturalist Charles Darwin's theory of **evolution by natural selection** was a revolutionary challenge to thousands of years of religious and philosophical thought regarding animals. Darwin's writings (*On the Origin of Species*, published in 1859, and *The Descent of Man*, published in 1871) challenged the notion that humans are special, and placed humans and animals into the same category: animals. Although animality had been defined as that which is beneath humanity, Darwin definitively affirmed that humans *are* animals. Darwin was not the first to claim that humans are animals; the eighteenth-century biologist Carolus Linnaeus classified humans in the primate order in 1735 in his *Systema Naturae*, which earned him a reprimand from his archbishop. Interestingly, Linnaeus included chimpanzees, which he called *Homo troglodytes*, in the same genus as humans, although in a different species. And in the seventeenth century, anatomist Edward Tyson demonstrated through dissection the similarities between humans and apes (Tyson 1966). But it was not until Darwin that there was a scientific theory to account for the relationship between the species: not only are humans and nonhuman animals similar, they also share these similarities due to common descent from a shared ancestor.

Not only that, but Darwin also argued that there are no fundamental differences between humans and the "higher animals" in terms of their mental abilities, again challenging the basic premise on which so much of the distinction rested. Darwin proved that not only are humans and all other animals related but also that we together feel pain, share emotions, and possess memory, reason, and imagination. Rather than seeing humans and animals as categorically different, Darwin showed that all animals, including humans, share a continuum of mental and emotional capacities. Of course, even though *On the Origin of Species* (Darwin's major work) was published in 1859, it would not be fully accepted for another hundred years. And it would not be until the late twentieth century and the rise of modern ethology that the idea of nonhuman animals experiencing emotions or possessing reason would be taken seriously.

Today, ethologists who study the minds and behaviors of nonhuman animals show that, such as Darwin theorized, there is no radical break between the emotional and mental capacities of humans and other animals; instead, there is a continuity of capacities. As we shall see in chapter 17, ethologists who work with great apes, dolphins, and parrots, as well as a wide variety of other animals, continue to find more and more examples of this continuity. We now know that many animals can feel and experience much of what we once considered to be "human" emotions, that they have self-recognition and self-awareness, that they can communicate with each other (and with us) through sophisticated communication systems and perhaps even languages; they can make and use tools, empathize with others, deceive others, joke, plan, and understand the past and the future.

Also since Darwin's time, the science of genetics has demonstrated the extent to which we share genetic material with other animals, and especially with the great apes; we share 98% of our genes with chimpanzees, for example, and chimpanzees are more closely related to humans than they are to gorillas. And finally, research in **paleoanthropology** has proven conclusively that humans share a common ancestor with chimpanzees that lived as recently as ten million years ago—a very short time in terms of evolutionary history.

In fact, we can say that since the nineteenth century the border between human and animal is actually narrowing. Through new discoveries in genetic science, paleoanthropology, neuropsychology, sociobiology, and ethology, we find that we are physically, behaviorally, and emotionally closer to other animals than we have ever been before. Where scientists, theologians, and philosophers of the past spent their time overemphasizing the differences between us and underestimating or ignoring the similarities, today's scientists have been closing the gap between the species.

As the border between human and nonhuman has continued to shift, patrolling that border remains ever more important for those who are invested in the idea that humans are not just separate from animals but that that separation entails superiority. As we shall see in this text, humans' "specialness" has been employed to justify virtually every practice engaged in by humans involving animals. Today, we keep redefining the criteria we use to differentiate humans from other animals as we discover bit by bit that animals are a lot cleverer, and a lot more *human*, than we thought.

What then, is an animal? Ultimately, we have not answered this question. But in chapter 3, we will turn toward a related one: how is it that we classify animals?

Suggested Additional Readings

Creager, Angela, ed. 2005. *The Animal/Human Boundary: Historical Perspectives.* Rochester, NY: University of Rochester Press.

Darwin, Charles. 1859/1985. *The Origin of Species.* New York: Penguin.

Fudge, Erica, Ruth Gilbert, and Susan Wiseman, eds. 1999. *At the Borders of the Human: Beasts, Bodies and Natural Philosophy in the Early Modern Period.* New York: St. Martin's Press.

Henninger-Voss, Mary. 2002. *Animals in Human Histories: The Mirror of Nature and Culture.* Rochester: University of Rochester Press.

Ingold, Tim, ed. 1988. *What Is an Animal?* London: Routledge.

Noske, Barbara. 1997. *Beyond Boundaries: Humans and Animals.* Montreal: Black Rose.

Salisbury, Joyce. 1994. *The Beast Within: Animals in the Middle Ages.* New York: Routledge.

Tattersall, Ian. 1998. *Becoming Human: Evolution and Human Uniqueness.* Oxford: Oxford University Press.

Thomas, Keith. 1983. *Man and the Natural World: A History of the Modern Sensibility.* New York: Pantheon.

3

The Social Construction of Animals

WHEN MY FRIEND AND fellow rabbit rescuer Susan Davis and I were researching our book, *Stories Rabbits Tell* (2003), we became familiar with the **Rainbow Bridge**—that special place that many animal lovers construe as a sort of heaven for pets, a place where pets (and sometimes their people) go after death. The Internet is full of thousands of sites where people post memorial photos, artwork, and poems and reminisce about their beloved companion animals. What struck us the most about this belief, however, was the realization that people who breed and kill rabbits for meat also recognize the Rainbow Bridge. These "rabbit growers" believe that their own dogs, cats, or birds will visit the bridge when they die, but not the rabbits they slaughter. Susan and I were dumbstruck: The community of rabbit lovers to which we belong wholeheartedly embraces the Rainbow Bridge concept. (Bunspace, the popular social networking site for rabbits, even has a small rainbow that is affixed to the profiles of rabbits that have "gone to the bridge.") But evidently not all people who are involved with rabbits share this sentiment.

The idea that the Rainbow Bridge is a heaven for only certain kinds of animals illustrates not only the odd nature of a belief such as this but also the ever shifting nature of how we classify animals. If there is such a thing as a heaven for animals, which ones get to go there? How does one get in? And what is the basis for the way that we classify animals? Some animals get not only gentle and loving treatment in life but a glorious afterlife as well, while other animals live short, brutal lives and then, in death, get nothing at all.

Figure 3.1. Jessie, a rabbit of the species *Oryctolagus cuniculus*, could be a pet rabbit, a lab rabbit, a meat rabbit, a fur rabbit, or—in Europe or Australia—a wild rabbit. (Photograph courtesy of Tracy Martin.)

What is the animal shown in figure 3.1? Is this a wild rabbit? A pet rabbit? A meat rabbit? A lab rabbit? An Easter rabbit? Biologically, it is all of the above. According to the **Linnaean taxonomy**, this creature is a member of the species *Oryctalygus cuniculus*, commonly known as the domesticated rabbit. This species of rabbit is found in the wild in Europe and in people's homes as pets, and it is raised for food, fur, and laboratory purposes. Domesticated rabbits, just like domesticated cats (and humans, for that matter), do not have subspecies; regardless of color or size, all rabbits share fundamentally the same genes.

How we classify rabbits—calling them pets, or meat, or lab animals—has to do with where we they live, and what they are used for. That tells us something important. How we use animals in society today defines at least in part how we classify them. And the reverse is true as well: How we classify them also impacts how they are treated. For instance, *Oryctalygus cuniculus* is affected by a different set of laws if it is classified as a pet, food, or a laboratory animal. In the United States, all of these "types" of rabbits are affected by different laws that impact their care and what can be done to them. If a rabbit is defined as a pet, killing it would be illegal under most states' animal cruelty laws; but put that rabbit into a laboratory or into a backyard rabbit farm and that killing is perfectly legal.

How do we classify animals? In chapter 2, we talked about how societies have constructed the border between human and other animals or as we commonly think of it, between humans and animals. In this chapter, we will look at how we classify *types* of animals. Understanding the various classification systems will help us to understand more about the social construction of animals and, perhaps more importantly, how we *use* animals in society today.

Biological Systems of Classification

Since 1735, when naturalist Carolus Linnaeus divided up the plant and animal kingdoms naming and classifying all that he saw, we have used the Linnaean taxonomy to classify elements of the "natural world." As we discussed in chapter 2, because Westerners have long considered animals and plants to be members of the natural world, we quickly accepted Linnaeus's plan. We define animals at least in part by their place in the Linnaean taxonomy (now modified).

Animals are divided into phyla, classes, orders, families, genera, and species, based on their possession of shared physical attributes. Since Linnaeus's time, we have given each animal species a unique "binomen," or two-part name. For instance, the wolf is of the phylum chordata and the subphylum vertebrata, the class mammalia, the order carnivore, the family canidae, the genus *Canis*, and the species *lupus*. These divisions are based on the fact that the wolf has a central nervous system, bears live young, and eats other animals. Since the rise of evolutionary theory, we now know that, generally speaking, traits shared by a group of animals are indicative of a shared ancestry. So mammals—all of whom give birth to live young and breast-feed those young—are classified together because they are considered to be more closely related from an evolutionary standpoint than, for example, animals that lay eggs.

But, although Linnaeus's system makes a great deal of sense and is the only system currently in use in the natural sciences, there have been other ways to classify animals. For example, prior to Linnaeus's time naturalists classified animals by their form of movement (walking, slithering, swimming, or flying) or by the habitat in which they lived (water, air, or ground).

The downside of a biological system of classification is a concept known as **biological determinism**. Biological determinism is the interpretation of

BOX 3.1

ENCOUNTERING GORILLAS

The first time that a human outside of sub-Saharan Africa encountered gorillas he thought they were a tribe of hairy, savage women. Hanno the Navigator, a Carthaginian explorer in the fifth century BCE, called this strange tribe "gorillai" after the word given to him by African informants. More than two thousand years later, an American missionary named Thomas Savage first described the species scientifically and used Hanno's ancient name in this new scientific classification: *Troglodytes gorilla.*

animal behavior from a strictly biological perspective that tends to exclude culture, social practices, and personality as factors in behavior. Perhaps because of the human fear of being considered *just* biological creatures, as we discussed in chapter 2, we continue to separate humans out from other animals. Even as all those trained in science must grant that humans and animals are alike because we are both members of the animal kingdom (and many, such as the great apes, also share our phylum, subphylum, class, order, and even family), there continues to be a form of **anthropodenial** which disallows our shared similarities. Humans may be animals, but we are *special* animals.

Other Systems of Classification

In medieval Europe, animals were generally classified according to Christian theology as we discussed in chapter 2. Even though all animals were seen as lower than human—and further from God—some were more elevated than others. For example, carnivores such as lions and eagles sat at the top of the hierarchy of animals, while vegetarians and domesticated animals sat at the bottom. The folklore of the time created animal heroes that exhibited characteristics such as bravery, cunning, and intelligence, and other animals that were seen as hapless, stupid, or weak. Animals could be noble, evil, or pure.

Another major way in which we have classified animals has to do with where they live, and whether or not they are part of human culture. In this scheme, animals are either wild (living outside the bounds of culture) or they are tame (living inside of human culture). In this view, whether or not an animal has been domesticated—selectively bred and controlled by humans—is the deciding factor as to whether that animal will be considered wild or not. But with this definition, there are problems as well. Animals that live in a zoo, for example, almost always come from the wild (no one would visit a zoo to see a dog, after all), but how wild are they, once they live behind a fence and are fed and raised by humans? Feral animals, or animals that were once domesticated but now live in the wild, provide another categorical problem. In New Mexico, where I live, the problem is not just semantic.

In my community, there is a herd of horses that local people and government agencies have been fighting over for years. Local observers have noted that these horses have lived free on this land since before the 1971 passage of the **Wild Free-Roaming Horses and Burros Act**, which mandates that "wild" horses and burros be allowed to live on public lands without harassment. But how are these wild horses to be defined? According to the law, wild horses and burros are "all unbranded and unclaimed horses and burros on public lands of the United States." That means that any horses living unbranded and unclaimed as of the passage of the law are, by definition, wild. Yet the Bureau of Land Management continues to round up these horses, which they claim have just wandered off from one of the local Native American reservations, and sell them at auction. In some cases, the animals have ended up at slaughterhouses in Mexico. To complicate matters further, some of the horses have had their blood analyzed, which advocates say prove that their ancestry goes back to the original horses brought to Mexico by the Spanish conquistadors in the sixteenth century. But because the horses tested only had 50 percent "Spanish blood," they are not qualified to live at a sanctuary set up for wild Spanish horses. So, in this example, whether the horses are considered wild, stray, Spanish, or domesticated means the difference between their being allowed to roam free or being caught and sold at auction to the highest bidder.

In all cultures, some animals are considered edible and others are considered inedible. Among Jews and Muslims, for example, pigs, shellfish, reptiles, and many other animals are considered inedible. In the United States, dogs, cats, and horses are inedible, while in some cultures, fish, insects, deer, or camels are not to be eaten. Related to the question of whether an animal is edible or not is whether the animal is fit to be sacrificed as part of a religious ritual. Usually, animals that are edible are often those that can be sacrificed. Even though in totemic societies there is generally a taboo regarding eating one's own totem animal, it is precisely that animal that *should* be sacrificed during rituals.

Totem animals are another example of a system of classification that includes animals. A totem is an animal that is considered to be spiritually related to a clan or a tribe, and is generally thought to be ancestral to a group of people. Today some people who are not part of a totemic culture have adopted the concept of a totem, which they see as a sort of spiritual helper or guardian. Totems are a way not just of classifying animals but of classifying the natural and cultural world: Who and what are related to each other and often who and what are considered edible.

In the West, a major distinction is made between animals that are edible, or food, and those that we consider family members, or pets. Generally, in the United States, an animal cannot be both food and pet. Although animals such as rabbits are food to some people and pets to others (and the U.S. Department of Agriculture labels them "mixed-use animals" because of this double classification) most people would never eat *their own* pet. One of the reasons for the rise of the pet/food distinction in the West is the occupational specialization that allows for very few people to be engaged in raising food. Most Americans have no connection whatsoever to the food that they eat, whether animal or vegetable. Because of this, we can enjoy the companionship and love of animals that we call pets, even though we eat animals that other people have had to raise—animals that we call food.

How Does One Become a Certain Type of Animal?

Today, use value—whether an animal is primarily used as food, milk, eggs, or whatever—is the main way in which humans in the West classify animals. But how does one *become* a pet, or livestock, or a laboratory animal?

A pet is defined as an animal that lives in a human household and is named. Naming an animal incorporates that creature into our social world. We use that name in a way that allows for interaction and emotional attachment. By talking about our animals to others—showing off their pictures or their blog or their videos on YouTube—the animal then gains a history that is meaningful to us. All of this remains even after the animal is dead. Many people continue to fondly remember and talk about their pet animals long after the pets have died. (Although most of us no longer talk *to* our animals once they are gone!)

Laboratory animals, meat animals, and fur animals are animals that are spatially separate from pet animals. These animals never live in a home. They also never get a name. They are objects, not subjects. They have no history, no biography, no intentions, and no emotions. They live in a space where they are crowded with others, reducing their individuality; they are handled rarely, when necessary, reducing an emotional connection; if necessary, they are given numbers that are used to refer to them (but not to address them); they do not feel pain so they are not anesthetized or killed with humane methods; and they have no agency, no ability to control their own lives. Ultimately, they are products.

Racing greyhounds, for example, do not live in a home, sleep in a bed, or spend time getting cuddled, talked to, or walked by their human caretakers. Instead, they live in a kennel with hundreds of other dogs. They are rarely spoken to and rarely touched at all. Any handling of a racing dog relates to his or her training and not to a relationship between dog and human. Racing greyhounds are known either by their number or their official name, which is used in the racing literature, but they are not called that name by humans who speak to them. For those greyhounds that are lucky enough to be rescued or retired from the racing industry and placed into a home, their lives change entirely—not because they themselves have changed, but because they are now subject to a new system of classification: that of pet.

In chapter 5, we will discuss the domestication of animals and, by looking at the history of that domestication, will see why some animals have ended up in one place and other animals someplace else. The dog, for example, was domesticated as a hunting partner to humans thousands of years before humans domesticated their first animal as food. Because of this, humans had thousands of years to build a partnership with dogs, which is probably a major reason why dogs are for the most part excluded as a food source and why they are among the most popular companion animals in the world.

But although history can explain some of why certain animals end up as meat when others end up as pets, it does not explain it entirely. Some animals, for example, have been able to cross category boundaries, moving from food to pet. Potbellied pigs, for example, have been raised for food in Vietnam for hundreds of years but today are considered to be a pet in the West.

The Sociozoologic Scale

All societies classify people but only stratified societies classify and rank people in such a way that vertical social hierarchies exist. These hierarchies are then naturalized, i.e., made to seem natural rather than human-made, such that the accompanying inequalities seem natural as well. Societies that are stratified on the basis of class, race, or caste organize humans on the basis of

Figure 3.2. A family of baby raccoons eating cat food while Pax, the family cat, watches. (Photograph courtesy of Robin Montgomery.)

arbitrary criteria and then allocate privileges and opportunities accordingly. Those at the top are given more privileges and those at the bottom are seen to have earned their poor place in society. Since the time of Aristotle, humans have always been ranked higher than animals, such as on the ancient great chain of being. The sociozoologic scale, a term coined by sociologists Arnold Arluke and Clinton Sanders (1996), does the same thing with animals: It categorizes and then ranks animals on the basis of their benefits to human society, which allows humans to define them, reinforce their position, and justify their interactions with other beings.

According to the sociozoologic scale, good animals are animals which provide some benefit to humans. They are both pets and tools; the latter includes meat animals, laboratory animals, and working animals. Because of the history of their domestication, animals allow us to use them, and thus they are nicely incorporated into human culture. Bad animals, on the other hand, are vermin and pests. Both stray from their proper place and resist being used by humans. Animals can cross boundaries and contexts from good to bad. A rat, for instance, can be a good animal when used in a laboratory context, and a bad animal when found lurking in an alley. These categories involve defining the

BOX 3.2

BECOMING A LAB RAT

The first time a rat was used in laboratory research was in 1828. Prior to that time, rats were seen as vermin because of their association with the plague. They were hunted in Europe by rat catchers who sold them as food and as bait for "rat baiting," a sport in which rats were placed into a pit and a terrier was released to kill them. The sport became so popular that participants began to breed the rats especially for baiting, producing differences in color and type, and ultimately producing the albino rat that would become the classic "lab rat."

Kenneth Shapiro (2002) has written about the social construction of the lab rat and how that rat is defined and used in the laboratory context. Laboratory rats are the opposite of the wild, dangerous rat. Instead, they are controllable, manipulable, standardized, and sterilized. In the United States, laboratory rats (as opposed to rats sold in stores as pets) are not even animals; according to the Animal Welfare Act, they are excluded from the definition of animals and deserve no legal protection because of this exclusion. They are instead "research tools" and data points. They are not individuals. They have no names, just numbers—and their life and death are recorded in the abstract language of science.

Today, rats and mice are the most popular of all animals used in labs; they make up as many as 95 percent of all animals used in biomedical research and product testing. Just two hundred years ago, they were considered filthy vermin.

animal so it fits the category; based on the category assigned, we then treat the animal a particular way. When we do this we understand, for example, that cow means food, and thus the cow is made to be killed and eaten. It is the cow's ability to be raised and eaten that makes it "good." The sociozoologic scale classifies and sets out the ways that we will use these animals.

Sociologist Leslie Irvine, in a book on animal-related disasters (2009), discusses the tragedies surrounding Hurricane Katrina, the massive hurricane that hit the Gulf Coast region of the United States in August 2005. Initially, many residents remained behind after mandatory evacuations were announced because they did not want to leave their pets. However, when these residents finally left their homes, they were forced to leave their animals behind because the government relief agencies would not allow pets to accompany evacuees. News reports were filled with stories and photos of weeping people and abandoned pets; many animals drowned or were lost forever. One such case was involved Snowball, a small white dog taken by police from a sobbing little boy as he and his family were boarding a bus at the Superdome. Snowball's story was featured in news accounts around the country, and countless people and groups spent time searching for him in the hopes of reuniting the dog and his boy. Yet Irvine points out the contradiction. Although people around the world donated money or volunteered to help rescue animals such as Snowball, few even stopped to consider that millions of chickens and countless cows, goats, and other farm animals lost their lives in the hurricane and its aftermath. Because dogs and chickens occupy very different places on the sociozoologic scale—dogs at or near the top, with chickens near the bottom—the public is able to spend millions of dollars and shed countless tears over the canine victims of disasters such as Katrina, but not a penny or a tear over the chickens.

Even many animal advocates when pressed find it difficult to feel empathy for animals that are not mammals or that fail to exhibit some of the charisma of other animals. For example, many people find insects to be repugnant and alien and cannot relate to them at all. This makes it hard for us to advocate on their behalf.

Ultimately, animals mean different things to different people. Although we share in the dominant system of categorization in our society—dogs are friends, chickens are dinner—some of us challenge these categories by keeping chickens, say, as pets. Once we have decided to allow a certain animal the title of pet—and the benefits that this definition provides—then we will

BOX 3.3

CHARISMATIC MEGAFAUNA

These animals rank high on the sociozoologic scale; they are cute or majestic and have characteristics that appeal to humans. Because of these characteristics, humans are drawn to them and want to save them.

Koalas
Pandas
Harp seals
Dolphins
Tigers
Whales

Figure 3.3. Egg-laying hens confined in battery cages. This photograph was taken at Quality Egg of New England. (Photograph courtesy of Mercy for Animals.)

not just treat them differently but will also be affected differently by them and will mourn their loss—whether they are chicken, dog, or pig.

In his essay, "Who Swims with the Tuna," David Quammen (2001) asks: Why do we worry about trapping dolphins in tuna nets and not worry about the tuna trapped in tuna nets? Quammen, a science and nature writer, notes that although the dolphin is intelligent, social, and appears to favor interactions with humans—even "allowing" humans to swim with them—the tuna is rarely thought of at all except in terms of its flavor as food. That is why we can find **dolphin-safe tuna** on grocery store shelves but not "tuna-safe dolphin." (Dolphins are often caught up in nets that are used to catch tuna; the "dolphin-safe" appellation is used to demonstrate that dolphin-safe methods were used when catching the tuna.) Quammen ends his essay with the question: Who swims with yellowfin tuna? The answer he gives is that dolphins do.

A New System of Classification

Ultimately, there are as many systems of classification as there are animals and cultures. Rabbits can be pet, food, or tool; they can be sacred or sacrificial; they can be separate from humans or can share with them a language and culture. But no matter the culture or the system, and even under the objective lens of biological science, what we know of animals is just a story—a story that we have made up about animals and ourselves. If these systems of classification that we have been discussing are all essentially just stories, can we create a new story that is more inclusive and humane? Can we conceive of one that does not rank animals on the basis of their importance to us, but on some other characteristic? Or one that does not rank them at all?

One possibility would be to try to flip things around: To try to see the world through animals' eyes rather than to try to see and understand animals through our own eyes. I think of the line spoken by the duck Ferdinand in the 1995 animated film *Babe*: "Christmas? Christmas means dinner, dinner means death! Death means carnage; Christmas means carnage!" To a European or an American, Christmas is one of the most beloved holidays of the year, a time when families get together, when people are more charitable toward others, and when Christians celebrate the birth of their savior. But for farm animals such as Ferdinand, Christmas means carnage. But although this is certainly the reality today for ducks, pigs, turkeys, and geese all around the world, this reality is itself socially constructed; there is nothing natural about it. In the next few chapters, we will address how these conditions came to be and how it is that we came to see them as normal and natural.

Suggested Additional Readings

Arluke, Arnold, and Clinton R. Sanders. 1996. *Regarding Animals*. Philadelphia. Temple University Press.

Sax, Boria. 2001. *The Mythical Zoo: An Encyclopedia of Animals in World Myth, Legend and Literature*. Santa Barbara: ABC-Clio.

Serpell, James. 1996. *In the Company of Animals: A Study of Human-Animal Relationships*. Cambridge, UK: Cambridge University Press.

Suggested Films

Cane Toads: An Unnatural History. VHS. Directed by Mark Lewis. New York: First Run, 1987.

Creature Comforts. DVD. Directed by Nick Park. Bristol, UK: Aardman Animations, 1990.

A Natural History of the Chicken. DVD. Directed by Mark Lewis. Washington, DC: Devillier Donegan Enterprises, 2000.

The Joy of Chickens

ANNIE POTTS
University of Canterbury

*My world has been delightfully filled with animals for as long as I can remember;
and chickens have played a particularly prominent and uplifting part in my life.
I am therefore privileged to advocate for these much maligned birds in both my
everyday life and through my scholarship and teaching of human-animal studies.*

*I grew up in a cold, rural city at the bottom of the South Island of New Zealand.
When my father returned from pilot training in Canada during WWII, he built
a house for my mother and himself on a quarter-acre section on the outskirts of
this town; he wanted to be able to raise chickens and grow his own vegetables
and fruit. So chickens were around from my conception. At age five I made the
connection between the white meat on my plate and the birds I was fond of in the
backyard. These feisty and life-loving "chooks" (as we popularly refer to chickens
in Australasia) did not often end up on the dinner plates at our house, but they
were to my mind no different from those who regularly did. Each time the family
was served chicken meat, I felt appalled that one of these curious and vibrant
birds had been killed for us to eat. So chicken flesh was the first meat I rejected on
the road to vegetarianism, and I owe* Gallus domesticus *for making me aware
of the animal* before *the food; and for jostling me into a career where I critique
meat culture and research vegetarianism in its many forms.*

*Chickens have also ensured I know my place in the pecking order: I am here
to advocate for them. When asked why he was vegetarian, novelist and Nobel
Prize winner Isaac Bashevis Singer commented: "I don't do it for the health of
myself; I do it for the health of the chickens." The strong influence that chickens
have on my work might be understood this way too, except for the fact that my
health is benefited far more than theirs: Despite writing about and educating
others on the plight of these birds and the need for rapid change in how we treat
chickens, millions of battery hens remain incarcerated on egg farms and billions
of obscenely overgrown broiler chicks hatch and die for the chicken meat industry. My aim is to expose the ways in which our contemporary denigration and
abuse of chickens is linked to erroneous assumptions about the emotional capacities and intelligence of these birds, assumptions which have been formed and
reinforced largely as a result of industrialization and intensive farming. Over
the past one hundred years or so, science, technology, and agribusiness have drastically modified how chickens are bred, raised, farmed, slaughtered, processed,
packaged, and consumed; inevitably the ways in which chickens are understood
and represented have also undergone radical change. Once chickens lived more
freely among people—we valued and appreciated them; roosters were respected*

for their bravery, vigilance, and loyalty (and admired for their beauty), while hens epitomized parental devotion and wisdom. Most people now have no relationship with chickens other than consuming their flesh or ova; these birds have been depersonalized, de-animalized, and even demonized. (I refer here to the recent avian flu panics.)

As a species, we tend to be strangely irrational when it comes to recognizing and respecting alternative forms of wisdom in other species. The greater the phylogenic (or physical) difference between humans and other living beings, the more likely we are to view those beings as "mindless," acting only on instinct. In common vernacular chickens are referred to as stupid—"bird-brained"—a prejudice that can be traced to the small-brain fallacy or the idea that "tiny things just can't be intelligent or aware." This assumption is reinforced by the belief that the cerebral cortex (the large, convoluted portion of the human brain) is the site of superior "intelligence" in all creatures. Because this anatomical structure is insubstantial in birds (the avian cortex is smooth), chickens, along with most other birds except perhaps ravens and parrots, are presumed to be unintelligent. However, this perspective on intelligence is decidedly anthropocentric. It fails to acknowledge that those abilities deemed to make humans superior are not as significant in the day-to-day lives and survival of many other species, and are therefore not appropriate measures of intelligence or adaptation for vastly diverse organisms. In fact, birds have a highly developed part in their brains called the hyperstriatum (an area not well evolved in human brains), which is associated with the ability to navigate (without instruments). When measuring intelligence according to this specialization of avian species, humans are notably deficient.

Although I personally believe that chickens have their own unique forms of wisdom we will never fathom, and that we perform a disservice to chickens whenever we try to measure their abilities according to our own assumptions and priorities, I nevertheless will list here some of the skills and aptitudes of chickens that have been scientifically demonstrated in laboratory and fieldwork studies on these birds. To begin, chicks form memories when still within their eggs. When they hatch into the world, their sensory systems are already so well developed they are able to recognize and learn from the mother hen within hours. As they grow, chickens realize the importance of the flock as a protective and stable social group; they learn where they fit in among flock members, and they come to memorize and recognize the faces of over one hundred other chickens. Able to differentiate among various visual, acoustic, olfactory, and tactile signals, they also demonstrate representational thinking; for example, chicken alarm calls convey semantic information to other birds, not only chickens, about the presence of predators and whether such foes are approaching by land or by air. Roosters tell hens if food they have discovered is novel or particularly enjoyable (and may even pretend *to have found delicious food in order to entice hens toward them for amorous reasons). Chickens have very acute hearing and precise color vision. They view the world panoramically at the same time as they are able to focus on an object directly in front of them, a visual experience we cannot really begin to comprehend. They grasp abstract concepts: young chicks recognize objects which have been partially obscured, a feat beyond the ability of small children. Chickens also anticipate the future, and thereby have the capacity to experience anxiety and frustration. As a trick at conferences, Australian avian expert Chris Evans lists many of the previously mentioned attributes and asks his audience to guess which species he is referring to. They invariably assume he is talking about monkeys.*

In much the same way we have been biased toward our own form of human intelligence, we have also formed erroneous assumptions about the emotional capacities of other species. More often than not it is inferred that only humans—and perhaps certain creatures we grow fond of and close to—are blessed with the capacity to really feel. Animals whose natural dispositions do not tend toward human company or affection are categorized differently from the "pets" who share our lives; thus farmed animals (cows, chickens, and sheep), certain wild animals (fish, rodents, mustelids), and wild birds are more easily dismissed as emotionless, or at least incapable of the level of feelings our companion animals show us. Because few people in industrialized and heavily urbanized countries get to know an individual chicken, let alone experience living with a flock of chickens, these birds are more easily dismissed as unfeeling and may even be misconceived as suffering less. One goal of mine is to educate others about the very real emotional lives of chickens. For example, wild and free-ranging chickens establish structured social relations and "friendships." They display tenderness, deception, altruism, and grief, and even suffer from post-traumatic stress disorder. The devotion of a mother hen toward her chicks is evidence enough of affection amongst chickens. Hens also form close relationships with particular members within their flock, often foraging alongside each other, showing each other tidbits, preparing dust-baths together, and nesting companionably when laying eggs. Chickens also form friendships with other species, including horses, goats, dogs, cats, ducks, and of course humans. In his sixteenth-century treatise on chickens, ornithologist Ulisse Aldrovandi described the devoted friendship he developed with a hen, to the extent that she would only go to sleep at night when in his company and surrounded by his books.

One story of chicken friendship concerns two hens that lived with my partner and me on our property in the port town of Lyttelton. One of these hens, Buffy, became ill at the end of her first year with us, and slowly deteriorated until she wanted only to sit under a cabbage tree in the garden, now and again sipping water as she faded away. Her friend Mecki was especially attentive during this time, choosing to sit with Buffy in her ground-nest despite the enticing activities of the other active and noisy hens. She would gently peck the ill hen around her face and on her back while uttering soft sounds, which Buffy responded to in kind. When Buffy died, Mecki retreated to the henhouse for some time, refusing to eat and disengaging from the routine activities of the flock.

Thus chickens grieve. And they also feel immense joy. I have had the pleasure of assisting factory-bred and raised chickens to adapt to the lives that were their birthright. This process can take several months, while chickens re-grow feathers and recover from viruses and other diseases resulting from overcrowding, learn to walk on weakened and deformed legs, and socialize appropriately with other free-ranging chickens. Former battery hens also need to adapt to natural sunlight, learn how to move on grass for the first time, forage for food, and take dust-baths. Their story is one of determination against the odds: These hens show immense courage and will to live despite suffering chronic pain and infirmity. It is also a story about discovering pleasure: Each hen experiences profound joy as she encounters a new world as part of a free-ranging flock.

Bearing witness to the emotional ups and downs of chickens has taught me how to step outside my own busy human life and enjoy moments with other beings whose experiences of the world are refreshingly different and uplifting. I spend time among the flock of chickens I live with when desiring company, solace, distraction, or fun. The antics of chickenkind also de-stress me after a human-focused day at the office. And it is to chicken advocacy I turn when seeking meaning in my work.

II

USING ANIMALS
Human-Animal Economies

4

Animals "in the Wild" and in Human Societies

A gaunt Wolf was almost dead with hunger when he happened
to meet a house dog who was passing by.

"Ah, Cousin," said the Dog.

"I knew how it would be; your irregular life will soon be the
ruin of you. Why do you not work steadily as I do, and get
your food regularly given to you?"

"I would have no objection," said the Wolf, "if I could only get
a place."

"I will easily arrange that for you," said the Dog; "come with me
to my master and you shall share my work."

So the Wolf and the Dog went toward the town together. On
the way there the Wolf noticed that the hair on a certain part
of the Dog's neck was very much worn away, so he asked
him how that had come about.

"Oh, it is nothing," said the Dog. "That is only the place where
the collar is put on at night to keep me chained up; it chafes
a bit, but one soon gets used to it."

"Is that all?" said the Wolf. "Then good-bye to you, Master
Dog."

—"THE DOG AND THE WOLF," *THE FABLES OF AESOP* (1894)

THE MORAL OF THIS TALE, told for thousands of years to children,
is that it is "better [to] starve free than [to] be a fat slave." But it is also an
elegant way of summing up the differences between wild and domesticated
animals. We discussed in chapter 3 the various systems of classification that
humans have used to categorize animals and how those categories then serve

as the justification for how we use and treat them. One of the most important categorical distinctions found in the West is that between "wild" and "domesticated" animals, which itself mirrors the nature/culture distinction so prominent in many modern societies. We have also talked about how during the Middle Ages, medieval Europeans were anxious about nature and saw it as something to be feared and controlled. It was only once Europeans stopped feeling at the mercy of nature, that nature—and wild animals—began to be viewed in a more positive light.

Today, wild animals are viewed by many people in industrial societies in the abstract—as a category worth saving—although we rarely give *individual* wild animals much thought. Wild animals are seen as a refuge from modern life; we visit zoos and marine mammal parks to see them and watch them on nature documentaries with pleasure. But this modern relationship with, and view of, wild animals is relatively recent. In this chapter, we will discuss the history of humans' relationships with wild animals.

Animals and Humans in the Paleolithic Era

The **Paleolithic era**, also known as the Old Stone Age, is the cultural age spanning the Pleistocene epoch, and dates from about 2.5 million years ago to about 15,000 years ago. It is named for the types of tools—stone—that were primarily used by our ancestors and other **hominin** species during this time. The Pleistocene was the era of the **glacial**, or ice age. During this period, a number of glacials extended over much of the earth's land mass causing not just massive cooling, but radical changes in plant and animal life. The glacial periods were separated by **interglacials** during which the ice caps would retreat, climates would warm, and plants and animals would once again adapt to those changes. Generally, each new glacial would cause animals to move south, away from the ice, leaving northern areas depopulated. Forests receded and grasslands and deserts developed. Interglacials, the periods in between the ice ages, allowed for animals to move northward again. Sometimes the climactic changes were so extreme that animal species became extinct. The late Pleistocene saw the extinction of a number of **megafauna**, such as mammoths, saber-toothed tigers, cave bears, and mastodons—perhaps from a combination of climate change and overhunting. Mammoths, for example, may have seen their population reduce in size due to climate change; then the remaining animals were hunted to extinction by humans. Those animals were replaced by smaller animals, cold-blooded animals, and migratory birds. These Pleistocene extinctions would have been the first example of major human effects on other animals.

What we know about animals in prehistoric times comes to us from the archaeological record. **Zoöarchaeology**, or the study of animal remains at archaeological sites, gives us a great deal of information about what kinds of animals ancient humans associated with, and what the nature of that association was. For example, evidence dating to about 14,000 years ago shows that in some areas dogs were carefully buried after death, indicating a definite relationship between dogs and humans. On the other hand, the most common type of animal remains found in human habitation sites were from those animals that were eaten by humans or by pre-humans, illustrating that predation was the first important form of human-animal interaction. Another major way to understand the relationship between animals and humans in prehistoric societies comes from the artifacts left behind by prehistoric peoples.

Based on that information, we know that our earliest pre-human ancestors probably had eating habitats very similar to those of modern chimpanzees— a diet that was primarily vegetarian, but supplemented by the occasional scavenged or hunted small animals. The stone tools left by creatures such as *Homo habilis*, who lived about 2.5 million years ago, were probably used for butchering animals but not for hunting. As our ancestors evolved, hunting became a more crucial subsistence strategy. *Homo erectus* lived from about 1.8 million years ago to approximately 300,000 years ago. With their larger bodies, larger brains, and more sophisticated tool use, they were most likely big game hunters. In fact, many anthropologists see cooperative hunting as the pivotal achievement of our ancestors that led to the development of the larger brain and more sophisticated culture. (Other scholars see language, cooperative parenting, or other social behavior as more critical in the development of our species.) As our ancestors evolved, their hunting tools and techniques continued to improve, leading to group hunting and, eventually, to the extinction of most of the megafauna on the planet. Toward the end of the Paleolithic, humans developed new tools such as fishhooks and harpoons, indicating the exploitation of new kinds of animal life, such as fish. And by the Upper Paleolithic, about 45,000 years ago, humans may have begun tracking the migration patterns of deer and other animals, allowing them to hunt more effectively. The appearance of arrowheads, spears, and nets in the fossil record indicated that people began devising new ways to hunt animals instead of chasing them down and clubbing them, which is presumably how humans initially hunted.

How much of the diet of ancient humans was made up of meat is not really known by scientists, and certainly would have varied by geographic region. It is probably the case that many early *Homo* species had diets made up of a significant amount of meat. However, by the Mesolithic (about

15,000 years ago), the end of the last glacial combined with overhunting caused the extinction of so many animals that humans began exploiting very different resources for food. At this time, our ancestors began relying more on small animals, fish, birds, and vegetable matter. Human populations had also spread into most regions of the earth, populating Africa, Europe, Asia, much of the Arctic, the islands of Oceania, and all of the Americas. People who lived in the far north exploited animals more heavily—either as hunters or as pastoralists—and southern cultures probably focused more on plant foods.

Modern *Homo sapiens*, who emerged at least as far back as 100,000 years ago, were not only hunting animals but using them in ritual life, as evidence from burials and cave paintings indicates. Clearly, animals were an important part of our ancestors' lives because so much of the art and ritual life of early humans was dedicated to their representation. By the Upper Paleolithic, evidence abounds that humans not only created an abundance of art—including jewelry, rock art, and body painting—but that much of it either involved animal parts (such as feathers, bone or skin) or included representations of animals. In Upper Paleolithic cave paintings, for example, animals are featured prominently, usually in hunting scenes. These scenes have often been interpreted as attempts by the painters to ensure a successful hunt via magic. Other interpretations focus on the half-human/half-animal creatures represented in the paintings, and suggest that these may be representations of shamans who may have dressed up as animals as part of their ritual practices. Animals not only may have been channeled by shamans, they also probably played a number of roles in ancient peoples' religious practices. They may have served as sacrificial gifts for the gods and were perhaps the focus of worship themselves. For instance, it has been suggested by some scholars that **Neanderthals**—a species of hominin living contemporaneously with modern humans in Europe for a small period of time—may have worshipped bears as part of a bear cult. Upper Paleolithic people living in South Africa may have worshipped pythons. Totemism, in which human groups take on an animal as a spiritual ancestor, may also have emerged during prehistory. In all of these cases, even though the particular animal may be worshipped, most likely it was also killed as part of the ritual surrounding it.

Subsistence Hunting and the Human-Animal Relationship

Modern hunter-gatherer, or foraging, societies depend on animals for much of their protein, but anthropologists have shown that most of the calories consumed by hunter-gatherers are derived from plant foods rather than animal foods. There are exceptions to this however: Native American and Arctic hunters tend to have more meat in their diet than do African hunter-gatherers, who eat meat less often. But even when animal meat does not make up the majority of the diet, it seems to always be the case that when meat is procured, it is celebrated. A successful hunt is often the excuse for a community celebration at which the meat will be shared. Meat is highly desired and hunting is a valued skill in hunter-gatherer societies.

What do hunter-gatherers think about the animals that they depend on, and with whom they coexist? Animals are not incorporated into hunter-gatherer societies in the way that they are in societies where animals are domesticated. They are still deeply important to those societies, but in very different ways. In hunter-gatherer societies, animals are hunted and eaten but not raised and controlled. They are predators to be feared but are sometimes kept as tame companions. Animals figure in religious systems and myths as well as in art and folklore. As we discussed in chapter 2, in many hunter-gatherer societies humans and animals are both part of nature. Humans do not live outside of, or as superior to, nature, as we do in the West. In foraging cultures, animals and humans coexist; in ritual, art, and mythology, they often transform themselves into each other.

As anthropologist Tim Ingold points out (1994), the relationship between animals and humans among traditional hunter-gatherers is often one of mutual trust in which the environment and its resources are shared by animals and people; animals that are hunted by humans are seen as equals. Hunter-gatherers tend to see animals as rational, intelligent creatures just like themselves, with the same spiritual importance as humans. However, animals are not thought of as *the same* as humans; they are brothers—in a sense—but still can be eaten. Unlike Westerners, who see companion animals as surrogate humans that cannot be eaten, traditional hunter-gatherers do not go quite that far.

Ingold writes about what he calls "the giving environment," where people share with one another and with the environment. Rather than attempting to control nature, hunters seek to maintain proper relations with animals. If an animal is not treated with respect, that animal's spirit can seek vengeance against those who mistreated it. Some cultures say that when the hunter is respectful, the animal will offer himself to the hunter. No matter how skilled the hunter, if the animal does not want to submit, he will not be caught. Other cultures feel that if an animal appears to a hunter that means that the spirits have sent it; therefore, the hunter *must* kill it.

Although we can certainly debate whether or not animals willingly give themselves to hunters, this viewpoint demonstrates the very different attitude taken by traditional hunter-gatherers to animals from that prevalent in society today where animals are seen as property, with no rights of their own. For hunters, animals were generally seen as equals to be treated with respect, even while being killed; in fact, many indigenous cultures see human hunters and their animal prey as being involved in an alliance. One concrete result of this attitude is that hunter-gatherers, for the most part, will attempt to avoid causing an animal to suffer when killing it. Once it is dead, all parts of the animal are used and shared; nothing is wasted. We may want to compare this attitude to Western methods of industrial animal production (discussed in chapter 7) that involve huge amounts of cruelty and suffering, and to food consumption habits in the United States in which huge amounts of meat (and other foods) are wasted.

With the rise of livestock domestication starting about 10,000 years ago, the human relationship with wild animals was transformed. Domestication changed this relationship into one of dominance and control as humans took on the role of master. Animals—no longer wild—became classified as property, and were considered items to be owned and exchanged. Today, domesticated animals can be bred, controlled, abused, or killed as the owner desires. Even the animal gods found in tribal cultures around the world have changed from being worshipped as part human/part animal to increasingly anthropomorphic deities with no animal qualities whatsoever.

With the spread of animal domestication around the world, this new relationship with animals—where animals are owned and controlled—has spread from domesticated animals to wild animals.

From Subsistence to Sport

Although hunting (or scavenging) animals has always been a critical part of the economy of humans throughout history, with the domestication of plants and animals in the Neolithic era, the exploitation of wild animals became far less important in many cultures. **Pastoralists**, for example, who herd domesticated animals, generally abhor hunting and see wild animals as a threat to their livestock. Some pastoralists do hunt but most only hunt predators that pose a threat to their animals. On the other hand, agricultural societies that successfully domesticated large animals stopped relying on wild animals for protein because of the easy presence of domesticated animals. Hunting, then, moved from an important part of the economy in hunter-gatherer societies to a sport for elites.

For example, in England during the medieval and Renaissance periods, hunting was an aristocratic tradition and commoners, for the most part, did not hunt. This is because the concept of "public lands" as we understand it today did not exist back then. Only wealthy landowners and their friends could hunt on private property, and virtually all property was private—owned by the crown, the lords, or the Church. Because hunting was primarily reserved for the wealthy, the public considered wild animals to be the property of the rich. Hunting, then, was a sign of prosperity and status; it demonstrated mastery over nature and the lower classes as well.

Unlike in England, hunting in the United States did not begin as an aristocratic tradition but as a democratic one. At least in the early colonial days, most land was considered "public." Early Americans did not consider Native Americans as having ownership rights to land. But by the turn of the twentieth century, most Americans—even those living in the West—no longer needed to hunt for their food. The food that Americans ate came increasingly from farms rather than from wild animals. With industrialism, more and more Americans moved from rural areas into cities and lost any contact with wild animals. Hunting in the United States turned into an elite sport, as it was in England, attracting a much smaller portion of the population and drawing the wealthiest participants. Rich Americans such as J. P. Morgan and Cornelius Vanderbilt began to develop private game reserves and duck-hunting clubs for themselves and their friends. This trend led working-class hunters to worry that the United States was going to end up like Europe, where hunting was restricted to the wealthy who could afford not only to hunt on private reserves but also to hire hunting guides to do all of the manual labor for them.

Today, sport hunting differs greatly from subsistence hunting, and even from the traditional way that hunting was practiced in the United States during early colonial times and through westward expansion. Sport hunting involves the killing of animals for recreation. Sometimes hunters consume the animal meat, but neither food consumption nor profit is the purpose of sport hunting. Instead, recreation and the acquisition of a **trophy** are the main goals.

Colonial Expansion and Animals

Wild and domestic animals both played a major role in European colonialism. Colonialism was motivated primarily by economics—the need to locate new resources and forms of labor in Asia, Africa, and the Americas. The colonial superpowers (primarily England, France, and Spain) were also seeking new markets for their goods, pursuing religious and strategic interests, and taking on what the poet Rudyard Kipling called the "white man's burden" of civilizing native peoples around the world (Kipling 2007). Colonialism would have incalculable repercussions felt in cultures around the world, and led to the rise of the modern global economy. But it also had important results for animals.

The Spanish, who were the first to arrive and settle the New World, brought with them domesticated horses, sheep, goats, cattle, and pigs. Horses played a major role in the conquistadors' domination of civilizations such as those of the Aztec and the Inca, and in the spread of Spanish culture from Mexico southward into South America and northward into North America. Spanish livestock were brought into each new settlement giving the Spanish an economic foothold in the region. The conquistadors used specially trained dogs such as mastiffs and greyhounds in wars against indigenous people.

Other European settlers brought animals with them to Australia and North America so they would have food sources after settling. One unintended result of the import of nonnative animals to the colonies was that in many cases, native animals and plants were displaced by the new species. These new animals also played a role in transforming the lives of many native populations. Plains Indians adopted the horse and began to use it in their own expansions while the Navajo began to herd sheep; sheep are still a central component of Navajo culture today.

Starting in the sixteenth century, as European superpowers colonized much of Africa and Asia, English aristocrats extended hunting to the colonial lands. English colonists and explorers hunted for sport and profit, killing large, dangerous animals whose bodies were then stuffed as trophies. European hunters also targeted animals whose hides or tusks could bring profit. Elephants fit both bills and were a major focal point of English hunting during the nineteenth and early twentieth centuries. Big game hunting and colonial warfare were intimately linked in Africa: Europeans wanted to control big game and native Africans, both of whom hunters could kill with little justification.

The presence of European hunters in Africa transformed the native Africans' views and uses of wildlife, which until that time had been important for subsistence as well as trade. After the Europeans began their trade in ivory, trophies, and other parts of animals, Africans also began to exploit these resources more intensively. In Africa as well as Asia, Europeans engaged native hunters to collect exotic live animals for display in Europe and later in the United States. This was the beginning of modern zoos and circuses.

Following the premise of **Manifest Destiny**—the God-given right and duty to expand westward across the American continent—settlers and miners moved westward from the original American colonies. They often encountered Native Americans along the way. Official government policy during the eighteenth and nineteenth centuries was to encourage Native Americans to sell their land to Euro-Americans and become civilized. Settlers expected Native Americans to eventually stop hunting animals and become farmers, freeing up vast amounts of land for whites to use for farming, mining, ranching, and other forms of development. Both the presence of Native Americans and the animals they traditionally hunted were an economic hindrance to these activities. Starting in the 1870s, the government developed a new policy to eliminate the bison from the plains, which would discourage the Native Americans from using that land.

The full-scale destruction of the American bison—thirty million were killed in just fifty years—served a number of purposes. Many American colonists saw both Native Americans and bison as savage beasts. By forcing Native Americans to stop hunting, it allowed for an easier process of displacement and confinement to reservations. The newly developed railroads benefited because the presence of bison on or around the tracks could impede the traffic of trains. White trappers and hunters flooded into open areas as native peoples were removed (or killed), and the newcomers benefited from killing thousands upon thousands of bison for their hides. Because most Americans were supportive of the goal to advance westward and to move Native Americans onto reservations, the elimination of this major food source was thought to be good policy. Even the soldiers got in on the act, killing bison as the U.S. Army went about its other business in the West. Not coincidentally, the eradication of the bison freed up the land for cattle grazing, helping beef become one of the major food sources for whites.

Controversies Surrounding Subsistence Hunting

Although very few hunter-gatherer cultures exist in the modern era, many traditional cultures still do use hunting as part of their subsistence strategies. In the United States and Canada, laws that protect certain animals from being overhunted often have exemptions for Native Americans, who are still allowed to practice their traditional hunting methods today.

Among the most controversial subsistence hunting practices by native peoples is the killing of marine mammals such as whales, dolphins, and seals. Native peoples who live in coastal areas have practiced whaling throughout history. Native Alaskans such as the Makah traditionally hunted whales by using a harpoon attached to an inflated sealskin; the whale would swim around until he either tired enough so that they could lance him to death, or he bled to death from internal injuries. Europeans, Euro-Americans, and the Japanese also used whales for their meat, blubber, oil, and bone. These whalers eventually developed modern methods of killing, which involved using large ships to track the whales as well as using the ship itself as the "drogue," or large item attached to the harpoon; later, they regularly used guns and explosives. By the twentieth century, whale population numbers had dropped due to overhunting by the large commercial whaling industries. This decline led to today's commercial bans on whale hunting.

The debate over indigenous hunting traditions has a number of complex factors. Animal protection advocates decry the killing—often quite brutal—of animals that are intelligent, social creatures such as whales and dolphins; they claim that the argument that these practices are traditional is no longer valid, as many native hunters now use modern weapons and sell the proceeds of the hunts for profit. Native communities argue that hunting is a practice that dates back hundreds or thousands of years, that it is protected by a number of international and national treaties, and that furthermore, bans on native hunting are a form of **neocolonialism** that threatens traditional culture. They also argue that animal advocates only want to protect the charismatic animals such as whales and baby seals. To complicate matters further, the public, and scholars too, often romanticize indigenous people's practices, approving of them if they are done in a traditional fashion but criticizing them if they are done for profit.

Modern Relationships with Wildlife: Hunting and Conservation

Today, most Americans have very little interaction with wild animals. Campers and other outdoor enthusiasts sometimes see wildlife when outdoors and certainly because of the expansion of cities and suburbs into once-wild lands, many Americans see wildlife in their yards, their neighborhoods, and on their roads. A decreasing number of Americans still hunt wild animals.

Today, only about 6 percent of the population in the United States (or 12.5 million Americans) hunts animals (U.S. Fish and Wildlife Service 2007). Yet hunting, even though practiced by a small minority of Americans, continues to carry an enormous amount of symbolic and emotional weight.

American hunters see themselves as inheritors of a great tradition—**the hunting heritage**. Because it is largely fathers or other male relatives who introduce most young hunters to the practice, participants see the activity as a sacred tradition embodying notions of family, history, and a love of the outdoors. Many hunting defenders call upon the history of the conservation movement and its connection to hunting as one of its most important justifications. Because turn-of-the-twentieth-century American hunters and conservationists such as Theodore Roosevelt and Aldo Leopold were instrumental in establishing game management policies and protections for wildlife and wildlife habitats, the history of hunting is tied to the rise of the American conservation movement.

Another strongly resonant idea about hunting has to do with the idea of self-reliance. Many hunters feel that they are reliving the pioneer lifestyle in which American settlers depended on their wits, skills, and perseverance to conquer nature and the American frontier. Of course, nearly all American hunters do not subsist on the animals that they kill. The reliance of hunters on high-tech equipment, guides to lead them to animals, and even **canned hunting** operations (where tame animals are brought directly to them to kill) largely negates this hardy pioneer image.

With respect to conservation, hunters definitely participate in wildlife conservation today. Hunting permit fees and taxes on hunting equipment fund wildlife management programs in every state in the country. Hunters also support a wide array of conservation projects aimed at ensuring that there exists not only wildlife to hunt but also habitats for game animals. Hunters also see themselves as conservationists in the sense that, through game management policies, hunting keeps populations in check that would otherwise spiral out of control and prevents starvation for some species. But opponents charge that hunters target healthy adult male animals rather than the sick or old that "nature" would otherwise take or the females whose killings would result in a drop in population. These healthy animals would not typically starve to death, say hunting opponents, and survivors likely will respond to the reduction in numbers by bearing more young anyway.

Because hunting is a practice typically passed on from father to son, many hunting advocates see it as an important, sacred family tradition, made more sacred because it takes place in the great outdoors. Because hunters traditionally have been from rural areas, many (following the tradition of Henry David Thoreau) see themselves as more connected to "nature" than urban dwellers—most of whom do not hunt and who show far less support of hunting than rural Americans.

But hunters are not all the same, even in the United States. Sport hunters, for example, do not hunt for food, and focus on big-game animals such as deer, bears, and moose; small animals such as rabbits, squirrels, foxes, and mink; predators such as mountain lions and coyotes; and birds such as pheasants, ducks, and geese. The term "trophy" refers to the antlers, horns, tusks, heads, or bodies of animals the hunters kill, and many American hunters enthusiastically seek them. In fact, the United States is now the chief market for "trophy animals." There are tens of thousands of animals from all around the world that hunters kill and bring back as trophies to the United States.

Because of this emphasis on bagging a trophy at all costs, trophy hunters have evolved into a somewhat different species than the traditional American hunter. Many trophy hunters now use baiting, dogs, game farms, canned hunting ranches, guides, and technology in order to ensure a trophy at the end of the day. Over the past two hundred years, Africa has been stripped of much of its population of large game animals. Those remaining have been permanently transformed as a result of hunting for trophies such as elephant tusks. Yet in Africa, as in America, many of the first conservation laws were passed by European hunters in order to protect and conserve game for the colonists.

BOX 4.1

CANNED HUNTS: WHAT DO YOU THINK?

Canned hunts are hunting operations that take place on private game farms or ranches that fence animals into a limited area so that hunters can more easily kill them. Canned hunting—also called high-fence hunting—is illegal or regulated in twenty states, and challenges the notion of "fair chase" that is central to many hunters in this country.

The animals used in canned hunting operations generally come from two types of backgrounds: Either they are raised on domestic game farms in order to be hunted, or they are purchased from animal dealers. Dealers buy the animals from a variety of sources. They include former exotic pets, animals legally or illegally imported from Africa or Asia, former circus animals, and "surplus" zoo animals. Not only do canned hunts completely eliminate any notion of fair chase, they also smack of the aristocratic style of hunting so hated by traditional American hunters. Indeed, wealthy hunters are allowed to kill whatever animals they like, at any cost, on private land where normal Americans cannot go, and with very little effort.

Human-Wildlife Conflicts

For the most part, wild animals are "managed" through the federal government by the U.S. Fish and Wildlife Service and by individual state agencies. These agencies function under the assumption that wild animals are a resource the public owns and the states manage. The states also handle some wild animals that are considered pests or problems. In either case, wild animals ostensibly exist to benefit the public, and the wide range of state agencies tasked with managing wildlife carry this basic philosophy with them. The mission of the U.S. Fish and Wildlife Service is to "conserve, protect, and enhance fish, wildlife, and plants and their habitats for the continuing benefit of the American people." The agency was established after the passage of the Fish and Wildlife Act of 1956, a comprehensive national fish, shellfish, and wildlife resources policy that emphasized oversight on the commercial fishing industry. In recent years, as many as eight million people per year have engaged in consumptive wildlife use, or hunting, trapping, and fishing in the national wildlife refuge system. This may be due in part to the passage of the National Wildlife Refuge System Improvement Act of 1997, which includes hunting and fishing as two of the six priority uses for the national wildlife refuge system.

State wildlife agencies—such as fish and game departments, conservation departments, and natural resources departments—regulate hunting and manage wildlife. The people who work for these agencies act as law enforcement officers as well as conservationists by ensuring the continuation of wildlife as a renewable resource. Additionally, some agencies exist to regulate conflict between humans and animals, ensuring that developers, ranchers, and others with economic interests do not have their interests impeded by the presence of wild animals. Although agencies use hunting as the solution to human-animal conflicts such as the presence of deer in suburban developments, hungry bears approaching campers, prairie dogs digging holes in city parks, or coyotes eating pet cats and dogs, the agenda of these agencies is very different. These animals are viewed as pests that must be eliminated because of human economic interests. Environmental sociologist Theresa Goedeke, for example, writes about how the social construction of wildlife by various stakeholders shapes policy decisions. In her article on the social construction of otters (2005), she shows how environmentalists see otters as playful angels but fishermen consider them to be "hungry little devils" that compete with them for fish. Not surprisingly, the more charismatic the animal, the more the public will tolerate "destructive" behavior; the less charismatic, the more likely that an animal will be defined by all parties as a pest and subject to stricter levels of control. Similarly, whether a mourning dove is considered to be a peaceful songbird or a drab

Figure 4.1. The prairie dog is seen as vermin by many people but is an important keystone species. (Photograph courtesy of Yvonne Boudreaux, Prairie Dog Pals.)

brown pigeon features heavily in whether or not it can be legally hunted. (People, too, can be socially constructed in wildlife law. The difference between "hunters" and "poachers" is artificial, and relates to whether a person is hunting the right animals in the right place using the right equipment at the right time—according to existing law.)

Congress passed the Animal Damage Control Act in 1931 to authorize the eradication of wolves, lions, coyotes, bobcats, prairie dogs, and other animals that pose a supposed threat to agriculture. The act created the animal damage control (ADC) program, which has at different times been implemented by a number of different federal agencies—most recently the U.S. Department of Agriculture's Animal and Plant Health Inspection Service (APHIS). The predecessor to ADC was the Bureau of Biological Survey. One of this agency's goals was the destruction of pests and predator animals such as wolves. Within two years, the professional hunters employed by the agency eliminated 1,800 wolves and 23,000 coyotes from national forests. Between 1915 and 1942, more than 24,000 wolves were killed (Isenberg 2002); in 1926, the bureau was responsible for the death of the last wolf in Yellowstone National Park. Even in these early days, the goals of the bureau were to protect livestock and to increase the population of animals for hunting. From a modern perspective, those results were predictable: The deer population did balloon after the loss of the wolves, but tens of thousands of deer died the following year because there was not enough forage to support them. After the federal government effectively wiped out the wolf population, the gray wolf was the first animal to be listed under the Endangered Species Act of 1973.

Originally, the Animal Damage Control Act focused on animal damage to public lands, but today it impacts public and private land. APHIS now administers ADC (now called Wildlife Services) often in conjunction with state fish and game departments. Wildlife Services kills predators in Arizona, California, Colorado, Idaho, Minnesota, Montana, Nebraska, Nevada, New Mexico, North Dakota, Oklahoma, Oregon, South Dakota, Texas, Utah, Washington, and Wyoming. It has been responsible for the near extermination of all nineteen of the largest mammals in the West, as well as a large number of small animals such as prairie dogs.

BOX 4.2

IMPORTANT LEGISLATION: ANIMAL DAMAGE CONTROL ACT (1931)

"This Act gives the Secretary of Agriculture broad authority to investigate and control certain predatory or wild animals and nuisance mammal and bird species. The Secretary is authorized to conduct investigations, experiments, and tests to determine the best methods of eradication, suppression, or bringing under control mountain lions, wolves, coyotes, bobcats, prairie dogs, gophers, ground squirrels, jack rabbits, brown tree snakes, and other animals injurious to agriculture, horticulture, forestry, animal husbandry, wild game animals, fur-bearing animals, and birds. Another purpose of these investigations is to protect stock and other domestic animals through the suppression of rabies and tularemia in predatory or other wild animals. The Secretary is also directed to conduct campaigns for the destruction or control of these animals."

Even though the goal of hunting for ADC programs is very different than the goal of hunting for sport, ADC hunting benefits hunters and hunting advocates because it provides another way for hunting to be perceived as a valuable public service. It also benefits a small group of people—wealthy ranchers, politicians, and appointed state and federal officials—who cash in on federal funds that help ranchers stay in business.

Animal damage control advocates often characterize *game animals* as *pest species*. Deer, for instance, do not kill farm animals but are blamed for destroying gardens, bringing disease, causing car accidents, and wreaking other forms of havoc in suburbs. So sport hunters are allowed to kill deer with public support—after all, no one wants to be involved in a collision with a deer. Unfortunately for deer, hunting does not necessarily control their populations. They can rebound soon after hunting season due to lessened competition for resources. And, of course, the animal damage control measures that wiped out much of their natural predators also play a role in their large numbers. There are numerous methods to prevent the damage that deer can cause, such as more responsible driving, speed limits, warning signs, roadside reflectors, as well as the use of fencing along roadways. Yet the fact remains that as long as developers continue to build on these areas, humans and wildlife will come into contact. Sadly, animal damage control programs have just one way of solving these problems—hunting.

BOX 4.3

WOLF ERADICATION EFFORTS

Looking at the history of wolf eradication in this country provides an interesting window into the logic of animal damage control. The government (or those in charge) has targeted various wolf species since the mid-seventeenth century. By 1800, the wolf had been largely eliminated from the East Coast because of the threat to the emerging cattle and sheep industries (which primarily grazed animals then, as now, on public lands). In the nineteenth and twentieth centuries, efforts to eradicate the wolf moved westward and picked up speed, with ranchers clamoring to have wolves killed in order to protect their cattle from predation. As European Americans wiped out the bison, they also slaughtered wolves. Sportsmen joined with ranchers in supporting this policy because they saw wolves as a threat to game animals.

Geographer Jody Emel (1995) points out that, beyond the economic reasons for slaughtering wolves, the ferocity with which hunters killed them had to do with the way they were characterized. People considered wolves wasteful, vicious, and unsportsmanlike in the way they killed farm animals. The public saw wolves as mean and treacherous animals that killed for pleasure, and would kill a human child if they could. Some of these perceptions date back to medieval Europe, when wolves were seen as not just dangerous but also evil. This was, by the way, very different from how many Native Americans saw wolves: They admired wolves for their endurance, strong family bonds, and bravery. A U.S. Fish and Wildlife program set up to reintroduce the Mexican gray wolf into Arizona and New Mexico after the twentieth century's eradication continues today. But, ironically, the same government agency is killing many of the wolves it has reintroduced in response to rancher complaints.

Wild animals are also threatened by the introduction of new diseases, the exotic pet trade, the loss of habitat due to commercial development, housing, oil, mining, and more. As an example, some western states still have small populations of wild horses. These horses are descendents of the highly trained Spanish mounts that were abandoned during the seventeenth and eighteenth centuries. These horses quickly reverted to their untrained state and became the "wild" mustangs (technically they are feral) that are still seen in states such as Colorado and New Mexico today. Sadly, thanks to the ever-encroaching demands of civilization, these beautiful horses are also increasingly rounded up and sold at auction regardless of

Figure 4.2. The BLM refuses to classify the wild horses of Placitas, New Mexico, as "wild," so they are currently unprotected by law. (Photograph courtesy of Robin Montgomery.)

their supposed protection under the Wild Free-Roaming Horses and Burros Act of 1971.

The plight of these western mustangs is a good example of the changes experienced by wild animals in this country. Even though wild horses are a potent symbol of freedom and the American West, they represent a breed that is unconquered and that must be brought under human control even if that means eliminating them entirely. Additionally, many people have the idea that animals are only good if they are used for something; without being trained or used for some personal or commercial purpose, these animals are viewed as not good for anything.

And so it is with the rest of this country's wildlife. Coyotes, wolves, bison, and prairie dogs are constructed by economic industries and state agencies as pests, not good for anything, and considered harmful to the economic interests of some people and industries. Thus, they must be eliminated. On the other hand, animals defined by hunters and by state game agencies as "game animals" are good for hunting, which is probably the one reason why many still exist today.

Suggested Additional Readings

Animal Studies Group. 2006. *Killing Animals.* Champaign: University of Illinois Press.

Brightman, Robert. 1993. *Grateful Prey: Rock Cree Human-Animal Relationships.* Berkeley: University of California Press.

Herda-Rapp, Anne and Theresa Goedeke. 2005. *Mad About Wildlife: Looking at Social Conflict over Wildlife.* Danvers, MA: Brill Academic Publishers.

Ingold Tim. 1980. *Hunters, Pastoralists and Ranchers.* Cambridge: Cambridge University Press.

Mighetto, Lisa. 1991. *Wild Animals and American Environmental Ethics.* Tucson: University of Arizona Press.

Nelson, Richard K. 1986. *Make Prayers to the Raven: A Koyukon View of the Northern Forest.* Chicago: University of Chicago Press.

Philo, Chris and Chris Wilbert, eds. 2000. *Animal Spaces, Beastly Places: New Geographies of Human-Animal Relations.* London: Routledge.

Proctor, Nicholas. 2002. *Bathed in Blood: Hunting and Mastery in the Old South.* Charlottesville, VA: University Press of Virginia.

Tober, James. 1981. *Who Owns the Wildlife: The Political Economy of Conservation in Nineteenth- Century America.* Westport, CT: Greenwood.

Wolch, Jennifer and Jody Emel, eds. 1998. *Animal Geographies: Place, Politics and Identity in the Nature–Culture Borderlands.* New York: Verso.

Suggested Films

Cost of Freedom. DVD. Directed by Vanessa Schulz. Bend, OR: 21st Paradigm, 2003.

Cull of the Wild: The Truth Behind Trapping. DVD. Produced by Camilla Fox. Sacramento, CA: Animal Protection Institute, 2002.

Wildlife for Sale: Dead or Alive. HGS. Directed by Italo Costa. Oley, PA: Bullfrog Films, 1998.

The Colonial Animal

WALTER PUTNAM
University of New Mexico

My "coming to animals" involved an unexpected eureka moment: While rocking my newborn daughter to sleep, I began wondering about the menagerie of stuffed animals that had taken up residence in her bedroom. Where did they come from? What did they mean? What was their relationship to their real counterparts? How did we manage to declaw, defang, desex, shrink, and stuff these otherwise ferocious animals in order to make them suitable companions for our cherished children? This commodification of animals set me wondering, a move which, as Aristotle commented, is the first step to philosophy. The answer, at least provisionally, turned out to be much more complicated than the rather simplistic urge to dismiss them as "cute and cuddly." Drawing on Marx, Freud, Baudrillard, and D.W. Winnicott, I argue that stuffed animals are a unique specimen of industrialized, adulterated wildlife originating from a colonial space and that they have infiltrated our psychological, physical, and cultural spheres in strange and alarming ways. These popular culture icons have been strapped to the fronts of New York City garbage trucks and used to smuggle drugs and bombs, and a recent study revealed that a quarter of traveling businessmen carry along a stuffed animal in their bags. Artists and collectors, adults and young children alike have reconfigured these natural species in order to make these transitional objects among the most common animal forms in global consumer culture.

The French anthropologist Claude Lévi-Strauss famously declared that animals were "good to think," thereby opening up the invitation not only to think about them but also to think with them. Much of my work revolves around questions of how animals figure into our human signifying systems and how they find their ways into the deepest recesses of our cultural landscapes. Culture is what is most often attributed to humans and denied to animals. My approach to studying animals is interdisciplinary and falls under the broad rubric of cultural studies. It is premised on the belief that human-animal relations span several academic disciplines and can only be grasped by transdisciplinary approaches that mutually enrich and inform their separate discourses. My early training in comparative literature dealt with British and French literary cosmopolitanism through authors, especially Joseph Conrad and André Gide, who wrote fictional accounts or travel narratives of colonial sub-Saharan Africa. This long-standing interest in tales of exploration and discovery—writings where animals appear frequently—has led me to an interdisciplinary project that I call "The Colonial Animal." It lies at the juncture of literature, science, philosophy, art, anthropology, and history in the evolving field of animal studies.

Africa—or a certain idea of Africa—became a site where European cultural tensions of race, gender, class, and nation could be staged. I would add "species" to this list as animals became the privileged signifiers of Western domination over the natural world and participated broadly in the construction of a colonial form of "mimetic circulation." In his Marvelous Possessions *(1992), Stephen Greenblatt has studied how the ability to create and circulate images across great distances resulted during early modern times in a modification of the social relation of production. By extending Greenblatt's theory, I wish to demonstrate how animals, whether depicted in media or displayed "in person," came to embody tropes of exoticism, savagery, wonder, and adventure which modified in profound and subtle ways the colonial relationship between Europe and Africa. Finally, I contend that many aspects of the current human relationship to nonhuman animals can be understood as "colonial" in its assumptions and practices.*

From antiquity to modern times, animals have figured prominently in the Western construction of the myth of a wild Africa. Africa became synonymous with subjugated nature in two ways: First, by the images conjured up by early explorers and settlers who, following Herodotus and Pliny, depicted Africa as "a land of wild beasts" and, subsequently, by the animals that were displaced to Europe and exhibited as spoils of empire. The spectacular display of lions, elephants, and giraffes in imperial Rome—often followed by their wholesale slaughter—came to represent the relations of power and authority between the capital and its far-flung empire. This pattern would be replicated in colonial times through a double movement of repulsion and attraction, grounded in fear and desire, which depended heavily on the European image of Africa as a place of great danger and immense seduction. From the earlier, romanticized trope of an Edenic, feminized Africa inviting European penetration, the continent progressively became a man's world where masculinity could be asserted by hunting the lions and elephants that required daring and courage in the face of real danger. Africa became synonymous with nature suspended in time and inhabited by creatures deprived of reason and judgment as opposed to the cultivated West engaged in progress through human endeavor. Animals became metonymic fragments of empire that cast Africa as a place both foreboding and inviting, resistant yet available to European expansion. My project charts how the animal presence contributed to the history of that myth and its impact on ideas and ideologies.

The colonial animal results from a long genealogy that cannot be recounted here in detail. It begins with the philosophical debate around animals that cast them as the earliest figures of alterity. They were encoded as "different" and "other" due to their purported lack of the defining human faculties of reason and language. Descartes' concept of the animal-machine, extended and amplified by Enlightenment ideals of self-determination aspiring to universal happiness, has governed many subsequent positions and assumptions regarding human and animal nature. Drawing on a long-standing metonymic relationship that associated African peoples and animals, racial discourse hypothesized a link—notoriously "missing" during the nineteenth century—that opposed primitive and advanced civilizations on the basis of the new biological and natural sciences. African animals have figured also in more contemporary literature. One point of focus has been the 1987 French novel by Marie Nimier, La girafe, which stages a modern, tragi-comic love story set in the Zoo de Vincennes against the background of the historical tale of the first giraffe to set hoof on French soil since Roman times. Through her main human character, Nimier weaves a contemporary tale of immigrant alienation and sexual aberration in parallel with the historical account of Zarafa's arrival in Paris. She thereby addresses the question of African migration to France in this story of human and nonhuman animal displacement and encounter. Large land mammals also figure prominently in the images and iconography inspired by the "discovery" of African wildlife. I conducted a research trip in summer 2008 to the Archives of the Overseas French Empire in Aix-en-Provence, which contain some 60,000 photos and over 45,000 postcards from the colonial period. I was especially interested in the use of African animals as ideological markers: For instance, many early maps depicted Africa as a "blank space" into which they inserted animals as placeholders in anticipation of European exploration and conquest. Finally, the colonial animal project deals with the exhibition and display of wild animals, especially in museums, jardins d'acclimatation, zoos, and colonial fairs. Not only did zoos organize the natural and colonial world by species and place of origin, they also became performance spaces for the reenactment of a European power play over lesser, colonized groups. The wholesale displacement of specimens of African wildlife to the metropolitan centers of the West involved logistical, political, and ideological efforts deployed largely for propaganda purposes. Much of the current debate over conservation and species protection mirrors colonial patterns that I try to lay bare in light of the vexed relationship that still exists between the West and the rest of the world as well as between human and nonhuman animals.

5

The Domestication of Animals

The earth trembled and a great rift appeared, separating the first man and woman from the rest of the animal kingdom. As the chasm grew deeper and wider, all other creatures, afraid for their lives, returned to the forest—except for the dog, who after much consideration, leapt the perilous rift to stay with the humans on the other side. His love for humanity was greater than his bond for other creatures, he explained, and he willingly forfeited his place in paradise to prove it.
—NATIVE AMERICAN FOLKTALE

THIS FOLKTALE IS ONE of many similar tales from around the world that purport to explain how the dog became domesticated. In all of such stories that I have found, whether the dog was coerced or manipulated into joining the land of humans (as is common), or whether he voluntarily joined human society, the end result is the same: He chooses to remain with humans, giving up his freedom, his wildness (and, according to this folktale, his place in paradise) for the privilege.

History of Domestication

As has been long noted by archaeologists and historians, animal domestication (as opposed to taming animals) first occurred during the Mesolithic period with the domestication of the dog. Domesticating animals involves more than simply taming them. Animals are considered to be domesticated when they are kept for a distinct purpose, humans control their breeding, their survival depends on humans, and they develop genetic traits that are not found in the wild.

Dogs were domesticated as hunting partners for humans at least 15,000 years ago (but perhaps as early as 30,000 years ago). Some scientists, using genetic evidence, suggest that the date may even be as far as 135,000 years ago. During this period, humans were still hunting and gathering and had no real control over the production of their food. Evidence suggests that dogs were first domesticated in East Asia and perhaps as many as three other independent locations; from there they may have spread around the world. Dogs have been found in human burials from around the world dating as far back as 14,000 years ago, demonstrating how important they were to Mesolithic humans.

All of the early breeds of dogs such as the pointer, the hound, and the retriever were bred for hunting. As humans migrated around the planet, a variety of dog breeds migrated with them. With agriculture came new specializations for the dog—from hunting to herding and protection, racing and pulling loads, and keeping rats away from the food supply. Today, there are more than 400 breeds of dog, all with specialized appearance and behavior, and all derived from the wolf.

Some researchers have speculated that the benefits dogs offer to humans may have contributed to the rapid expansion of humans into the New World. Archaeological and historical evidence show that the earliest inhabitants of North America were selectively breeding dogs for traits that would be useful for hunting; they used dogs as beasts of burden, tying wooden

---

BOX 5.1

WHEN WAS THE DOG DOMESTICATED?

Archaeologists have long said that dogs were first domesticated 15,000 years ago, based on evidence from burials in which dogs were buried along with humans. The oldest of these burials dates to 14,000 years ago in Bonn, Germany. However, that date is being challenged on at least two fronts. Archaeologists have located what they believe is a domesticated dog skull in a cave called Goyet in Belgium that they date to 31,700 years ago. Another cave called Chauvet, in France, contains fossilized dog skeletons dating to 26,000 years ago. In addition, new DNA evidence attained using mitochondrial DNA shows the split between wolves and dogs could go as far back as 100,000 years, meaning that humans may have been living with dogs for much longer than we thought.

BOX 5.2

DOMESTICATED ANIMALS

- Sheep – Middle East – 9,000–11,000 BCE
- Cats – Near East – 8500 BCE
- Goats – Middle East – 10,000 BCE
- Pigs – Near East, China – 7000 BCE
- Cattle – India, Middle East, Africa – 8000 BCE
- Chickens – India, Southeast Asia – 6000 BCE
- Guinea pigs – Andes Mountains – 5000 BCE
- Donkeys – Northeast Africa – 4000 BCE
- Ducks – China – 4000 BCE
- Horses – Kazakhstan – 3600 BCE
- Llamas – Peru – 3500 BCE
- Bactrian camels – Southern Russia – 3000 BCE
- Dromedary camels – Saudi Arabia – 3000 BCE
- Honey bees – Egypt – 3000 BCE
- Water buffalos – Pakistan – 2500 BCE
- Yaks – Tibet – 2500 BCE
- Alpacas – Peru – 1500 BCE
- Turkeys – Mexico – 1000 BCE

sleds called travois to them, on which they loaded firewood and other necessities. The Clallam Indians of Puget Sound kept long-haired dogs whose hair they spun into clothing and blankets; other native groups most likely used dogs for protection as well.

With the domestication of the "livestock" animals as well as the domestication of plants in the Neolithic era, starting about 10,000 years ago, the primary economic activity of our ancestors moved from food collection to food production. This change is among the most monumental in human—and animal—history.

After the dog, the first animals to be domesticated were the goat, sheep, pig, and cow. All were domesticated for their milk, meat, and wool or skin. Horses, which were initially hunted as food animals, were the first true beast of burden, being used to pull ploughs in farming communities perhaps 5,000 years ago; 2,000 years later, horse riding developed. Horses were tremendously important to humans, allowing for travel, warfare, trade, and work, and, later, entertainment such as horse racing and sports such as the rodeo. Horses' importance to humans only declined with the development of the internal combustion engine in the early nineteenth century. However, by that time, horses had become so valuable to human society that they served as an important source of companionship in many cultures.

Horses were followed by cats, chickens, llamas, alpacas, and camels. Finally, less than 2,000 years ago, smaller animals such as rabbits were domesticated.

As has been demonstrated by a number of scholars, of the fourteen or so large animals to have been *truly* domesticated, most can be defined by a number of behavioral traits, including a tendency toward scavenging, rapid maturity rate, reasonable size, calm disposition, ability to be bred in captivity, gregarious nature and willingness to live with others in close

BOX 5.3

UNSUITABLE FOR DOMESTICATION

- Ferocity (zebras, rhinos, hippos)
- Eats high on the food chain (lions, tigers, and other carnivores)
- Picky diet (pandas, koalas)
- Slow growth (elephants)
- Territoriality and solitary habits (deer, antelope)
- Reclusive breeding or elaborate courtship (cheetahs)
- Tendency to panic (gazelles)

quarters, and hierarchical social life. The cat is a notable exception. All of these traits make animals such as the dog or horse amenable to living with humans, in exchange for feeding and care. Of course, the animals themselves must have had something to offer to humans as well, such as food, clothing, an ability to work as hunters or beasts of burden, and, later, the promise of companionship.

Although the species that were ultimately domesticated fit these criteria, making them natural choices for domestication, the process of domestication itself was a result of natural and cultural evolution. First, specific behavioral and physical traits of individual animals that scavenged or hung around human encampments were favored by natural selection in the process of domestication. For example, those animals that demonstrated less fear and more curiosity would be among the first to approach human societies; after reproducing, those traits would continue in their offspring. As each generation was born, the "flight distance" of those animals was shorter than in the preceding generation until the domesticates were comfortable living around humans. Humans would have made choices as well, such as selecting woolly animals from among wild sheep (that are not normally woolly), and thus acquiring livestock better suited to lowland heat and from which to obtain wool. Animals would have been selected for smaller size so that they could be more easily handled, or larger size in order to produce more meat, or for different colors or textures of wool, fur, or hair.

Next, humans most likely adapted their own behavior to that of the animals, incorporating them into human social and economic structures and, later, manipulating the physiology and behavior of the animals themselves. To be a domesticated animal is to be—at the very minimum—owned and controlled by humans in a human cultural environment. This is a profoundly different human-animal relationship than what was seen in nonagricultural societies, where humans could not conceive of "owning" animals in this way. As we will discuss, domestication in the twenty-first century has moved from natural selection to artificial and has been shaped almost entirely by human hands.

Results of Domestication

The domestication of animals was a truly revolutionary stage in the development of human civilization. It allowed humans to have a steady food supply, new sources of labor, and new forms of companionship and protection. It also provided resources for new forms of religious worship. In pastoral societies, in which people raised ruminants but did not farm, animals allowed for the feeding of people in unstable environments and in climates as diverse as Siberia, Mongolia, and the deserts of Africa. In these cultures, animals are used for their milk, fur, skin, blood, dung, and rarely for meat. They are also valued for bride-price and gift exchanges; much of the status of the people is based on animals. Although animals are often lavished with attention and affection, they are also eaten on occasion.

In agricultural societies, when the domestication of large ruminants is combined with plant domestication, animals could be used to plough fields. Their manure could be used to fertilize the fields; they could eat the stubble off the fields after harvest and be used to transport people and goods. They would have also have enabled trade, migration, and warfare and assisted in producing a surplus of goods. These factors allowed for the development of the early city-states with their complex division of labor and high degrees of inequality and, eventually, the political and economic dominance of a handful of European and Asian societies over much of the rest of the world. Other results from animal domestication included greater protein consumption and the introduction of communicable diseases such as measles, mumps, and the plague, which effectively wiped out much of Native America. More recently, diseases such as SARS, avian flu, mad cow disease, and AIDS have also resulted from domestication. Today, with the heavy emphasis on meat consumption and industrialized methods of meat production, we also have seen new diseases of the circulatory system as well as greater degrees of environmental degradation.

But the results of domestication for the human-animal relationship and for the animals themselves were no less revolutionary. As anthropologist Tim Ingold points out (1994, 1998), the relationship between animals and humans among traditional hunter-gatherers was often one of mutual trust in which the environment and its resources were shared by animals and people; animals that were hunted by humans were seen as equals. Domestication changed this relationship into one of dominance and control; humans took on the role of master and animals that of property. Animals became items to be owned and exchanged.

Domestication also had long-range consequences for the animals themselves; the very nature of the animals changed throughout the process—typically not in the animals' favor. Through domestication, once-wild animals become increasingly more dependent on humans, physically and emotionally. Because a handful of traits (such as curiosity, lack of fear, willingness to try new things, food begging, submissiveness, etc.) found among the juveniles of a species are those selected in domestication, the physical traits of the young (shorter faces, excess fat, smaller brains, smaller teeth, etc.) will also be selected. This leads to modern domesticates that are physically and behaviorally unable to live independently and that are, in fact, perpetual juveniles (a condition known as **neoteny**). Once humans began selectively breeding their animal charges to emphasize or discourage certain physical or behavioral traits (and killing both those offspring that did not fit the bill, as well as wild competitors), the animals changed even further. Today, domesticates are, for the most part, smaller (yet fleshier), more brightly colored, with shorter faces, rounder skulls, and more variations in fur and hair type as well as ear and tail appearance. In addition, domestication has resulted in a permanent loss of genetic diversity within the species.

Researchers believe that years of domestication have led to dogs losing the problem-solving skills they once had in the wild. In behavioral studies, pet dogs fail basic intelligence tests that wolves and wild dogs pass with ease. The findings (Smith and Litchfield 2010) suggest dogs are now so dependent on people that they are "stupid" versions of their forefathers. Another way to look at this is to say that as dogs became dependent on people, they got better at communicating with humans and using us as tools. Rather than opening doors themselves, domesticated dogs just wait until we open them for them!

According to some scholars (Budiansky 1997), the process of domestication was one of mutual benefit in which the early domesticates and our human ancestors had something to gain from joining forces. The question we must ask ourselves now, 10,000 years later, is this: If nonhuman animals did in fact benefit from aligning themselves with human societies, gaining protection from predators, easy access to food, and shelter from the weather, do modern domesticated animals experience the same benefits?

Indeed, a survey of the variety of venues in which domesticated animals are kept and used today reveals a startling dichotomy in terms of the way that we relate to domesticated animals. If we start with the most recent and certainly most mutually beneficial of the forms of domestication—companion animals—we see animals that are loved and treated as family members, lavished with gourmet food, glamorous clothing, an enormous variety of toys and entertainment, and even exotic travel opportunities. But

Figure 5.1. Pepe, a seven-pound Chihuahua, illustrates some of the most extreme characteristics of neoteny; he cannot survive without human care. (Photograph courtesy of the author.)

we also see animals that spend their lives confined in a fish bowl or cage or at the end of a rope or chain, animals that get little or no medical care, the most meager of provisions, and no shelter, love, or nurturance at all. And even though the most cherished of our companions live and die surrounded by love, millions of others are bred for profit and die unmourned. Animals used for food rarely receive even a fraction of the positive treatment lavished on some of our companions and, in fact, generally face a much shorter, harder life, and a much more brutal end.

Altering the Animal Body

The story of animal domestication did not end with the creation of today's major domesticated animals. Domestication continues to this day, imposing new shapes and traits upon animals, finding new uses for these "improved" creatures, and creating new benefits and profits for humans.

As we noted, since the first animals were domesticated for food, labor, and skin, domesticated animals have changed in a whole host of ways, behaviorally and physically. Natural selection favored those traits that made individual species, and individual animals good prospects for domestication, making the earliest domesticates look and behave differently from their wild relations.

As farmers, and later show breeders, learned more about the inheritance of traits, animal breeders began selectively breeding their animals for more specific characteristics, such as overall size, fur and wool color or texture, ear and tail shape, and more. Termed **artificial selection** by Darwin, selective breeding has led to the creation of hundreds of breeds of dogs, one of the most intensively bred animals in the world. Using dogs as an example, breeds were created in order to fulfill human desires. Some breeds were created to retrieve ducks during a hunt, others were created to herd sheep, and still others were created to race.

With the advent of industrial methods of food production in the twentieth century, changes in livestock breeds accelerated. To produce the most meat in the shortest amount of time, animal agribusiness companies breed farm animals such as pigs and chickens to grow at unnaturally rapid rates. These changes have been encouraged by new developments in agricultural science, aimed at improving the productivity of food animals. For example, American beef cattle are routinely administered hormones to stimulate growth. And to increase milk yield, producers often inject dairy cows with hormones.

Since the early part of the twentieth century, farmers have been experimenting with creating new livestock breeds via careful cross-breeding in order to maximize size, fat composition, productivity, or other traits. Since the development of artificial insemination and the ability to freeze semen, cattle farmers are able to more selectively breed their prized bulls and cows to replicate the traits of the parents.

The pet and show industries also rely on artificial selection (and today, following the livestock industry, artificial insemination) to create breeds of animals with traits favorable to humans. Recent years have seen an escalation in the varieties of dogs, cats, and other companion animals being developed in order to appeal to discriminating consumers.

And although the early breeds of dogs were created to highlight working traits, recent breeds have been geared more toward aesthetics. On the other hand, because cats are not working animals, most cat breeds have been created for aesthetic purposes, with an eye toward color, size, fur type, tail, ear, and body type. The result is hundreds of breeds of dogs, and dozens of breeds of cats, rabbits, and other species, all bred by large and small breeders to sell through the pet industry. Another result is a whole host of health problems associated with these breeds. Dogs in particular are at risk of problems associated with the odd proportions in body, legs, and head that are bred into many of the breeds. Another form of artificial selection refers to breeders' emphasizing deleterious traits in the breeding process. Japanese Bobtails (cats with a genetic mutation resulting in a bobbed tail), hairless cats, and Scottish Folds (that have folded-down ears) are examples of this type of breeding.

Genetic manipulation of animals represents a new scientific development that has irreversibly changed animal bodies. Because pigs, beef cows, and chickens are created for one purpose—food consumption—their genes have been altered in a whole host of ways to suit that purpose. For example, pigs have been engineered to have leaner meat, tailor-made to suit a more health-conscious consumer. The biotech firm Gentech, for example, has discovered a gene marker that will allow scientists to breed pigs with leaner meat; the Meat Animal Research Center has been able to select for a gene that produces "double muscling" in cattle, producing more lean meat per cow (and a variety of health complications). As American tastes change, animals themselves change in order to conform to dietary and culinary trends.

Genetically engineered animals are also becoming more popular among scientists who experiment on or test with animals. Genetically modified mice and rats are especially popular, allowing researchers to study the ways that genes are expressed and how they mutate. Genetic engineering has even found its way into the pet world with the production of a new hypoaller-genic cat (selling for $12,000–$28,000), created by manipulating the genes that produce allergens.

In terms of reproduction, cloning animals is the wave of the future allowing humans the greatest level of control over animal bodies. Thus far, the livestock industry has been most active in the use of cloning. One example is the cloning of prized breeder animals in order to ensure higher yields (in meat, wool, etc.) by copying only very productive animals. But cloning is found in the vivi-section and pet industries as well. Laboratory scientists are also cloning mice, rabbits, and other laboratory animals in order to ensure that the animals used in research are genetically identical and to control for any "imperfections." In the pet world, cloning has been less successful in part because of the enormous number of animals that are either born with terrible deformities or that are "sacrificed" in order to produce the cloned animal. However, a handful of companies today either offer companion animal-cloning or tissue-freezing services for those animals that cannot yet be cloned.

Another way that animal bodies have been changed is through surgical procedures. Because the control of animal reproduction is critical to domesticating animals, castration has been used for thousands of years to ensure that undesirable animals cannot breed or to increase the size or control the temperament of certain animals. In the twentieth century, with companion animals

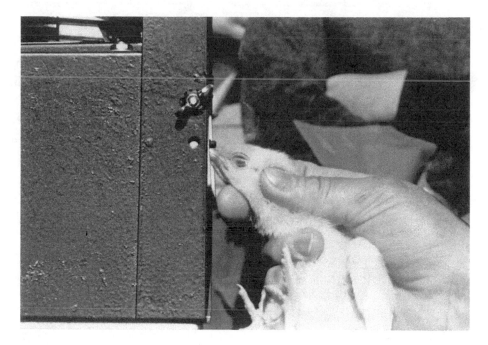

Figure 5.2. In a process called *debeaking*, baby chicks have part of their beaks sliced off with a heated blade to prevent them from pecking each other when crowded together into battery cages. (Photograph courtesy of Mercy for Animals.)

rising in popularity, surgical techniques to remove the uterus and ovaries of female animals were developed, and spaying is now an extremely common surgery for companion animals although it is very rarely performed on livestock.

Other forms of surgical modification have also been common for years, particularly in livestock and purebred companion animals. In the last century, the close confinement necessitated by factory farm production has resulted in a number of procedures being performed on livestock. The **debeaking** of hens (amputating, without anesthesia, the front of the chicken's beak) is common in the egg industry, where chickens are so intensively confined in tiny cages that they may attack each other out of stress and overcrowding. Even in situations where livestock are not as closely confined, farmers often remove body parts. One mutilation that is increasing in popularity is **tail-docking** of dairy cows. Producers amputate up to two-thirds of the tail, usually without painkillers. Cattle are often de-horned and sheep may have their tails removed (usually via banding, without anesthesia).

In the pet-breeding world, companion animals undergo surgical procedures in order to make them conform to the artificial requirements of the breed. Breed standards demand that certain dogs, for example, must have their tails docked, their ears cropped, or both. In addition, many companion animals today experience surgical procedures that are used to control behavior unwanted by humans. Some people, for example, have their dogs debarked (by cutting their vocal cords) in order to reduce barking, and many cat owners elect to have their cats declawed (which involves amputating the front portions of a cat's toes) in order to prevent harm to their furniture.

All of these uses of the animal body can be compared to French philosopher Michel Foucault's (1998) notion of **biopower**: the ways in which the modern state controls and regulates their citizens' bodies. With respect to human bodies, the use of branding to mark human slaves and convicts— a practice found in ancient Egypt and Rome that endured in antebellum America—illustrates the power that the state wields over bodies. When it comes to animals, it is easy to see how society's needs and desires have shaped the changes to the animal body discussed in this chapter.

Is Domestication Good or Bad?

Human civilization would not be what it is today if it were not for animal domestication. Without the assistance of working animals that pull ploughs, carts, sleds, and carriages, carry goods or people, assist humans with hunting, and herd other animals, and without the use of animals as food and fiber, it is difficult to imagine where human societies would be today. But what of animals?

Certainly many companion animals gain a great deal by living with humans, and many working animals may live better lives with humans than without. Even though the mutual dependence/mutual benefit theory of domestication that we outlined earlier suggests that animals chose to be domesticated because they recognized that life is better with humans, this does not necessarily mean, especially given the modern evolution of the agricultural animal, that their current lives must be better than what they would be if they still lived "in the wild." According to ecologist Paul Shepard, "[T]he benefit to animals of being domestic is fictitious, for they are slaves, however coddled, becoming more demented and attenuated as the years pass" (1995:267). In any case, the point is moot because, with the possible exception of the mixed-breed cat, today's domesticated animals are so highly bred and engineered for human benefit that they could never again survive on their own. In short, we care for them because they could not live without our care, and they live with and obey us because they no longer have a choice in the matter.

But we also care for domestic animals because we have grown dependent on them. Although this is perhaps more the case in pre-industrial societies where animals' labor is and was a critical part of the economy, it continues to be the case today in modern society's dependence upon meat and other products taken from the bodies of animals as well as in our dependence on pets for companionship, love, and affection. But whereas the dependency of most domestic animals on humans is irreversible—even animal rights advocates do not foresee a day when domesticated chickens, pigs and cows, much less Chihuahuas and Persian cats, can be "wild" again—that is not necessarily the case for human dependency on animals.

Does this unwillingness to release ourselves from our dependency on domestic animals stem from love, greed, selfishness, or a desire to dominate others? Just as we have bred dependency into the domestic human-animal relationship, we have also bred its corollary—dominance—into that same relationship. Not only do we dominate farm animals through every level of control exercised over their minds and bodies, but we dominate albeit with affection our pets as well (from whom we demand unconditional love and absolute obedience). Or does our dependence stem from a human need to stay connected to animals and to the wild from which they came, and by extension, to our own roots?

Suggested Additional Readings

Budiansky, Steven. 1992. *The Covenant of the Wild: Why Animals Chose Domestication.* New Haven, CT: Yale University Press.

Clutton-Brock, Juliet. 1987. *A Natural History of Domesticated Mammals.* Cambridge: Cambridge University Press.

Diamond, Jared. 1999. *Guns, Germs and Steel: The Fate of Human Societies.* New York: W.W. Norton & Co.

Henninger-Voss, Mary, ed. 2002. *Animals in Human Histories.* Rochester, NY: University of Rochester Press.

Manning, Aubrey and James Serpell, eds. 1994. *Animals and Human Society: Changing Perspectives.* New York: Routledge.

Serpell, James. 1986. *In the Company of Animals: A Study of Human-Animal Relationships.* Cambridge: Cambridge University Press.

Shepard, Paul. 1995. *The Others: How Animals Made Us Human.* Washington, DC: Island Press.

Suggested Films

Dogs That Changed the World. DVD. Directed by Corinna Faith; narrated by F. Murray Abraham. New York: Thirteen/WNET New York, 2007.

Holy Cow. VHS. Directed by Harry Marshall. New York: Thirteen/WNET New York, 2004.

Coming to Animals

MOLLY MULLIN
Albion College

I grew up on a farm in north-central Florida. It was a lonely place for a child—no sidewalks, no other kids, and hardly any neighbors at all. As is true for a lot of American kids, animals helped relieve my loneliness. Animals also provided one of the few interests and sources of pleasure shared by members of my family, a family whose very identity as a family remained uncertain.

By the time I went off to college, I had become more interested in understanding humans than animals. Humans for me had become profoundly confusing and problematic, and therefore fascinating. I majored in history and took a single anthropology course—on gender—my senior year. But it was anthropology where I felt I could best pursue my curiosity about humans—their troubling and often surprising similarities and differences. In 1985, I entered a graduate program in anthropology at Duke University.

Duke's anthropology department, in the mid-1980s, included both cultural and biological anthropologists, most of them primatologists. For the most part, there was little coordination or collaboration between the biological program and the cultural program. Many of us, myself included, felt we had enough to learn from other disciplines entirely and that left little time for engaging with the other subfields of anthropology. But there was also an underlying concern that our perspectives about humans and our approaches to studying them were so different that attempts to bridge the divides might only result in hostility.

Eventually, in the mid-1990s, the university divided the anthropology department into separate departments, with cultural anthropology remaining in the existing space, some of the biological anthropologists moving to the medical school, and others to Duke's Primate Center. But their initial presence in the same building inspired me to continue thinking about the relationship between primatology and cultural anthropology and about relationships between humans and animals, culture and nature.

An interest in humans' relationships with animals was not, in the late 1980s and early 1990s, particularly encouraged by my professors or the other graduate students with whom I worked. During an oral preliminary exam, I became flustered when I found myself rambling on about the ending of Foucault's The Order of Things, *where he suggests that "man," the focus of "the human sciences," will soon be "erased, like a face drawn in sand at the edge of the sea" (1970:387). I interpreted his statement as a call for anthropologists to recognize the arbitrariness of human-animal boundaries, just as we had recognized the arbitrariness of boundaries between genders, and boundaries that our discipline had been* built upon. I was embarrassed because I realized that in no way did I have any experience talking about human-animal boundaries and their significance for anthropology.

I began acquiring that experience toward the very end of my work on a doctoral dissertation that examined relationships among gender, class, and concepts of culture in the patronage of indigenous art in the American Southwest. Some of the art patrons I studied had various interests in animals. The White sisters, renowned patrons of anthropology and Native American art, bred Irish wolfhounds in the 1930s and 1940s. When eventually I thought to research this fact, I was surprised to find that in the art patrons' minds, their dog breeding had much in common with their art patronage. Both endeavors allowed them to negotiate relationships between past and future and to translate seemingly domestic skills into public influence (Mullin 2001). Animals, I began to think, were actually quite central to understanding anthropology and its history—and important for anyone interested in understanding humans.

Encouraged and inspired by the work of scholars such as Donna Haraway, Sarah Franklin, Garry Marvin, and Rebecca Cassidy, I continued to explore relationships between culture and nature and human-animal relationships. At the small college where I arrived to teach anthropology in 1995, I designed a new anthropology and environmental studies course on the cultural politics of animals. While I was designing my course and writing a chapter for the Annual Review of Anthropology about recent work in human-animal studies (Mullin 1999), I contacted many scholars in the United States, Europe, Australia, and New Zealand. Encouraged by their enthusiasm and generosity, I coorganized two conference sessions at annual meetings of the American Anthropological Association with Sarah Franklin, an anthropologist known for her research on biotechnology and her research on Dolly, the first successfully cloned animal (Franklin 2007).

In 2004, Rebecca Cassidy and I organized an international symposium funded by the Wenner-Gren Foundation for Anthropological Research, Inc. We chose our topic, domestication, carefully, in the hope of encouraging productive discussion and collaboration across anthropology's disparate "four fields" and across the humanities-natural sciences divide. No project is without its challenges, and the symposium and its resulting volume, Where the Wild Things Are Now (Cassidy and Mullin 2007), proved no exception. But the topic of animals, and humans' relationships with them, proved capable of uniting, at least temporarily, scholars of diverse backgrounds and perspectives for a productive exchange of perspectives and discussion.

My research today continues to focus on domestication. I've studied dog breeders and the history and politics of the pet food industry. Most recently, I have *started an ethnographic research project on "Coop Loops and Cow Shares: De-industrial Domestications." I'm studying two related but different phenomena: the backyard chicken movement and the selling of "cow shares," a strategy to get around the ban on some states, including the one where I live, of sales of unpasteurized milk. I'm considering both trends in relation to the context of de-industrialization.*

Growing up, animals helped me to make sense of my confusing human family. In my research and teaching in human-animal studies, I continue to search for points of connection—with human and nonhuman animals, among scholars in the humanities and natural sciences, and among farmers, hunters, vegans, and conservationists. I encourage my students to value diverse perspectives, human and animal, and to recognize that no one has all the answers to the many urgent social, environmental, and political problems that we face. Collaboration across subfields and disciplines is necessary, as is negotiation and compromise.

6

Display, Performance, and Sport

A circus tiger mauled and killed his trainer.
"I wonder what set him off," said the commentator.
I don't know. How would you feel if separated from your family,
You were shipped to different cities in a cage no less,
Bound of life, with pain/pleasure techniques,
And complete humility for performance under duress,
A whip no less.
If you were a tiger would you do it?
Would you break away,
Think of escape and if desperate,
Kill and avow your infinite humiliation and guaranteed
Death?
Do you do it, now, as a human?
If not, then I understand why you were not sure
What set the tiger off, Mr. Commentator.
—SERJ TANKIAN, "CIRCUS TIGER" (2002)

Why Do We Watch Animals?

Americans love watching animals. We love watching them eat, play, inter-act with each other, and even sleep. We also love touching them and being as close as possible to them. If we are not watching our own animals, we are birding, whale watching, photographing wildlife, scuba diving, snorkel-ing, or watching Animal Planet and Webcam footage from zoos and animal sanctuaries.

One reason modern Americans are so captivated by animals today is the disappearance of animals from our lives. In our post-industrial world, com-panion animals remain the only form of physical connection that Americans have with animals. Since animal agriculture now takes place behind the closed doors of huge factories and most Americans live either in cities or suburbs, interacting with non-companion animals has effectively become a thing of the past for the great majority of us. Thus, the proliferation of animals in films and on television allows many of us to view animals that we would never generally get the chance to see. But long before television and film, Westerners had devised ways of seeing and connecting with wild animals through zoos and circuses that brought wild animals into domestic enclosures for urban dwellers to see.

Having a fascination and a desire to watch animals—especially wild ones—is not enough to explain the other ways people use and have used animals as entertainment, such as circuses, marine mammal parks, dog or horse racing, animal fighting, and rodeos. What is it that draws so many people to entertainment venues in which animals are not just present but are forced to perform sometimes dangerous stunts for our pleasure? What is the pleasure in seeing large wild animals such elephants, chimpanzees, and tigers dressed up like children and performing tricks?

Whether it is through the billions of tourist dollars spent at zoos and wildlife parks per year, the proliferation of the modern ecotourism and wildlife safari industry, whale watching, birding, or the enormous popularity of Animal Planet television shows such as *Animal Cops* and *Meerkat Manor*, it becomes obvious that the public craves seeing animals. Whether we observe animals from the comfort of our couches during *The Dog Whisperer* or whether we travel thousands of miles to Australia, Africa, or Antarctica to watch koalas, lions, or penguins, it is clear that this trend will continue. One question that we could ask is: Is all this attention good or bad for the animals? Certainly documentaries that focus on the plight of wild animals and their loss of habitat seem good for animals—educating the public on the ways in which animals are imperiled and what we can do about it seems to only have positive consequences. Likewise, shows that focus on animal rescuers, animal trainers who use animal-friendly methods, and many of the other ways that we can watch animals on television and in film seem to be beneficial.

On the other hand, many of the other ways in which humans can see and even interact with dolphins, elephants, bulls, horses, or other wild and domesticated animals do not necessarily provide those animals with any benefits. Circuses, rodeos, and marine mammal parks not only force animals to perform in ways that are unnatural for them, they also keep many of their animals confined in conditions that are sometimes very intensive and do not seem to benefit the "stars" of the shows at all.

In this chapter, we will look at a variety of venues where humans can watch animals and observe their living conditions, the possible benefit or harm to the animals, and the motivation of the people who derive pleasure from watching them. We focus on wild as well as domesticated animals, as both are used for entertainment in a number of ways. Typically, the wilder and more exotic the animals, the more pleasure most of us get from simply watching them or observing their performances or tricks in circuses and marine parks. In some ways, watching a very wild or exotic animal act like a human is even more interesting than watching a domesticated animal— already much like us—do humanlike tricks. But domestic animals of their proximity to many of us typically must perform—whether racing, fighting, or participating in a rodeo—since watching a domesticated horse, rooster, or cow in a zoo would not fit most people's idea of entertainment.

Ultimately animals, whether domesticated or wild, are heavily featured in human entertainment because of our pleasure in watching them and because animals, as sociologist Adrian Franklin points out (1999), are both like us and different from us; they can be interpreted by us in a variety of ways, to represent difference and otherness and to represent sameness and family.

Zoos

People have kept animals in captivity for thousands of years, long before the concept of "zoo" ever existed—as creatures of worship, as part of gladiatorial contests in the Roman Empire, for activities such as bearbaiting and bullfighting in medieval Europe, and more. Wealthy elites in ancient Egypt, Greece, China, and Rome, and later in medieval and Renaissance Europe, also kept exotic animals. The keeping of these animals—such as giraffes, monkeys, elephants, and lions—in these early collections demonstrated either the wealth of the individuals or the wealth of the empire as well as a mastery of nature through the ability to contain "ferocious" animals. The animals themselves were often gifts from the leaders of other kingdoms or states. By the late seventeenth century, private **menageries**, as they were called, were status symbols for wealthy Europeans and commoners were not able to view them.

Ancient and early European exotic animal collections were not meant for public observation; the idea of zoos for the public to visit and see wild animals and the idea of zoos as anything more than oddities or amusements really did not develop until the eighteenth century. One way that commoners were able to see these animals was in the form of traveling entertainers who, in the mid-nineteenth century, also offered minstrel acts such as juggling, singing, poetry recitals, and human oddities shows. The purpose of these shows was spectacle—animals lived in small cages (when they were not held by chains) alongside human oddities and native peoples captured from colonial lands.

The first **zoological garden**, the Ménagerie du Jardin des Plantes, opened in Paris in 1794. Zoological gardens were different from private menageries and traveling shows. Here, animals were available for viewing as people walked and looked about (rather than just standing in one spot and observing them in cages). In addition, where menageries were disordered groupings of random animals, zoos were ordered collections often organized by continent or taxonomic label. Like modern zoos today, these early zoos competed for the best and most exotic animals and displays. And like the ancient collections, they were a place to put all the animals African and Asian rulers sent as gifts.

The early French zoos were the first to propose the idea that zoos could advance the study of natural history, and they slowly began to emphasize education. This new development corresponded with the rise of scientific ideas about animal nature that, not coincidentally, supported practices of animal confinement. At this time, the idea that menageries could further scientific knowledge became popular and spread throughout Europe.

Nineteenth-century zoos were still focused on the upper classes, charging entry fees that the poor could not afford. Instead, the poor satisfied themselves with animal attractions such as animal fighting, bearbaiting, and racing, all of which Northern European countries later prohibited—not because of concerns about cruelty to animals but as a way to control the poor. Even though the zoos of this time period promoted an educational message, they were still largely about entertainment; a chimpanzee tea party was the main attraction at the London Zoo from the 1920s to the 1970s, while the Bronx Zoo held tea parties for the orangutans.

The history of American zoos is somewhat different. The first American zoos were the Central Park Zoo in New York, founded in 1860, and the Philadelphia Zoo, opened in 1874. These zoos, as well as other early American zoos, developed during the time when the first public parks were being devised and constructed and indeed many of the early zoos were built in parks. Attendance was free to the public as a way of drawing in the middle class and the poor and providing them with an educational, uplifting experience. Whether American or European, by the nineteenth century all the major cities had to have a public zoo.

What animals can be placed in a zoo? The zoo maintains the primary distinction between "wild" and "domestic" animal, and only wild animals can be placed in a zoo. Furthermore, the conventional wisdom says that it is not a zoo if it does not have an elephant. For hundreds of years, zoos procured their animals by paying hunters or traders to catch live animals in the wild. During the growth of zoos in the nineteenth century, thanks to the stress of not just capture but transport across Africa or India to Europe, anywhere from one- to two-thirds of the animals died en route. The collectors—themselves big-game hunters—wrote extensively about their excursions. Many left "kill diaries" in which they boasted in excruciating detail of their kills and of the baby animals that mourned at the sides of their dead mothers until they were snatched away, put into cages, or tied or chained up, and transported to Europe. Because most social animals such as gorillas, chimpanzees, elephants, and hippos guard their young, collectors had to kill the adults (sometimes the females, but often the entire herd) when capturing their babies.

After the end of the colonial era, animals for zoos continued to be captured in the wild. It was not until 1973 with the signing of the Convention on International Trade in Endangered Species of Wild Flora and Fauna and the passage of the Endangered Species Act that wild imports began to decline in the United States. However, then as now, unscrupulous dealers can fake certificates of entry to say that the animals are captive bred, and customs officials are not trained in the identification of exotic animals. In addition, until the modern period, zoos used to take in animals from the public—either "donations" from well-meaning individuals or else no-longer-wanted exotic pets. In other cases, community groups would often fundraise in order to buy zoos a special animal that they felt would complete the zoo's collection. Even today, it is often a matter of civic pride that a local zoo is able to have a prestigious animal such as an elephant or, the most prestigious of all, a panda.

Zoos ship animals around the country many times throughout their lives. They manage populations in a way that is cost effective, keeps the right balance of animals for the zoo's mission, brings in visitors, and provides for breeding opportunities. This means that zoos will buy, sell, and borrow animals, sometimes temporarily and sometimes permanently, and remove animals from familiar environments and animals. So what do zoos do when they have "surplus" animals? In the best cases, zoos find sanctuaries or other places where animals will be "retired." Sometimes zoos euthanize animals. And sometimes zoos sell animals through dealers and brokers to a variety of locations, including roadside zoos, private homes, exotic meat farms, research laboratories, the entertainment industry, and canned hunting operations. Thankfully, the American Zoo and Aquarium Association, the major U.S.

zoo accrediting agency, now prohibits member zoos from selling surplus animals to canned hunting operations or any other non-accredited facilities. Unfortunately, this means that with fewer places to use to dispose of surplus animals, many zoos resort to killing them.

Early European zoos, like the traveling displays, showed their animals in small, barred cages. Some zoos, however, created elaborate displays such as Bristol Zoo's Monkey Temple, an open-air "Indian temple" that humans had supposedly abandoned and the jungle had taken over. Monkey Temple was an early attempt to confine animals—in this case, rhesus macaques—in a "naturalistic" enclosure without bars or wires, where visitors could enjoy the animals behaving in a way that they thought was "natural." The design was also based on a motif common to zoos in the nineteenth and early twentieth centuries that echoed not only the jungle environments from which the animals supposedly came but also the exotic cultures with which the animals were associated. (The connection among exotic cultures and peoples and exotic animals was made explicitly clear when Carl Hagenbeck, an animal dealer and collector who supplied European zoos in addition to having famous clients such as P. T. Barnum, imported two young Cameroonian boys to keep a captured baby gorilla company at his zoo in Austria in 1910—the implication being that African boys and African apes were closely connected. In 1906, the Bronx Zoo exhibited an African pygmy man named Ota Benga with the chimpanzees until the city's African American community lodged a protest.)

It was not until the early twentieth century that the new style of animal enclosure was fully developed using concrete moats, sunken fences, and other design aspects to confine animals but also allow the public to have an unimpeded view of them. Hagenback was the innovator behind this new style of zoo enclosure when he introduced it in Hamburg's Tierpark in 1907. Hagenbeck created sophisticated panoramas, which utilized concrete rocks and other substitutes for natural materials that were designed to make it appear as if the animals were living in the wild. These new, easy-to-replicate, "naturalistic" designs soon became the standard zoo enclosure style. (Large numbers of premature animal deaths had plagued early European zoos due to the stresses of the animals' capture and transport as well as the inappropriate conditions.) But the ultimate focus of these changes was not to improve the animals' lives. The new enclosures were created to make the public enjoy visiting zoos more, since zoos now had to compete with more modern forms of entertainment and needed to do something to attract more customers. In fact, eliminating the bars often made the animals' lives even worse; the bars were at least a feature of their enclosure that they could climb. Many of the newer enclosures that featured moats or sunken pits without bars left the animals in a barren environment where there was absolutely nothing for them to interact with.

Even today, the modern approach to zoo enclosure design known as **landscape immersion** replicates the animals' environments as closely as possible by using concrete forms to simulate rocks and other natural objects and to connect visitors with the habitat. The goal is to make visitors happier—indeed, studies show that visitors do not like seeing animals behind bars because it reduces their own viewing pleasure. As anthropologists Bob Mullan and Garry Marvin (1987) point out, by improving living conditions or by providing them "with a better stage and with more complex scenery and props to make the illusion more satisfactory" (159), visitors will feel satisfied with the treatment of the animals and thus feel that the captivity of animals in zoos is morally acceptable. A recent study (Melvin, McCormick, and Gibbs 2004) demonstrated that zoogoers view naturalistic enclosures as providing the best welfare for the animals.

Starting in the 1970s, a growing debate about animal protection meant that people started to question the ethics of exhibiting wild animals in what many regarded as restrictive, cruel enclosures. At the same time, Congress expanded the **Animal Welfare Act** to include standards of care for animals in exhibits. More recently, a 1995 poll by the Roper Center for Public Opinion Research showed that 69 percent of Americans are concerned about the treatment of animals at zoos, aquariums, and wild animal parks.

Many zoos now use **environmental enrichment** to reduce boredom, stress, and a condition known as **zoochosis**—psychosis caused by captivity—and to increase species-specific behaviors. For some zoos, this means environmental complexity (such as trees, structures, jungle gyms, etc.) that can increase the ability of the animals to exercise; occupational or feeding enrichment in which the animals have to work for their food; physical and sensory enrichment, which involves the addition of new objects for the animals to investigate, smell, touch, or play with; and social housing. But sometimes this only means giving a gorilla a tire or a ball, or putting a group of unrelated animals together into an enclosure rather than keeping animals with their kin groups during their lifetimes. (The latter is particularly difficult to do when zoos need to manage their populations by selling off animals.) Studies have shown that environmental enrichment has documented positive effects on zoo animals' psychological well being. In recent years, a number of American zoos have closed their elephant exhibits, sending their elephants to live out the rest of their lives at one of a handful of elephant sanctuaries in the United States. This is viewed as a positive development by animal welfare advocates, who decry the small spaces that zoo elephants live in that shorten their lifetime in captivity—17 years for Asian elephants and 19 for African elephants, compared to 42 years for wild Asian elephants and 56 years for wild African elephants (Clubb et al. 2008).

Among the reasons people visit zoos are to escape from urban and suburban environments and to be able to view and even interact with wild animals, something that has long been missing from the urban or suburban lifestyle. That is why even reputable zoos complement their educational message with exhibits and events that allow the public to ride, touch, feed, or get very close to animals. Visitors also like to see animals move. They become bored when animals are sleeping, even when they are nocturnal and should not be awake in the daytime. This leads to zoo patrons yelling at animals or pounding or tapping on enclosure windows. And even when zoo visitors do not react negatively to the animals, studies have shown that their presence is associated with behavioral changes in zoo animals—especially primates—that are indicative of stress.

The message that zoos promote most today is conservation. Some zoos use the term "arks" to emphasize their role in conserving species whose habitats were destroyed or that are on the brink of extinction thanks to overhunting or other problems. In addition, many zoos play a part in breeding rare and endangered species—this is known as *ex situ* **conservation**, as opposed to *in situ* **conservation**, which refers to conservation programs in the wild. Others, such as the San Diego Zoo, create research programs that focus on creating sustainable populations, conserving wildlife habitats, improving animal health, and even collecting endangered species' DNA. Some zoos have released zoo-raised endangered animals into the wild. All of these are worthy causes.

But it is difficult to imagine that zoos, with a hundred-year history of wildlife destruction in order to acquire animals that would then live for only a couple of years in captivity, should be society's institution responsible for preserving thousands of species. Even habitat conservation and reintroduction programs can do only so much when thousand acres of rain forest are being paved over or burned every day. Habitats continue to disappear to make room for development, cattle grazing, cropland, and more. Sport hunters continue to kill rare and endangered animals for trophies. And as humans continue to threaten habitats and entire species, it makes little sense to invest much hope in **captive breeding programs** and the like when the root problems of species extinction continue to flourish. Perhaps most important to note, the majority of animals in zoos are not even endangered.

Most zoos are not good models for captive breeding and species conservation. Those that have captive breeding programs often use them to create more zoo animals; they play no role in ensuring that wild animals can survive in their native habitats. Sadly, people around the world are confronting the very real problems of habitat loss and species extinction on a scale unparalleled in human history. If this trend continues, a few remaining individuals in captivity will be but a pathetic reminder of what once existed.

Ecotourism and the creation of wildlife preserves could be two ways to preserve habitats and to allow animals to live unmolested in their natural environments. By allowing limited ecotourism, visitors (although sadly only those who can afford it) could visit these places and their expenditures could help fund efforts. It is worth noting that ecotourism can be hard on the environment, given how many more resources Westerners are accustomed to consuming. In addition, some preliminary research is beginning to emerge on whether or not animals can be harmed by ecotourism. For example, one recent study (Matheson et al 2006) found increased levels of aggression in Tibetan macaques that interact frequently with tourists; scholars think that the feeding of the animals may be the cause of the aggression.

Marine Mammal Parks

Marine mammal parks and swim-with-dolphin programs differ from most zoos in that they make animals perform for the public rather than promoting their observation in enclosures. Similar to the zoo industry, marine mammal park advocates state that keeping marine mammals such as whales, dolphins, seals, and sea lions captive provides education to the public, allows scientists to gain information about the animals, and aids world conservation efforts. Unlike zoos, marine mammal parks are relatively recent inventions: The first park, Marine Studios, opened in 1938 in St. Augustine, Florida.

Marine mammal parks, like zoos and other venues in which people can view wildlife, believe that entertaining the public with wild animals and educating them about wild animals' lives make the public care more for those animals. And, if the public cares, perhaps they will also support conservation efforts. By including lectures and exhibits on the lives of marine mammals and their natural habitats, as well as the importance of marine conservation, the parks mix educational messages with the fun of watching these playful animals. Dolphins, like pandas or other especially cute animals, are often effective at conveying this message because of their perceived friendliness, playfulness, and even the way that their jawline looks like a human smile.

ANIMALS AND SOCIETY 108

Although zoos and marine parks encourage visitors to form an emotional connection to animals by making them behave in humanlike ways, they also attempt to turn the public's fuzzy connection to these animals into an interest in conservation and an obligation to ensure that these animals survive in the wild. A 1995 poll by the Roper Center for Public Opinion Research reported that 90 percent of respondents believe that public display facilities provide a valuable means of educating the public. However, no studies have measured the influence of these "educational programs" on people's behavior, especially as it relates to conservation.

Marine mammal parks also focus on learning about the animals themselves, employing scientists of all types to study the behavior, biology, and anatomy of the animals in the hopes of using that knowledge to extend the animals' lives in the wild. Of course, that knowledge may not be totally applicable to wild animals since the subjects are captive animals that often exhibit **stereotypic behaviors** associated with captivity. Like many zoos' captive breeding programs, this is more of a stopgap measure given the massive human threats to dolphins, whales, and other animals from legal and illegal hunting, pollution, habitat loss, and more.

Finally, like zoos, marine mammal parks that have captive breeding programs promote their efforts to ensure the continuation of endangered species. Certainly the captive breeding of dolphins, for instance, has reduced the need to remove dolphins from the wild for use in marine mammal parks, but so far no captive-born dolphins have been released to help wild populations sustain themselves. Less than 10 percent of zoos and marine parks are involved in conservation programs.

Animal advocates worry about the animals living in marine mammal parks, and especially worry about the small tanks in which these animals live. In the wild, many species of marine mammals travel as many as one hundred miles per day and live in large, complicated social groups that fish or hunt for their own food and dive to extremely deep depths. None of these conditions can be met in a concrete pool, no matter how expansive. As a result of the stresses and boredom of confinement, marine mammals can exhibit stereotypic behaviors such as aggressiveness, repetitive motions, obsessive chewing, and more. Their health suffers, too. They can suffer and die from reactions to chemicals such as chlorine in the water, poor water quality, bacterial infections, pneumonia, cardiac arrest, lesions, eye problems, ulcers, abscesses, and more.

Though some marine mammals can live longer in captivity now that conditions have improved, many do not; like zoo animals, many die of avoidable causes including poisoning, consuming foreign objects, transit stress,

BOX 6.1
_____

FAMOUS ANIMALS: KEIKO

Keiko was an orca that garnered international fame thanks to his role in the film *Free Willy*. He was captured in 1979 in Iceland and spent years in three different marine mammal parks in Iceland, Canada, and Mexico. In 1993, he was featured in *Free Willy*, in which a young boy befriends a whale kept at a marine mammal park and later helps Willy to escape and rejoin his family in the ocean.

After the release of the film, many people began clamoring for Keiko himself to gain his freedom. The Free Willy Keiko Foundation was established in 1995 in order to raise funds to buy Keiko and to transport him from Mexico to the Oregon Coast Aquarium where he would be nursed back to health and ultimately released. He was transported by air to Oregon where he lived for a year and gained back thousands of pounds. In 1998, he was finally flown to Iceland where he underwent training to once again live in the wild. Unfortunately, Keiko was lost during one of his training sessions. He ended up in the Atlantic Ocean off the coast of Norway where he was found to be suffering from hunger and was attempting to interact with humans. He died in 2003 from pneumonia and since his death scientists have concluded that releasing an animal that lived so long in captivity was probably not a good idea.

and capture shock. Also like zoo animals, most marine mammals in captivity were wild caught. Catching marine mammals involves killing many other animals. Prior to 1989, the National Marine Fishery Service, which approves permits for wild capture in the United States, approved the vast majority of all permits. Other nations—most notably Japan—continue to catch dolphins during bloody hunts for scientific research as well as for display in local marine parks.

The Public Reaction to Zoos and Marine Mammal Parks

Surveys taken on behalf of zoos show that the vast majority of zoogoers want zoos to play a role in saving wild animals and believe that zoos perform this function. In addition, the vast majority of them think children will learn more about wild animals than in school or on TV, and will develop concern

BOX 6.2

ANIMALS IN THE NEWS

In February 2010, marine mammals kept in captivity once again made the news when animal trainer Dawn Brancheau was killed by Tilikum, a performing whale kept at SeaWorld in Orlando. As expected, the animal welfare community and a surprising number of supporters from outside of this community recommend ceasing the practice of keeping marine mammals as entertainment even though representatives of marine mammal parks and zoos advocate keeping captive wild animals. They argue that presentations such as the Shamu show at SeaWorld are less about entertainment and more about education and conservation.

For much of the public, it is difficult to see the harm in keeping wild animals captive, when entertainment venues such as circuses, marine mammal parks, and even zoos hide their morally unpleasant dealings behind a façade of glitzy performances or even conservational rhetoric. What is wrong with visiting the zoo, or the circus, or a marine mammal park?

In my adopted state of New Mexico, residents were recently horrified to hear that Kashka, a "beloved" sixteen-year old giraffe kept at the Rio Grande Zoo, was dumped in a zoo dumpster and carted off to the landfill after being euthanized in 2010. What was the outrage about? Were people horrified at the callous treatment of an animal that brought profits to the local zoo and pleasure to local residents?

It turns out that dumping dead zoo animals in the landfill is standard procedure after an animal has died. However, Kashka's body should have been driven directly to the landfill rather than placed into the dumpster for pickup with the rest of the zoo trash. At the time of this writing, a worker was under investigation for this breach in protocol.

Apparently no one cared about the fact that Kashka, a 2,200-pound animal that in Africa would roam with her family over a range that extends up to 100 square miles, and could run as fast as 35 miles per hour, was kept in an enclosure at the zoo that was a tiny fraction of her natural habitat. Kashka should have been living in Africa with her kin, traveling and mating and socializing with her fellow giraffes, foraging for food, and even dying in the wild. Why was she removed from that life and forced to live in a tiny space, to give birth to babies that will eventually be sold to other zoos, all to entertain and "educate" the public? And although she certainly should not have been dumped in a dumpster after her death, the reality is that that sad ending was only the final sad coda to a sad life.

for wild animals by visiting zoos. Other research, however, demonstrates that this is not the case. Research has shown that the average visitor spends thirty seconds to two minutes per enclosure—for example, forty-four seconds is the average time spent in front of a reptile enclosure at the National Zoo in Washington, DC (Mullan and Marvin 1999). Most visitors do not read the labels attached to these enclosures, which indicates that there is very little educational information being conveyed. Social ecologist Stephen Kellert's research (1979, 1997) has indicated that zoogoers remain poorly educated about animals and their plight. In fact, according to Kellert's research, after visiting the zoo the major message for many is that humans are superior to other animals.

In a recent study of Chicago's Lincoln Park Zoo ape house visitors, researchers discovered that people ignored signs, complained when apes were resting, and fabricated answers to children's questions (Janega 2007). In 2007, the American Zoo and Aquarium Association conducted a study which, according to the AZA, demonstrated that zoos promote conservation messages among zoogoers. Yet Lori Marino and her colleagues evaluated the study and, based on the flaws that they found as well as other research into zoogoers' attitudes, determined that "there is no compelling or even particularly suggestive evidence for the claim that zoos and aquariums promote attitude change, education, and interest in conservation in visitors" (Marino et al. 2010:137).

The concept of zoos is full of contradictions or, as human-animal studies scholar Jonathan Burt puts it, they "are often places out of joint" (2002:259). Zoos offer visitors a chance to escape the city and journey into "nature" yet there is nothing natural about keeping penguins, tigers, or elephants in a city zoo. The animals are often housed indoors, with climate control to keep the animals alive, whether in a small, barred enclosure or in a large naturalistic setting made of fiberglass and concrete rocks. And as philosopher Keekok Lee asks (2006), are the animals kept in zoos really wild animals, or just shadows of wild animals? Zoos today focus on conservation, yet this is a recent and somewhat profit-driven change. The history of Western zoos is tied to the destruction of wildlife around the world, and is still involved in the capture of wild animals and the deaths of surplus animals—all in the name of conservation. Zoos do focus on education, yet some animals have more educational value than others. Pandas and other animals that are easy to anthropomorphize, and have the round furry bodies and big round eyes that draw people to them, are the most highly sought after and receive the most visitors. People visit zoos because they love animals, yet even while they feel guilty about the conditions in which the animals live, self-interest

Figure 6.1. Among several in residence, these giraffes are being fed at the Living Desert Zoo and Botanical Gardens, Palm Springs, CA. (Photograph courtesy of the author.)

(the desire to see or touch the animals) wins out, keeping zoos in perpetual business. As philosopher Ralph Acampora points out (2005), no matter how authentic the enclosure, the whole point of a zoo is to bring humans into at least viewing contact with animals—a situation that very rarely occurs in the wild—thereby rendering the animals' behavior unnatural.

It is strange that our love of animals and our ability to anthropomorphize at least some of them do not then allow us to empathize with them and end their captivity. Yet as Nigel Rothfels (2002) puts it, zoos still disappoint. People do not just want to *see* animals; they also want to *connect* with them, a condition that is impossible given the structural limitations of the zoo. So even though zoos are for people and not for animals, we are still left unsatisfied.

Circuses

Although circuses as we know them today are an American invention, they have their roots in two different historical phenomena: Roman public exhibitions and medieval European traveling shows.

Ancient Romans enjoyed attending a variety of public games and festivals, including horse and chariot races, gladiator competitions, and other human-animal events held in open-air arenas. Roman animal entertainment venues included the Circus Maximus and the Flavian Amphitheater (later known as the Colosseum); when the latter was dedicated in the first century CE, 9,000 animals—elephants, lions, tigers, and other exotic and dangerous creatures—were slaughtered over a period of 100 days. Greek naturalist and philosopher Pliny the Elder, writing of the suffering of elephants in the Colosseum, wrote:

> When the elephants in the exhibition given by Pompeius had lost all hopes of escaping, they implored the compassion of the multitude by attitudes which surpass all description, and with a kind of lamentation bewailed their unhappy fate. So greatly were the people affected by the scene, that, forgetting the general altogether, and the munificence which had been at such pains to do them honour, the whole assembly rose up in tears, and showered curses on Pompeius, of which he soon afterwards became the victim.
> BOSTOCK AND RILEY (1890:253–254)

But these attractions although popular did not migrate to other parts of Europe. Europeans instead enjoyed traveling acts featuring wild animals, performers, and human oddities. At the end of the eighteenth century, they moved to the United States in the form of **dime museums** and were run by the likes of P. T. Barnum. These early dime museums exhibited animals alongside people with disabilities, tattooed people, native people, and manufactured fakes such as the **Fijian mermaid**, a mummified creation made up of the parts of multiple animals, which was intended to resemble a mythological creature. By the 1840s, the dime museum part of the circus finally became the circus sideshow, and P. T. Barnum, founder of the American Museum, went on to found P. T. Barnum's Museum, Menagerie, and Circus.

Like zoo animals, circus animals were caught as babies by animal collectors and hunters. Jumbo, the famous Barnum circus elephant, was caught by a German hunter who killed Jumbo's mother and wrote, "She collapsed in the rear and gave me the opportunity to jump quickly sideways and bring to bear a deadly shot, after which she immediately died. Obeying the laws of nature, the young animal remained standing beside its mother. . . . Until my men arrived, I observed how the pitiful little baby continuously ran about its mother while hitting her with his trunk as if he wanted to wake her and make their escape" (Rothfels 2002:64).

Most circuses still contain a mix of human and animal acts, although circuses no longer showcase human oddities. Animal acts include old-fashioned equestrian events, wild animal acts, and a variety of trained elephant acts, which continue to remain the biggest crowd pleasers and the biggest money makers for modern circuses. Circus-goers can watch the trainer demonstrate his control over a dangerous wild animal through stunts by placing his head in a lion's mouth or by wrestling with a 350-pound tiger.

We know that, during the early days of the circus, trainers threatened, whipped, and beat animals in order to get them to perform. Animal rights activists claim that is still the case in many circuses today. Former circus employees and undercover videos shot by animal rights groups show that circus personnel may use food deprivation, intimidation, and various forms of physical and emotional punishment to train animals to perform tricks. Circuses claim that their training methods are based on a loving bond between animal and trainer, yet undercover video footage shows elephants being whipped and shoved with bull hooks—a fireplace poker-like tool used to control behavior—and electric prods. When not performing, elephants and other circus animals are caged or "picketed" (one front and one rear leg are chained to a cable) for most of their lives. Many respond to life in captivity by demonstrating stereotypic behaviors such as weaving and rocking, which are associated with captivity-related stress.

BOX 6.3

ORGANIZATIONAL FOCUS: PERFORMING ANIMAL WELFARE SOCIETY

The Performing Animal Welfare Society (PAWS), founded in 1984 by former Hollywood animal trainer Pat Derby, is a captive wildlife sanctuary where abandoned, abused, or retired performing animals and victims of the exotic animal trade can live in peace and dignity. Derby trained the animals on the shows *Lassie*, *Flipper*, and *Gunsmoke*, as well as those appearing in a number of animal films and television commercials. She was dismayed to see many of the abusive techniques used to train animals in entertainment. After leaving the industry, she realized that the plight of these animals, especially after they were no longer needed in Hollywood, was dismal and decided to do something. PAWS not only takes in animals from the entertainment industry, it also works with public officials to create policy and legislation that protect animal actors from abuse.

Figure 6.2. "What are you in for?" (Cartoon by Dan Piraro. Courtesy of http://www.bizarro.com.)

Unfortunately, this stress often results in attacks where elephants have lashed out at trainers or caregivers, sometimes killing them. In the United States alone, there have been twenty-eight deaths due to elephant attacks since 1983. These attacks virtually always result in the elephants being killed.

Historically, all of the animals found in circus shows were captured from the wild. Even though this is no longer the case for many animals today, there are still older elephants and other animals in modern circuses that were born in the wild to parents that were shot to death in order to catch their young. In 1995, Ringling Brothers opened the Center for Elephant Conservation, an elephant captive breeding program, in Florida. This center has so far bred twenty Asian elephants, all of whom are used to supply Ringling's two touring units with elephant performers, of whom there are currently sixty-one.

Animal Racing

Animal racing has been around for thousands of years. The ancient Greeks and Romans, for instance, famously held chariot races using horses. Greyhound racing, the oldest and most popular of dog races, has its origins in **coursing**, an ancient activity in which hunters used hounds to chase and bring down animals such as hares, rabbits, foxes, and deer. Betting has accompanied dog and horse racing for thousands of years, thus providing profit not only to those people involved in breeding, training, or racing the animals but also to the general public. People attend horse races around the country and dog races in fifteen states, for the thrill of the race as well as the hope of winning money.

Horse racing is most commonly practiced in the form of thoroughbred racing, where wealthy owners of finely bred horses hire trainers and jockeys to race them. Horse races are held in the United States on a variety of tracks and at a variety of distances, usually from 900 meters to a mile and a half per race.

Thoroughbred breeding and horse racing is a multibillion-dollar industry. Training a thoroughbred can cost $22,000 per year; stud fees for top stallions can start at $25,000 and go up to $500,000, and top racing horses can cost up to $10 million. Add to those amounts the money made from gambling bets, tourist expenditures, the "purse" awarded to the owner of the winning horse (which now exceeds $1 million for the top races), and other related income, and horse racing is big money.

Even though handlers devote exacting attention to ensure the well being of the top-ranked horses, the same cannot be said for those horses that are not so successful. Minimum living and training standards must certainly be met, but the practice of racing itself is rife with dangers to the horse: It is common for them to fall or fracture their bones while racing or training, which are often fatal conditions for horses. Sometimes, injured horses are drugged so that they will race despite an injury. Other common injuries are those to ligaments or muscles as well as joint sprains, and many race-horses are also susceptible to a disease called exercise-induced pulmonary hemorrhage.

What happens when a racing horse's career is over? A winning stallion such as Barbaro will usually be put up to stud when he is retired. The most successful horses can hope to live a life of leisure once retired, but being a success is certainly no golden ticket to a happy retirement. Thousands of horses, most of whom will never win a race, are bred each year. Most horses will see their careers end after just a season or two and will be sold at auction— sometimes to people who want them as pets, sometimes to businesses such as horseback-riding outfits, and sometimes to slaughter. Indeed, the win-ner of the 1986 Kentucky Derby, Ferdinand, was reportedly slaughtered in Japan for pet food. In 2007, the last of the American horse slaughterhouses were closed. (Prior to that time, more than 100,000 horses were slaughtered annually in the United States.) Currently, there is a movement afoot in the United States to reopen them. Until that happens, many horses are shipped to Mexico or Canada for slaughter.

Like horse racing, greyhound racing is associated with gambling and takes place at the same tracks on which horses are raced and where the infrastruc-ture is in place for gambling. It is a less lucrative industry than horse racing, but it has the potential to bring in big dollars: Millions of fans bet $3.5 billion in 1992, the year the sport was at its peak. Since then, attendance has been down as a result of the sport's unsavory reputation and wide public concern about mistreatment of the dogs, but revenue still tops $1 billion a year.

In a world where millions of companion dogs are still euthanized every year for no other reason than there are too many of them, the breeding in

Figure 6.3. Revenues from greyhound racing have topped the billion-dollar mark. (Photograph courtesy of Jan Eduard, Wikimedia Commons.)

the greyhound industry is a cause for concern among animal welfare advocates. More than 1,500 breeding farms produce nearly 30,000 dogs every year for this sport. Breeding greyhounds live stacked in kennels either outdoors or in barns, with no exercise, no toys, no love, and no life outside of the cage. Even racing dogs live in small kennels during their life off the track; sometimes as many as a thousand dogs live at each track. There is so little regulation about the care of racing dogs that most racetracks have their own rules regarding dog welfare.

Because greyhounds are so much cheaper to breed and train than thoroughbreds, they are much more expendable. In addition to heart attacks, injuries such as broken legs and necks are rampant in the industry. Some dogs are drugged in order to perform, and kennel cough is common due to the close living conditions. Veterinary care is minimal for animals that are expected to live for at most only a couple of years.

Perhaps nothing is sadder in the greyhound industry than what happens to a dog when his or her career is over. Although the dogs' lifespan is more than twelve years, they stop racing after three to five years. Some dogs are returned to the breeding farms where they were born, spending the rest of their lives in small kennels as breeders. Owners or track operators kill some dogs outright. In 2002 the bodies of 3,000 greyhounds were found at the home of a former Alabama racetrack security guard, who was paid ten dollars apiece to "retire" them. Just a few years later, in 2006, it was discovered that British trainers paid a man ten pounds apiece to dispose of all their old and surplus dogs. Over fifteen years, he electrocuted or shot over 10,000 dogs and used their bones in his garden.

Animal Fighting

Blood sports were popular in ancient Rome, Imperial Japan, and China, and in Southeast Asia; these included everything from cricket fighting in China and cockfighting in Southeast Asia to gladiator events in which animals were pitted against animals, people against people, and animals against people. Europe from the Middle Ages to the modern era saw a variety of blood sports, including dogfighting, bearbaiting, and bullfighting. Many of these practices were banned during the Victorian era in Europe, in some cases because of growing concerns about animal welfare, but in other cases because of concerns about the impact of such practices on their practitioners.

Dogfighting involves placing two viciously trained dogs in an enclosure where they fight either until one is too injured to continue and quits the fight due to extreme pain or severe exhaustion or until one dies. Some dogfighters use performance-enhancing drugs as well. At a dogfight, more than a hundred people may place bets up to $50,000. The underground dogfighting industry is huge, with millions of dollars involved in selling, breeding, training, fighting, and betting on the dogs. Some top dogs are worth tens of thousands of dollars.

Photos of dogs that have survived dogfights show pit bulls with faces so badly scarred they often cannot see. Injuries include ripped ears, ripped mouths and noses, crushed sinuses, tissue damage, and broken bones. Deaths are usually attributable to these injuries, massive blood loss, and sometimes exhaustion. Even though rules generally state that a fight ends when a dog gives up or refuses to engage the other dog, some fights can last hours and end only when one dog dies an agonizing death. Seriously injured survivors may die days later from blood loss or infection, or their owners may kill them if they do not wish to keep a dog that loses. Rhonda Evans, DeAnn Gauthier, and Craig J. Forsyth, in their study of white Southern dogfighters or dog men (2007), suggest that the dogs symbolize the men's masculinity; when the dog loses, he or she must be killed in order to restore the man's honor and virility. Though dogfighting is illegal in every state, it is growing in popularity. About 250,000 dogs—mostly pit bulls—are victims every year. It is estimated that at least 40,000 people across the country either own or breed pit bulls for fighting, and it is difficult for police to monitor criminal activity because dogfighters are so secretive.

Cockfighting involves placing two specially bred and trained "gamecocks" together in a pit and betting on the winner. Cockfighters, or "cockers," usually attach razor-sharp knives or ice pick-like gaffs to the birds' legs so they can injure and mutilate opponents. Birds often suffer from lacerations, eye injuries, punctured lungs, and broken bones. Like dogfights, cockfights often end in death, although some animals are forced to fight again and again. Hundreds of people can attend a fight, and violence can erupt even outside of the pit. High-stakes betting and weapons are commonplace at cockfights and, like dogfighting, the practice is connected to the illicit drug trade. Arguments at cockfights can result in human injuries, and fatal shootings are not uncommon. As with dogfighters, cockfighters identify with their birds and even feel strong attachments to them, mourning when the animals lose (and die) and feeling pride when they win.

Alternative Ways of Watching Animals

Americans crave animals in their lives. For many people, beloved companion animals do not completely fill that need. As the continued existence of zoos and circuses and the use of animals in rodeos, animal racing, and more demonstrate, many of us seek out wild (and domesticated) animals even when that means we will support industries that cause harm to those very same animals.

Other activities that allow people to see, and sometimes interact with, wild animals are gaining in popularity as well. Whale watching is one such activity, as is ecotourism. Both are driven by some of the same impulses that drive us to watch wild animals in zoos or marine mammal parks: what some scholars call the quest for wildness, often to fill in a gap that is missing in Western industrialized lives. Even in some of these venues, however, the quest for wildness that drives the tourists may be a fabrication: In order to allow visitors access to wild animals, the animals must be, in fact, contained in some way. Even when we visit animals in large wild animal parks, there must be some form of containment for the animals; otherwise, we would not be able to see them. Women's studies scholar Chilla Bulbeck has studied ecotourism sites (2005) and has interviewed attendees; she has found that many visitors experience some guilt about visiting these sites knowing that the presence of humans is not good for the animals. Ultimately, though, self-interest (the desire to see or touch the animals) wins out, even for the more conservation-minded of the tourists. The irony is that the more wild the site, the less the animals' movements and behaviors are controlled but the more that the visitors' activities are constrained. The animals' freedom is increased (including their freedom to not be present), but for many visitors the pleasure is decreased. Research on the benefits of ecotourism indicates that there are a number of negative environmental costs, such as the high costs in water, food, and energy to the host countries because of Westerners' much higher consumption of resources than indigenous populations. Other costs include the displacement of indigenous peoples from tourism sites, death and injuries to tourists, and stress and behavior change for the animals. Still, many advocates argue that the benefits of ecotourism for the human tourist and for the animals and their environment outweigh the costs.

Whale watching is another activity that brings people in contact with "wildness" but which has, perhaps, less of an impact on the animals being watched. Every year, millions of people sign up for organized whale-watching trips through commercial ventures and many thousands more go out in kayaks or small boats to watch, and sometimes interact with, whales as well as dolphins and porpoises. In these encounters, people report having an authentic, or even spiritual, feeling, and many return with an interest in conserving the habitats of the creatures that they just saw. And although there can be negative repercussions to the animals from whale-watching excursions (such as from the intrusion of motorized vessels into a whale pod's migratory pattern), dolphins and whales can adapt to such changes. In addition, in many cases, the animals actually appear to be watching the humans as well. Often times, especially in the case of dolphins, the animals will spontaneously perform for, or approach, the viewing people. One interesting side note is that in countries such as Japan, which have thriving whale-hunting (and dolphin-hunting) industries, the opportunity to go on whale-watching trips may play a role in changing the attitudes of many Japanese about the killing of whales and dolphins.

Another way that we can see animals is through television and film. Today, there is a huge variety of documentary programs and films available that

Figure 6.4. "Show Business." (Cartoon by Dan Piraro. Courtesy of http://www.bizarro.com.)

show wild animals in their own habitats—such as the poignant and remarkable *March of the Penguins*—as well as the incredibly realistic computer-generated animation and animatronics that allow us to view all manner of animals without interfering in an animal's life.

In 2009 alone, animal films such as *Marley and Me, Hotel for Dogs, Bolt,* and *Space Buddies* grabbed moviegoers' attention with animal-friendly messages. At the same time, the number of Americans who enjoy watching wildlife in animals' natural habitats reached more than seventy million, demonstrating a growing willingness to connect with animals on their own turf, and on their own terms. We will discuss the use of animals in film more in chapter 16.

Suggested Additional Readings

Acampora, Ralph. 2005. "Zoos and Eyes: Contesting Captivity and Seeking Successor Practices." *Society & Animals,* 13: 69–88.

Acampora, Ralph, ed. 2010. *Metamorphoses of the Zoo: Animal Encounter after Noah.* Lanham, MD: Lexington Books.

Berger, John. 1977. "Why Zoos Disappoint." *New Society* 40: 122–123.

Berger, John. 1980. *About Looking.* New York: Pantheon.

Bulbeck, Chilla. 2005. *Facing the Wild: Ecotourism, Conservation, and Animal Encounters.* London: Earthscan.

Hanson, Elizabeth. 2002. *Animal Attractions: Nature on Display in American Zoos.* Princeton, NJ: Princeton University Press.

Lawrence, Elizabeth. 1985. *Hoofbeats and Society: Studies of Human-Horse Interactions.* Bloomington: Indiana University Press.

Lee, Keekok. 2006. *Zoos: A Philosophical Tour.* New York: Palgrave MacMillan.

Malamud, Randy. 1998. *Reading Zoos: Representations of Animals and Captivity.* New York: New York University Press.

Mullan, B. and G. Marvin. 1997. *Zoo Culture: The Book about Watching People Watch Animals.* 2nd ed. Chicago: University of Illinois Press.

Rothfels, Nigel. 2002. *Savages and Beasts: The Birth of the Modern Zoo.* Baltimore: Johns Hopkins University Press.

Warkentin, T. and L. Fawcett. 2010. "Whale and Human Agency in World-Making: Decolonizing Whale-Human Encounters." *Metamorphoses of the Zoo: Animal Encounter after Noah*. Ralph Acampora, ed. Lanham, MD: Lexington Books.

Suggested Films

The Cove. DVD. Directed by Louie Psihoyos. Boulder, CO: Oceanic Preservation Society, 2009.

A Life Sentence: The Sad and Dangerous Realities of Exotic Animals in Entertainment. VHS. Sacramento, CA: Animal Protection Institute, 2006.

Lolita: Slave to Entertainment. DVD. Directed by Timothy Michael Gorski. Blackwood, NJ: Rattle the Cage Productions, 1993.

March of the Penguins. DVD. Directed by Luc Jacquet. Los Angeles: Warner Independent Films, 2005.

The Urban Elephant. DVD. Directed by Nigel Cole/Allison Argo. New York: Thirteen/ WNET New York, 2000.

A Whale of a Business. VHS. Directed by Neil Docherty. Melbourne, FL: PBS Frontline, 1997.

Wildlife for Sale: Dead or Alive. VHS. Directed by Italo Costa. Oley, PA: Bullfrog Films, 1998.

Working from Within: An Ethnographer in Human-Animal Worlds

GARRY MARVIN
Roehampton University

In 1996, I returned to the academic world after ten years working in television documentary making, and I quickly needed to work out where my discipline, social anthropology, had moved on to since I was last teaching and researching. However, what engaged my attention and interest were not so much new theoretical perspectives in social anthropology but rather the emergence and development of a multidisciplinary field in which humans and their relations with other animals were being brought from the margins of academic interest to the fore. I had previously written about human-animal relations, but was this human-animal studies?

This field was being shaped and developed, in the main, by scholars from the humanities in disciplines such as history, literary studies, philosophy, performance studies, the visual arts, and gender studies. What might anthropological studies of human-animal relations contribute to this new field?

The significance of animals in human cultures—in hunting, pastoralism, and agriculture, as beasts of burden and transport, and in cosmological systems and religious practices—had been recognized in much anthropological work from the beginnings of the discipline, but was there something specific in modern anthropological studies of humans and animals that could respond to issues at the heart of human-animal studies? Attending conferences with, and reading the publications of scholars who were writing in this field, I felt that what anthropology had to offer was not so much a particular or specific subject matter but rather the nature of the studies generated out of a particular kind of research process.

Scholars in the humanities are immersed in and engaged with philosophical and other theoretical arguments, documents, historical texts, literary texts, and works of art. The materials of their research are complex in and of themselves but are necessarily at least one remove from the human-animal relations to which they refer. A key research method of social anthropology is a different form of immersion and engagement—that of participant observation—being with the subjects of their research, sharing in their everyday lives for an extended period of time. What I work with is the immediacy, the presentness, the rawness of the relationships between people and animals, and such a research approach engages with a different complexity—that of the contingent nature of such relationships as they emerge, happen, and end.

An important aspect, perhaps an issue pervading this field, is a concern with the acceptability or unacceptability of the uses and treatment of animals by humans. This was immediately troubling for me because my research, firstly with bullfighting and now with hunting, has centered on relationships between humans and animals that result in the deaths of animals. However, issues of ethics were not central to my research. That such practices are criticized and condemned as unacceptable in the modern world is not surprising but, as an anthropologist, my task is different. My anthropological project is to understand them, to understand how human and animal lives and deaths are configured within such events, how they are experienced (or thought to be experienced in the case of animals), what meanings they have for those who participate, and what can be said about their social and cultural construction.

During my preparatory reading of literature for my PhD research on bullfighting in Spain, my early thoughts were that I would need to understand what might be the significance of killing bulls in a highly ritualized manner in a public arena. I initially thought that the event might be some remnant of a sacrificial ritual and that I would need to explain the purpose and meaning of bull killing/sacrifice in modern Spain. It was only ethnographic fieldwork—engaging with bull breeders, matadors, and aficionados—that revealed to me that although bulls are killed in a bullfight, focusing on the fact that they are killed could not explain the complexity of the event, the experiences of it, or the responses to it. I came to understand that the central concern of the bullfight, for those who participate, is not that bulls are killed but rather with how those deaths are brought about. It was the cultural sense of that "how" that I needed to understand and explain. In conversations with me, bull breeders spoke of their admiration, respect, and even love for the animals they raised to perform and ultimately die in the arena. Matadors spoke in similar ways. Bulls, they hoped, would be partners or collaborators in a work of art; they too expressed a love for bulls. Such thoughts, emotions, and experiences cannot be understood from outside the event; they can only be understood, in their cultural context, from within. It is such understanding that anthropological fieldwork offers. It is generated out of months of close engagement with those who finally come together in the arena for the bullfight. I needed to spend time with those who worked with bulls, to observe matadors as they trained and were taught the craft of their profession, to discuss their hopes in hotel dressing rooms before a performance, and to spend many, many, hours in bars with aficionados as they discussed their passion for, and the aesthetics of, los toros. Simply watching a bullfight is not enough to generate an understanding of the event, how it is shaped, how it is experienced, and the nature of its meaning for those who perform and for those who witness the performance. Indeed, without understanding all that comes before a bullfight, an outside observer would not, and could not, comprehend the event at all.

In my work on bullfighting, I now think I dwelt too much on meaning— understood as purpose, intention, significance. I was too concerned to answer what it meant to raise bulls, what it meant to be a matador, and what was the cultural meaning of the bullfight. I now feel I should have paid more attention to experience. Not just what it meant *to be a matador but also* how *it was to be a matador or a bull breeder or an aficionado. This would have allowed for a richer account of what is going on between men and bulls in this event.*

In my attempt to understand hunting, I am now much more attuned to experience—how it is for a hunter to hunt, how it is to be in hunting mode, the relationships between hunter and hunted, and what experiences are sought and generated in hunting. The meanings of hunting (again as purpose, intention, and significance) are generated before and after hunting, not while it is happening. It is understanding and interpreting what happens and how it happens that comprise my anthropological project. When I approached hunting, I was already alert to the possibility that the killing of animals might not be the end (in the sense of both its purpose and finality) that defines the activity. From outside hunting, especially if one has a critical view of the event, it is easy, perhaps inevitable, to focus on the fact that animals are killed. However, as with bullfighting, such a focus is too narrow for understanding the nature of hunting and its experiential and cultural complexities; complexities that can only be explained by hunters themselves. As with the bullfight, for hunters it is not that animals are killed which is significant but rather how those deaths are brought about. What I have learned is that hunting, for the hunter, is all that which occurs before a shot is taken. What constitutes hunting is the engagement with the landscape, the slow tracking and stalking, or the silent waiting, that might allow the hunter to close in on their potential prey. The hunters with whom I work are very clear on this. Shooting is simply shooting and it is the experience of hunting, not shooting, that they seek. Understanding what hunting is for them, how it is for them to hunt, can only come through immersing myself in their world—listening to the hunting stories they tell to fellow hunters, preparing to go out hunting with them, being with them in the woods at dawn or dusk as they hunt, and relaxing with them after a hunt as they recount and reflect on the day's activities and experiences. Also, importantly, I need to be attentive and receptive to them teaching me about who they are, how they are, and what it is they are doing. My responsibility is to learn this as well as I can and then to represent and interpret their practices as completely and fairly as possible in my academic work.

Ethnographic fieldwork involves the researcher in at least four ways of being during research—being there, being with, being open to the unexpected, and, crucially, being open, in a non-normative way, to the ways of life of those whom we seek to study. I believe that the results of such research practices make valuable contributions to the field of human-animal studies in that they are grounded in detailed specificity—studies that attend to complexities of the relationships that particular people have with particular animals, at particular times, and in particular places.

7

The Making and Consumption of Meat

The special sounds good, but can I substitute the pork chop for a fried chunk of your left buttock?

Figure 7.1. "Pig Buttocks." (Cartoon by Dan Piraro. Courtesy of http://www.bizarro.com.)

WHY DO WE CONSIDER it totally normal to eat pig buttocks, yet totally ridiculous to eat human buttocks? In this chapter, we will discuss what "meat" is, how animals are made into meat, and why only some animals can be made into meat.

For most people in the United States, the only interaction we have with the animals that became our dinner is the preparation and consumption of them. People today eat "meat," not "animals." Separated from the production process by geography and the behind-the-scenes nature of meat production, Americans consume billions of animals each year, without even really recognizing it. Our supermarket meat, neatly covered in plastic wrap on its Styrofoam tray, bears little or no resemblance to a dead animal, let alone a living one. We do not see eating meat as contact with animals; we see it as contact with "food" and are conditioned to see such consumption as "natural." Yet the relationship that modern Americans now have to the meat that they consume is not natural at all. It is also very different from the relationship enacted in traditional societies—whether hunting and gathering, pastoral, or agricultural—and even in our own society until about a hundred years ago.

Meat Taboos

Technically, any animal can become meat, but every society has social rules regarding which animals are edible and which are not. Although meat is highly prized in societies around the world, certain kinds of meat are forbidden by those same societies. Those rules have nothing to do with the animal itself, and everything to do with the meanings surrounding animals and food in those societies.

In general, scholars who have tried to understand the reasons behind food taboos—which overwhelmingly feature meat—have tended to focus on two types of explanations: functional explanations and symbolic ones. Functional explanations tend to focus on the utility of a particular animal— whether it is more valuable alive or dead. Other functional frameworks focus on whether it is cost-efficient to eat certain kinds of animals, or whether restricting the consumption of certain animals will preserve resources in a particular environment, or whether eating certain animals can cause health problems in humans. Symbolic explanations, on the other hand, emphasize the meanings found in the animal itself. For example, totemic societies are societies in which people are said to be descended from animal totems. Typically, the totem is forbidden as a food source (except on very rare ritual occasions). In this case, those animals are not consumed because people are considered to be related to them, and humans do not eat their own kin.

In India, for example, cows are taboo to eat because they are considered sacred to Hindus. Anthropologists generally explain the taboo on cow flesh from an economic standpoint: Cows are worth more alive (or "on the hoof") than dead. Plough-pulling cows are extremely valuable in India for farming; their dung is used in building and for fertilizer and fuel, and their milk is consumed daily. Even the urine of cows is valuable, and is used in **ayurvedic medicine**. Because of their economic value, cows are exalted and protected. The sacred status of cattle also stems from the history of animal sacrifice in Indian society. When India was still a hunting culture, it was very common to regularly sacrifice large and small animals during religious rituals. After India domesticated animals and began to rely on them for more than just meat—and especially after the introduction of the plough in the eighth century BCE, killing cows in this way became untenable. The Hindu ban on killing cattle, which was influenced by Buddhism and Jainism, could thus be seen as a religious response to an economic problem. Not only is it taboo for Hindus to eat cows, it is also illegal to slaughter a cow in most of India, even though Indian Muslims and Christians both eat cows. (Many Hindus—especially Brahmins, the highest caste of Hindus—are vegetarians.) All Indians, however, consume dairy products (especially from cows), and milk and milk products are used during Hindu rituals. Finally, the Hindu ban on cattle slaughter and consumption also helps to maintain the Indian **caste** system—because untouchables (who are considered to be unclean) are allowed to eat beef. Thus, the food differences reinforce the status differences among the populations.

Among Jews and Muslims, pigs are considered taboo (*kashrut* to Jews and *haram* to Muslims). Some scholars have explained this by again focusing on the practicalities of pig production: Raising pigs in the hot, dry Middle East makes little sense because of pigs' need for moisture and shade. In addition, because pigs are omnivores and do not graze, they would need to be fed human food in order to survive; this makes little economic sense. Anthropologist Mary Douglas (1975), on the other hand, explained the biblical taboo on not just pigs but also on all of the animals said to be "abominable" in the Book of Leviticus with an argument that focused on the purity and impurity of certain animals. She argued for a systemic explanation of food taboos by focusing on the ancient Hebrews' symbolic system, which was based on the concept of holiness. Dietary rules, for Douglas, exemplify the metaphor of holiness. Specifically, she argued that animals that are unclean, or abominable, do not conform to their proper class—animals that fly must have feathers and eat seeds, animals that live in the water must have fins and scales, and animals that live on the ground must chew the cud and have cloven hooves. All animals that do not conform to expectations—such as shellfish, amphibians, reptiles, carnivores, bats, and birds of prey—are therefore abominable and thus taboo. Douglas's explanation is a symbolic and a practical one: Because the ancient Hebrews were pastoralists and raised ruminants such as sheep and goats, the prototypical land animal would have matched the characteristics of the animals that they already raised. Although Muslims do not share all of the same food taboos with Jews—for instance, they eat camels, which are prohibited by Jews—they do share many, including the prohibition on birds of prey. Of the two major Muslim sects, Shia share most of the Jewish prohibitions, and Sunnis only share some. Seventh Day Adventists also share the Jewish food taboos and many are also vegetarians.

THE MAKING AND CONSUMPTION OF MEAT   129

In the United States and the West in general, dogs are inedible because they were initially domesticated as a hunting partner, and not a food animal; this ultimately led to them attaining the status of pet. Once an animal is defined as a pet rather than a food, it becomes very difficult for that animal to be consumed. To be a "pet" is to be considered, at least in part, family, and eating a member of the family (even an animal) is a symbolic form of cannibalism. In addition, from an economic perspective, it does not make sense to eat an animal that must be fed other animals first.

However, dogs *are* eaten in China, Vietnam, and Korea, as well as in some Pacific Island cultures. Again, anthropologists generally explain this contradiction from an economic perspective. In cultures such as that of the United States, where there is an abundance of animals for consumption, dogs are more valuable as hunting partners, and for security and companionship. In theory, they can be consumed in cultures where either there are few other animal resources or their other services are not highly valued. This explanation fails, however, because there are indeed plenty of other protein sources in China, and dogs do serve other functions there. Yet anywhere from ten to twenty million dogs are still raised for slaughter in that country. Another explanation might lie in their symbolic value. In China and the Philippines, dogs, when they are consumed, are thought to "warm the body," a highly prized characteristic.

Even though dogs, cows, and pigs are the most well-known of all the tabooed animals, most cultures have at least one food taboo, and most of those relate to meat. Scavenger animals such as vultures are often prohibited, for example, because of their association with death and disease, as are rats and mice. Sometimes animals, or parts of animals, are taboo because they are associated with the poor or with famine. In the United States, organ meats, pigeons, and squirrels are associated with the poor, and are thus not highly valued by middle- or upper-class Americans. Cats are rarely raised as food, but during desperate economic times cats have been eaten in China, Russia, and Europe.

How Animals Become Meat

Ultimately, it has to make economic and symbolic sense for particular animals to be consumed as food in a given society. But how does an animal become meat?

In order for an animal to become meat, the animal has to be considered "edible" on the basis of that culture's mode of production as well as on the

Figure 7.2. Processed meat products on display at a U.S. grocery store. (Photograph courtesy of Blair Butterfield, Wikipedia Commons.)

basis of its symbolic system. Then the animal must be *defined* as meat. In English-speaking countries, for example, those animals considered edible are known as "livestock"—literally "supply" or "money" that is alive. It is interesting to note that the term does not actually mean "meat animal." Instead, it refers to animals that served as a form of currency or wealth, indicating that the edible component of livestock was at one time not as important as the other functions of such animals. Another element in making an animal edible is turning the animal from a sentient creature into an edible object. How is subjectivity removed from an animal? Not naming animals that are to be eaten is one way; for the most part, we do not eat those with whom we have a personal relationship.

Once an animal has been defined as one that can be consumed, an animal becomes meat only upon being slaughtered and then butchered. Animals must be transformed from living creatures into edible foodstuff. Hunting cultures, pastoral societies, and farming societies all have specific methods of killing and butchering an animal. Butchering transforms the whole animal into individual parts, which then become known as "meat." Meat, then, is really just a disassembled or deconstructed animal.

In the United States, another factor that determines the creation of meat relates to how the animal is raised. In the industrial West, animals are born and raised in order to *be* meat. They are seen as products ("stock") from their birth until their death, and their importance is based only and entirely on their economic value as meat. So one way that an animal is transformed into meat is through its *production* in a factory whose final product is meat.

In the West, after the animal is butchered comes packaging. Pieces of animal flesh are, for most Americans, consumed only after they have been packaged in Styrofoam and plastic and purchased at a grocery store. This packaging further distances the live animal from the final product and the consumers from the reality of what they are eating.

Meat Consumption in the Past

Our pre-human ancestors, the *Australopithecines* and *Homo habilis*, were most likely scavengers and gatherers and may have even been hunted themselves by wild animals. For instance, paleoanthropologists have found skulls of *Australopthecines* with holes that were probably caused by saber tooth tigers, eagles, and other predators. Primarily, they would have been vegetarians who supplemented their diet with the occasional dead animal that they could scavenge.

With the evolution of *Homo erectus* around a million and a half years ago, our ancestors became hunters, eating big game animals while still consuming vegetable matter. Many anthropologists feel that the bigger brain of *Homo erectus*, combined with the development of more sophisticated tools, may be related to the rise of cooperative hunting as an economic strategy at this time. As archaic *Homo sapiens* and anatomically modern *Homo sapiens* arose a couple of hundred thousand years later, our species made their living primarily through hunting of large animals and gathering plants.

About 15,000 years ago, during what is known as the **Mesolithic revolution**, the most recent glacier began to retreat leading to overall climate warming. As the earth's climate warmed up, many large herd animals moved north and people living in the south began to adopt a more generalized economic strategy. They focused less on large animals and more on small animals, birds, and fish, as well as a variety of grasses, beans, peas, and cereals. As time passed, many of the large megafauna ultimately became extinct thanks to overhunting. Also during this time, the dog was domesticated as a hunting partner, allowing humans to more efficiently hunt small animals. Hunting and gathering remained the primary economic activity of all humans until the Neolithic Revolution, which began about 10,000 years ago, when humans first domesticated plants and animals. But even after that time, many human populations around the world never domesticated animals and remained hunter-gatherers. Anthropologists who have studied these cultures have shown that in most cases, these cultures primarily ate vegetable food with meat as only a small part of the diet. Some cultures, however, are exceptions to that rule. The Inuit, for example, survive on a diet primarily made up of animal flesh.

With the Neolithic Revolution, the first food animals were domesticated in the Middle East and later in Asia and Africa. This led to the first civilizations in which animals were purpose-bred, and raised and slaughtered for food. In traditional agricultural and pastoral societies, livestock have now been raised for thousands of years—years in which the animals were not fed, but were allowed to graze on pasture, a system that was simple, economically efficient, and good for the environment. Animals were slaughtered for food only rarely—often for ritual purposes. The only cultures that ate meat on a daily basis were those who subsisted largely on fish, and populations such as the Inuit who live in environments with very little plant food.

As animal domestication became more intensive, and as the great civilizations arose in ancient times, meat eating became more common. Even then, most people did not eat meat on a daily basis. Only the wealthy consumed huge amounts of meat; elites not only ate meat at every meal, but threw away large amounts of meat as well. For the Greeks, as in many cultures, the wealthy used the consumption of animals as a marker of their elevated status. Even today, the very wealthy demonstrate their wealth and status by consuming exotic, expensive, and even endangered animals.

Modern Meat Production

It was not until the nineteenth and twentieth centuries that meat consumption became a daily activity in the United States, and this is only because of major changes in how livestock were raised and how meat was produced. Changes in production patterns resulted in vast changes in consumption; rising rates of consumption in turn fueled the drive to find more efficient ways of increasing production. This new demand for animal products has increased to a point where it can no longer be satiated by the family farm system that emerged thousands of years ago. **Factory farms** now produce the overwhelming majority of meat, dairy, and eggs today.

The first major innovations with respect to modernizing livestock production in the United States were the expansion of the railroad into the South and West and the development of the refrigerated railroad car. Prior to that time, in the nineteenth century, cattle that were raised in western states such as Texas and California could not easily be converted to meat and brought to market in the large cities of the northeast. The expansion of the railroad into the South and West allowed cattle to be transported to Chicago, where large stockyards were constructed to house the animals before slaughter. After slaughtering, the refrigerated railcar allowed the fresh meat (as well as dairy products and eggs) to be shipped east, increasing meat consumption among Americans. Prior to the innovation of the refrigerated railcar, the most popular meat in America was pork. Small farmers could raise their own pigs, but cattle had to be raised by big cattle ranchers out west. With the railcar, beef became widely available and quickly was established as the single most popular meat in the country. (Beef's continuing popularity is also linked to the large amounts of undeveloped land available in the United States on which to graze cattle and raise feed. Very few other countries have as much land as we have, and thus have far lower rates of meat consumption.)

The next development with respect to modern animal-raising techniques was the introduction of methods drawn from industrialization, which can be summed up as large-scale, centralized production and intensive animal rearing. Animals confined in small spaces and with regulated food, water, and temperatures enabled easier health monitoring, and controlled what the industry calls "unnecessary" and "inefficient" animal movements and that describes in any industry increased production and increased consumption. Automation turned human workers and animals into cogs on the production line. Ironically, although industrial methods of animal production have been borrowed from the assembly line model pioneered by Henry Ford in the auto industry, Ford himself was inspired by the "disassembly lines" of the Chicago slaughterhouses when developing his own automobile production model.

Today, livestock are housed in large facilities known as **confined animal feeding operations**, where all aspects of the animals' lives are completely controlled and human-made: no outside air, no dirt, no sunlight, and no capacity for natural movement or activities such as grooming, play, exercise, unaided reproduction, or the like. Ironically, the same social behaviors that allowed livestock to be domesticated in the first place are eliminated because the animals' social structure must be subverted in favor of total confinement—either alone or crowded together into non-kin groupings. In these systems, animals are no longer seen as sentient beings; instead, they are industrial products.

Because of the close confinement necessitated by factory farm production, a number of new agricultural practices have emerged. **Debeaking** (amputating, without anesthesia, the front of the chicken's beak) is common in the egg industry, where chickens are so intensively confined in tiny cages that they may attack each other due to stress and overcrowding. One increasingly common practice is tail docking (without anesthesia, usually via banding) of pigs, to keep intensively confined pigs from chewing each other's tails. Dehorning cattle is also becoming increasingly popular; dehorned cattle require less feeding space at the trough, are easier to handle, and cannot injure other cattle. Injuries result in bruised meat that cannot be profitably sold.

Over 90 percent of eggs worldwide are now produced in **battery** conditions, where tightly packed and stacked cages of birds are kept in a large facility in which light, temperature, food, and water are strictly controlled and a steady stream of antibiotics is used to keep the birds healthy. This system is so efficient that one person can care for as many as 30,000 birds, and a single egg-laying operation can have as many as five million birds at one time.

Traditional pasture systems for pigs have been replaced in the West by confinement in large warehouses. Sows are subjected to the greatest degree of control and are kept pregnant almost full time in gestation stalls, which completely restrict movement. Antibiotics again are an important part of the diet. Veal calves (the offspring of dairy cows, from whom they are separated immediately after birth) undergo the strictest confinement. They are kept in veal crates where they cannot turn or walk, and are fed a milk substitute lacking iron in order to produce the anemic flesh prized by veal lovers. Beef and dairy cattle production has also been intensified in the past fifty years, with feedlots gradually replacing pasture, grains replacing grasses, artificial milking machines used to increase milk production, and weight maximization achieved through hormones and antibiotics.

Although animals have been raised for food for thousands of years, those were years in which the animals were not fed by man but allowed to graze in a pasture. This system was simple, economically efficient, and good for the environment. Today, however, that system has been replaced by one that consumes vast amounts of resources (water to feed the animals and to clean up the waste, chemicals to pump into the animals to keep them healthy, oil to run the factories and power the trucks that transport the animals, and grain that would normally be eaten by humans but is now fed to the animals). The principles of modern farming are to keep costs down and bring productivity up. Costs are brought down by reducing human care of the animals, by cramming animals into the smallest possible spaces, by cleaning the factories only once the animals are killed, and by killing the animals as soon as possible (such as the male layer chicks that are crushed to death soon after

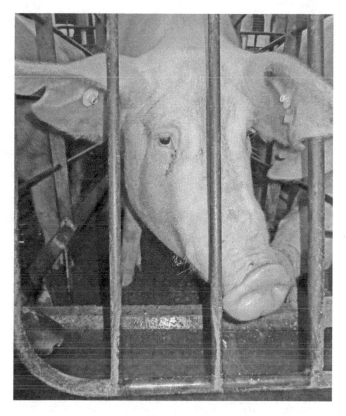

Figure 7.3. Sows like this one are confined to gestation crates throughout their lives. Because of the lack of mental stimulation and boredom, they often resort to bar biting. (Photograph courtesy of Mercy for Animals.)

birth). Productivity is increased through the use of antibiotics, hormones, and genetic modification, as well as via practices such as manipulating the light in animal barns or removing food. American beef cattle, for example, are routinely administered hormones to stimulate growth; to increase milk yield, producers inject dairy cows with hormones. As we discussed in chapter 5, farmers have also been experimenting with genetic manipulation to create new livestock breeds or modify existing ones.

Even the methods in which the animals are killed are as efficient as possible. Although it is not yet possible to mechanize slaughter entirely—a worker still has to be there to shoot the bolt into the animals' heads, followed by other workers who slit their throats, and still other workers who disassemble the animals—the process is constantly being refined to make it as efficient as possible. Animal scientist Temple Grandin, who says that her autism makes her think and feel like an animal, has pioneered a number of new methods for bringing animals to slaughter; these are intended to lessen the animals' fear and thus make the job of killing them easier. Before Grandin started working in this area, animals such as cows and sheep were so terrorized by the sights, sounds, and smells of the slaughterhouse that they had to be forced into slaughter; later, "decoy animals" were used to make the animals feel that it was safe to proceed ahead, where they were killed. By using Grandin's innovations, such as the "Stairway to Heaven" on which cows walk to their deaths, Grandin and her supporters believe that coercion and decoys are no longer needed. Being slaughtered is still nasty business, however. Although the Humane Methods of Slaughter Act mandates that animals be stunned prior to slaughter, the law excludes from the stunning process certain animals such as rabbits and birds, and is poorly enforced by regulators.

Why We Eat Meat: The Political Economy of Agribusiness

If you ask someone why they eat meat, most people will tell you that it is because it is healthy, because it tastes good, or because they have always eaten meat and everyone in their family—and, indeed, in our whole culture—eats meat. But why we eat meat, and eat meat in such large quantities in the United States, has much to do with the meat industry and how meat eating is promoted and supported by the agricultural industry and the U.S. government.

The meat industry is one of the nation's most powerful businesses, supported by huge profits as well as by subsidies from the government. In the past several decades, virtually every aspect of the meat industry has become increasingly consolidated, with a very small number of companies controlling the markets for eggs, dairy, and milk. According to the United States Department of Agriculture (USDA), the largest 2 percent of factory farms produce more than 40 percent of all farm animals. Today, just four companies—Pilgrim's Pride, Tyson, Perdue, and Sanderson Farms—produce 58.5 percent of chickens used for meat. A small number of powerful chicken, pork, and beef producers and processors dominate the U.S. market, edging out not just the small family farms but the medium-sized farms as well. These corporate giants such as Cargill, Tyson Foods, IBP, and ConAgra are now "vertically integrated," owning the facilities that produce the animals, the feedlots to fatten them, and the meatpacking facilities to slaughter the animals and package their meat. This type of integration allows massive companies such as these to control every aspect of production, making it

BOX 7.1

VERTICAL INTEGRATION IN THE CHICKEN INDUSTRY

Pilgrim's Pride is the fifth largest poultry producer in the country. The company owns the breeder farms, the egg hatcheries, and the grow-out farms, as well as the trucks that ship the chickens to the processing plants, and the plants themselves—where the poultry is killed and processed. It also owns the feed mills that provide the food for the chickens, as well as separate egg farms that produce eggs. This vertical integration allows Pilgrim's Pride to control much of the market for poultry.

nearly impossible for smaller farms to compete and creating huge profits for those who succeed.

As one of the most economically powerful industries in the United States and abroad, animal agribusiness enjoys close ties with politicians. The industry often wields its economic clout in the form of campaign contributions to specific candidates; once elected, these candidates are more likely to support agribusiness lobbying interests. The National Cattlemen's Beef Association is one of the most powerful such groups in the country and spent more than $2 million lobbying the federal government from 1998 to 2004.

On the local level, factory farms are usually situated in states where regulations are weak or in poor communities where residents possess little political power to refuse construction of facilities. Corporate animal interest groups also spend a tremendous amount on lobbying at state and local levels. In 2008, for example, Californians passed a ballot initiative that would prohibit a number of factory farm practices such as veal crates and battery cages for hens, but the agribusiness industry spent almost $9 million to fight the campaign. If states do implement new regulations, factory farms can simply move to other locations. In some areas where factory farms are located, a high percentage of politicians enjoy close ties with the industry or are industry insiders themselves.

Often, regulatory agencies are staffed with former representatives from the very industries they are charged with regulating. Recent top-ranking USDA officials have included the former public relations director and chief lobbyist of the National Cattlemen's Beef Association, a former president of the National Pork Producers Council, and other former meat industry leaders. Thanks to the cozy relationship between the government and the

---

BOX 7.2

ORGANIZATIONAL FOCUS: FARM SANCTUARY

Farm Sanctuary, founded by Gene and Lorri Bauston in 1986, was the first sanctuary in the United States dedicated exclusively to rescued farmed animals. The Baustons rescued their animals from slaughterhouses, stockyards, and farms, and often got their animals from cruelty investigations. Today, Farm Sanctuary (now headed by Gene Baur, who changed his name after the Baustons divorced) operates two locations—in upstate New York and northern California—and houses approximately 1,200 cows, turkeys, chickens, pigs, sheep, goats, and rabbits. Farm Sanctuary is also actively involved in educating the public about the realities of factory farming and in encouraging a vegan lifestyle. The organization also works to see that federal and state legislation is passed that protects farmed animals. For instance, they were involved in Florida's campaign to ban gestation crates for pigs, in Arizona's campaign to ban gestation crates and veal crates, and in California's Proposition 2, which mandates that egg-laying hens, veal calves, and pregnant and nursing pigs all be housed in larger, more humane conditions.

---

meat industry, the industry benefits from federal subsidies in areas such as meat research, inspections, and predator control (through Wildlife Services, a branch of the USDA). Other benefits include buyouts of surplus meat products, and incentives for farmers to grow huge amounts of corn and soy—which are primarily used to feed livestock.

Another way in which meat eating is encouraged is via advertising, much of it subsidized by the government. For instance, the "Got Milk" campaign is produced by the National Milk Processor Board, which was established by the USDA's Fluid Milk Promotion Act of 1990 to promote milk drinking.

Sociologist Melanie Joy (2009) coined the term **carnism** to refer to the belief system which supports meat eating. According to Joy, because carnism became the dominant ideology surrounding food, it is the unquestioned default—meat eaters do not question why they eat meat; it is simply a given. On the other hand, vegetarianism is seen as a strange alternative, well outside of the mainstream, that is questioned and must be defended by its adherents. Because (until now) carnism was unnamed, it was invisible, and thus most people never even think about it. Once named, however, it becomes visible; consumers may begin to question some of the choices they make with respect to eating.

Slaughterhouse Workers

One of the reasons that meat is so heavily consumed in American society is because it is produced so cheaply. One reason why meat costs so little in the United States is the factory conditions under which meat animals are raised. Another reason is the poor treatment of the workers who toil in the industry. There are approximately 5,700 slaughterhouses and processing plants in the United States, and the industry employs approximately 527,000 workers. The pay is not good. According to the U.S. Department of Labor, the median annual income of a slaughterhouse employee in 2004 was $21,440, and the median annual salary of a meat trimmer was only $18,660 (U.S. Department of Labor 2007). In addition to low pay, slaughterhouse work is among the most dangerous in the country.

Slaughterhouse workers spend long days doing repetitive work at rapid speeds using dangerous equipment and sharp tools. They are injured in a number of ways: they slip and fall in the blood, feces, and other fluids that cover the floors; they are kicked and cut by animals struggling for their lives; they are cut by knives that disembowel and disassemble animals; and they endure painful and chronic repetitive motion injuries. The industry's ever-increasing line speeds intensify the risk of being cut, bruised, burned, stabbed, blinded, dismembered, disfigured, and worse.

According to the U.S. Department of Labor, more than 13 percent of slaughterhouse workers are injured or fall ill each year due to work conditions. Not only is this one of the highest rates in the entire private sector, but slaughterhouses also have the highest rates of injury and illness in the food-manufacturing industry—an industry already notorious for having one of the highest incident rates.

Slaughterhouse line speeds are constantly accelerating; for example, in chicken slaughterhouses, as many as fifty birds per minute can roll past workers. This means that employees must shackle, kill, or cut apart multiple animals every minute, for eight hours or more every day—often without breaks to check equipment, sharpen their knives, or rest for a few minutes. The noise level is high, and temperatures can soar to 120 degrees on the killing floor or drop below subzero temperatures in the refrigeration units. Because all birds and many pigs and cows are conscious as the workers shackle them, they are terrified—thrashing, kicking, or flapping as they try to escape.

The cleaners have one of the most dangerous jobs. With duties that include working at night in dense steam and fog, climbing onto equipment, hosing off machines, using harsh chemicals and pressurized water, and trying to retain footing on slippery bodily fluids, cleaners are at high risk for injury or death. Line workers repeat the same motions thousands of times in a single shift—cutting and lifting heavy, live animals, carcasses, and more—and their likelihood of developing crippling, painful repetitive strain injuries is high. The air inside slaughterhouses is filled with dust, dirt, and airborne feces and blood particulates. As a result, workers can become infected with a number of illnesses. Often many animals on the line are sick. When they defecate or vomit on the workers, they can spread disease-causing bacteria such as *E. coli*, *Campylobacter*, and *Listeria*. Animals treated with antibiotics may become infected with bacterial illnesses that are resistant to antibiotics, and workers can become sickened by the same dangerous strains.

Worker turnover can exceed 100 percent in a year. High turnover means that workers often do not accrue sick or vacation time, nor are they at one job long enough to be covered by insurance. Management often pressures workers to continue working and to ignore injuries. Because management may fire workers who take sick leave, workers may continue working despite being in pain. Managers may also force injured workers to quit, especially because those whose staff has lost the fewest workdays may receive bonuses. The meat industry is also notorious for union busting and for retaliating against employees who demand basic improvements in working conditions and pay.

Meat companies recruit Mexican immigrants who are not aware of their rights and will accept lower wages. More than one-fourth of slaughterhouse workers are foreign-born noncitizens. Thirty-eight percent of the 304,000 production and sanitation workers in the meat industry are foreign-born noncitizens. Many of them, unable to speak English and fearful of losing their jobs or being deported, are easy targets for intimidation and manipulation. Workers may not be aware of their protection under workers' compensation laws; the language barrier can prevent workers who are sick or injured from communicating with management.

Cultural Implications of Modern Meat Production and Consumption

One focus of human-animal studies has been to examine the links between meat production and the exploitation of people. From a political economy or conflict theory perspective, we might say that all devalued and oppressed groups are victims of the same overwhelming forces of capitalist society: profit over humane treatment of animals, equitable treatment of workers, and long-term, sustainable agriculture and environmental policy.

Conflict theorists would argue that oppression—whether aimed at humans or other animals—results from institutional and economic forces, not individual attitudes or practices. Because profit is the goal in a capitalist economy, humans and animals may be harmed if doing so improves profits.

American agribusiness gives very little consideration to the quality of life for animals—that are referred to as products, the lives of the workers—whose labor is exploited, or to the environment—that suffers from the pollution caused by the industry. However, if animal welfare considerations are seen to impact profits, then the corporations will sometimes undertake changes. For example, thanks to the activism of U.S. organizations such as People for the Ethical Treatment of Animals and the Humane Society of the United States, more Americans are now aware of the level of animal suffering in commercial agriculture; many citizens are demanding more humane ways of raising animals. Large companies such as McDonald's and Wendy's have responded by changing the conditions under which some of the animals that produce their meat are raised. In addition, certain types of treatment of animals—such as beating—can also affect profits because they result in bruising of the meat, which cannot be sold for as much money.

Realistically, when taking into account the environmental damage from factory farms (the USDA estimates that animals raised for meat produce 1.5 billion tons of waste each year, which pollutes water and air and is a major contributor to global warming), the health costs associated with a meat-based diet, and the harm to workers, it does not make rational sense to raise animals in factory farm conditions. It takes hundreds of gallons more water to produce meat than vegetables, and many pounds of grain (that humans could eat) to produce one pound of meat. As the world's population grows, people will need more food. Because most of the world is shifting to an American-style diet, people will want more meat and more resources will be needed to produce this meat than to produce grains or vegetables.

How will we satisfy this demand? And do we even want to? If we continue on our current path, we will meet this demand by constructing more factory farms and diverting more land and resources to raise livestock in the conditions outlined in this chapter. Yet, interestingly, human hunger is exacerbated even while we consume more meat. Meat production causes hunger by diverting food, water, and land away from vegetable and grain production to meat production.

Many scholars have pointed out the links between the consumption of meat and power. As Plutarch (1874) noted, most people in the ancient world rarely ate meat; elites not only ate huge amounts of it, but wasted huge amounts of it as well. Feminist scholar Carol Adams (1991) also notes that people with power have always eaten meat and meat eating in particular tends to be associated with masculinity. Women, children, the poor, and minorities often eat what is thought to be second-class food: vegetables, grains, and fruit. Although women often do not eat meat, or eat it rarely, they are expected to prepare it for their husbands, sons, and fathers. In past centuries, the Chinese and the Japanese were thought of as "rice eaters" and the Irish were potato eaters—these characterizations seemingly made them inferior to the English, and justified their conquering. Even today, in cultures around the world, if the meat supply is plentiful all people have access to it; but if it is limited, only elites get it.

Ethics and Meat Eating

Despite the fact that nearly nine billion land animals (eight billion of which are poultry) are raised and killed for food each year in the United States (USDA National Agricultural Statistics Service 2009), there are virtually no laws that protect them. Most state cruelty codes exempt common agricultural practices, so cruelty that would result in criminal prosecution if the victim were a dog or a cat is not defined as cruelty when it pertains to farming.

There is ample reason why most people prefer their meat to come prebutchered and wrapped in plastic, or increasingly, precooked and prepared. We want to maintain a comfortable distance and we do not want to think about the animals we are eating or how they lived and died. And when we do in fact consider where our food comes from, we imagine Old MacDonald's farm. There is comfort in the myth that the animals at least had a life worth living, and that small farms still provide us with our food. But family farmers have become nearly extinct in this country, their ranks comprising less than 2 percent of the population. The reality is that the vast majority of farmed animals in this country live a fraction of their normal life span; their lives are filled with suffering that is nearly impossible for us to even imagine. Brought into existence by corporations that consider them mere units of production, these animals endure routine, institutional—yet terrible—abuses.

One way that we avoid thinking about the meat that we consume is by our naming practices. Many forms of meat have names that conceal the animal that they come from: pork and not pig; beef and not cow. As

BOX 7.3

FAMOUS ANIMALS: CINCINNATI FREEDOM

Cincinnati Freedom was a Charolais beef cow that escaped from a slaughterhouse in Cincinnati, Ohio, in 2002, and evaded police and animal control officers for eleven days until she was finally captured. Because of the publicity surrounding her escape and capture (she had become something of a local hero in the press), officials decided not to return her to the slaughterhouse and attempted to find a home for her. Artist and animal activist Peter Max donated some of his paintings to the Cincinnati SPCA in exchange for custody of the cow, whom he later brought to Farm Sanctuary in upstate New York. She lived at Farm Sanctuary for six years, spending time with other cows that had also famously escaped slaughter—such as Queenie, Annie Dodge, and Maxine—and died of spinal cancer in 2008. From the Farm Sanctuary website:

> When it came time to say goodbye to Cinci, the herd gathered close around her. One of the eldest steers, Kevin, stepped forward to lick her face, while Iris, an older female, licked her back, soothing and keeping her calm up until she took her final breath. After our beautiful girl passed, every member of the herd approached to say goodbye, each one sharing with Cinci one last moment of affection.

feminist scholar Carol Adams points out, the animal is the **absent referent** in meat: Without the animal there is no meat, yet they are absent from meat because they have been transformed—via slaughtering, butchering, and marketing—into food. The animal is absent because it is dead; it is absent because we talk about animals differently when we eat them, and it is absent because animals become metaphors for describing something else. When we say things such as "I felt like a piece of meat," we are not *actually* thinking about meat, and we are certainly not thinking about the animal that made that piece of meat.

Even though meat consumption has been practiced—and highly valued—in virtually every human society, most societies express some form of cultural ambivalence about the practice, which is expressed a variety of ways. One example is the meat taboos that we discussed earlier in this chapter. Another example has to do with rules and rituals surrounding the killing of animals.

In many of the ancient civilizations of the Middle East, animals had to be sacrificed to the gods by priests before they could be consumed; in many cases, the animals were expected to "assent" to their killing. Even in ancient civilizations, most people did not kill the animals they ate, and they certainly did not eat the animals raw; instead, they transformed them via butchering and cooking them, demonstrating that people even then were perhaps not quite that comfortable with the concept. Some scientists have noted that humans may have an inbred aversion to meat. We are much more likely to express disgust about meat than we are about any other item of food; children are notoriously averse to eating new kinds of meat. Meat, then, is highly desired and highly tabooed.

Today, because of the intensive confinement associated with modern factory farming and the cruelty inherent in modern agricultural practices, millions of people around the world have adopted a vegetarian or vegan diet; millions more try to purchase meat from animals that were raised organically, sustainably, or humanely. In fact, the market research firm Mintel has found that the number of vegetarian or vegan menu items at U.S. restaurants increased 26 percent from 2008 to 2010. In addition, cage-free eggs have become a popular alternative for people who want to eat eggs but do not want to participate in the suffering of intensively raised hens. Another modern movement in many U.S. cities is the **slow food** or **local food** movement, in which meat is consumed from animals that were raised, slaughtered, and butchered locally, in much more humane conditions than on factory farms. Many urban Americans are also raising their own chickens and rabbits—small animals that can be easily raised at home and that because of their exemption from the "livestock" definition in the Humane Methods of Slaughter Act can be slaughtered at home, too.

Suggested Additional Readings

Davis, Karen. 2001. *More Than a Meal: The Turkey in History, Myth, Ritual and Reality*. New York: Lantern Books.

Eisnitz, Gail. 1997. *Slaughterhouse: The Shocking Story of Greed, Neglect and Inhumane Treatment Inside the U.S. Meat Industry*. Buffalo, NY: Prometheus Books.

Joy, Melanie. 2009. *Why We Love Dogs, Eat Pigs, and Wear Cows: An Introduction to Carnism*. Newburyport, MA: Conari Press.

Mason, Jim and Peter Singer. 1990. *Animal Factories*. 2nd ed. New York: Harmony Books.

Sapontzis, S. F. 2004. *Food for Thought: The Debate Over Eating Meat*. Amherst, NY: Prometheus Books.

Williams, Erin and Margo DeMello. 2007. *Why Animals Matter: The Case for Animal Protection*. Amherst, NY: Prometheus Books.

Suggested Films

Animal Appetites. VHS. Directed by Michael Cho. New York: Third World Newsreel, 1991.

A Cow at My Table. VHS. Directed by Jennifer Abbot. Galiano Island, BC, Canada: Flying Eye Productions, 1998.

Death on a Factory Farm. DVD. Produced by Tom Simon and Sarah Teale. Hastings-on-Hudson, NY: Working Dog Productions, 2009.

45 Days: The Life and Death of a Broiler Chicken. VHS/DVD. Washington, DC: Compassion Over Killing, 2007.

Life Behind Bars: The Sad Truth About Factory Farming. DVD. Watkins Glen, NY: Farm Sanctuary, 2002.

Meat. DVD. Directed by Frederick Wiseman. Cambridge, MA: Zipporah Films, 1976.

Meet Your Meat Collection. DVD. Directed by Bruce Friedrich. Norfolk, VA: PETA, 2003.

8

The Pet Animal

THERE HAVE BEEN A number of stories in the news recently that have captured the public's attention regarding animals. Many of them involve interesting cases of cross-species friendship—the elephant Tara (that lives at the Elephant Sanctuary in Hohenwald, Tennessee) that befriended Bella the dog; the deer Bambi that befriended Thumper the rabbit; and Owen the baby hippo that befriended Mzee, the giant tortoise, in the aftermath of the tsunami that devastated the coast Indonesia in 2004. Cross-species friendships, many of which cross the wild/domestic animal border, demonstrate that even animals that would normally have a predator/prey relationship with each other can find happiness together.

But perhaps even more extraordinary are the stories of animals that keep an animal of another species as a pet. Koko, the signing gorilla that lives at the Gorilla Foundation in California, is a notable example. Most people know the story of Koko and her beloved kitten All Ball, and Koko's grief at All Ball's death. But what does it mean to keep an animal as a pet? Typically, pet keeping is defined as keeping an animal from another species for enjoyment, rather than utility; it has generally been thought to be something that only humans do. Psychologist Hal Herzog (2010), for example, points out that only animals living in a captive or semicaptive environment in which food is provided have ever "kept pets," and that truly wild animals, for whom food provision is a difficult job, never do. Even if that were the case (and new stories are emerging about truly wild primates that have been seen keeping either other primates as pets or, in one case, a kitten), would that really negate the idea of animals keeping animals as pets? We shall see, for example, that because the urge to keep animals for enjoyment is universal in human societies, pet keeping has only been truly popular in cultures where resources are plentiful and people were not struggling for survival. It may be that among mammals, or at least among primates, the desire to share a close bond with other animals not of one's own species is universal, and that freedom from hunger allows that desire to be realized.

For urbanized Westerners, the human-pet relationship is the only real relationship—other than through the consumption of meat—that most of us ever have with nonhuman animals. In 2009, according to the American Pet Products Manufacturing Association (APPMA), more Americans lived with a companion animal than lived with children (62 percent, up from 46 percent in 2008). According to the APPMA, Americans spent more than $45 billion on pet food, toys, clothing, travel paraphernalia, and other pet-related items in 2009. That same year, almost 412 million animals lived as pets in 71 million American homes. That is 100 million more pets than people in the United States! But what allows some animals to have such favored treatment while other animals—billions of them per year—live such short, cruel, and agonizing lives?

What Makes a Pet a Pet?

Throughout this text, we have been discussing how animals are socially constructed, and the different categories to which we assign different animals. In chapter 7, for example, we discussed meat and the making of meat. In order for an animal to be consumed by humans, it must first be defined as a "meat animal." One cannot find and kill and eat any animal—not only would that break a number of laws in a country such as the United States, our symbolic and classificatory system would not allow it.

Another extremely important animal class, from a functional and symbolic perspective, is the pet. A pet, or companion animal, is an animal that is defined by its close relationship to human beings. As with so many other animal categories—meat, livestock, working animal, and laboratory animal—there is nothing distinctive about the animals that we consider to be pets, other than the fact that they have been chosen by humans and turned into pets. Think about the many types of animals considered to be pets in this country. Although many are domesticated and have long lived in close confines with humans—dogs and cats come most quickly to mind—many others are not domesticated at all and in fact are still wild animals. Birds, fish, and the many animals considered **exotic pets** such as turtles, snakes, frogs, hedgehogs, spiders, or primates are wild animals that have been moved into a domestic environment. Historian Keith Thomas (1991), discussing pet keeping in England between 1400 and 1800, wrote that a pet was an animal that was named, allowed into the house, and never eaten. These criteria fit many, but not all, definitions of pets around the world.

Most pets become pets in part because of how they are born. Domestic animals, including companion animals, are the result of controlled breeding for human purposes. This process goes back many thousands of years and has impacted everything from the size, shape, and color of the animals that live with us to their temperaments and even their relationships with us. Even though food animals are "produced" by the agribusiness industry, more and more companion animals are being bred in factory farm conditions for pet breeders to sell or by pet owners (intentionally or unintentionally).

When intentionally bred by breeders, pets may be born either in small facilities or homes or in bigger facilities called puppy mills or kitten mills. These are large, often unlicensed, and usually filthy facilities typically found in rural areas, in which puppies, kittens, bunnies, and today even birds are bred in large numbers, usually to be sold to pet stores via brokers. Breeding animals living in these facilities spend their entire existence in wire cages. They are bred over and over again, producing litter after litter, until they literally are worn out. The puppies, kittens, and bunnies born here often develop health conditions due to a lack of medical care and proper treatment. Nearly all animals available at pet stores in North America come from these commercial breeding facilities. Other types of breeders include hobby, show, and backyard breeders and "accidental" breeders, who typically sell their animals directly to the public via newspaper ads, the Internet, or similar means. Commercial breeding facilities usually sell their animals wholesale to brokers or dealers, who transport them to the pet stores that will ultimately sell the animals to the public. Because of the vast numbers of animals bred every year by breeders, and the large number of animals abandoned by their human families (a topic to be discussed later in this chapter), many pets find their way to animal shelters and private rescue groups, where the lucky ones are "recycled" and once again enter human households as pets.

A pet is an animal that lives in a human household. (Many pets do not live inside of the home but reside outdoors, in the backyard, or—in the case of horses and other "livestock" that are also considered pets—in a barn.) Another major criteron of the pet is that it is named. In the West, at least, one cannot be a pet and not have a name. Naming an animal incorporates

Figure 8.1. Ruby the rabbit visits with children at a juvenile detention facility in California. (Photograph courtesy of Suzi Hibbard.)

him or her into the human social world and allows us to use their name as a term of address and a term of reference. We can speak to them as we do our family and friends, and we can speak about them as we do about others that are important to us. In both cases, naming allows for interaction and emotional attachment. In addition, by talking about our pets to other people, the animal's history and personality become clear.

Pets, then, are animals that are generally *purpose-bred* to become pets, are kept in or near a human household, are relatively controllable and cared for by humans, and are either domesticated or at least tame. The most popular pets have personalities amenable to being with humans. Many animals, such as dogs, seem to genuinely like being with humans. Ideally, pets should (at least in the minds of animal lovers) enjoy a life of love and attention. The term "pet," after all, was a fifteenth-century English term meaning "spoiled child." This word probably derived from the French term *petit,* or "little," and grew to mean anything or anyone that was spoiled or indulged. However, there are many exceptions. Exotic pets are wild animals caught and sold through the (often illegal) exotic pet industry. They are neither domesticated nor purpose bred nor, in many cases, even tame. In addition, although we often define pets by their primary role as companions, many pets are kept outside either in a cage or on a chain, and thus provide no companionship for the family members at all. Others are kept primarily for purposes such as security or economics (as with fancy pet breeders and those who show animals), and still others, such as snakes or fish, show no companionable tendencies whatsoever.

The Rise of Pet Keeping

Although pet keeping has been practiced for millennia in societies around the world, it has only been in the last hundred or so years that pet keeping in the West has exploded, creating multibillion-dollar industries focused on producing, feeding, caring for, medically treating, and even disposing of millions of animals per year.

People have long kept animals as companions; even hunter-gatherers that had no domesticated animals kept pets. The animals kept by such groups were tamed wild animals caught as babies and used for companionship. Aboriginal Australians, for example, once kept dingoes and wallabies as pets; Polynesians kept dogs and parrots; South American tribes kept monkeys, parrots, and even wild cats; and Native Americans kept deer, dogs, and crows. Even though these same animals were hunted and consumed in these cultures, generally once an animal was adopted as a pet, it was exempt from consumption. In some tribes, women nursed orphan animals at their own breasts. In primitive cultures around the world, women (and children) had the closest relationships with pet animals, as is the case in modern Western cultures.

After animal domestication began about 8,000 years ago, it would have been much easier for people to keep animals as companions because they could choose from animals that had already been domesticated and were thus tame and used to humans. Archaeologists have found the presence of pets in ancient civilizations going back at least 5,000 years. For much of history, animals in small communities served multiple purposes through-out their lives—as sources of eggs, milk, or fertilizer or perhaps as working animals, providing companionship and, often, ultimately being slaughtered for food. Because keeping (and feeding) animals solely for companionship was quite a luxury, it is likely that the animals which were *only* pets were a luxury for elites.

Which animals were the first pets? Domesticated at least 15,000 years ago as hunting partners, dogs were almost certainly the first pets and were among the only animals not domesticated for specifically for food. It is logical to assume that hunters developed close relationships with some of their canine partners. The first pets, then, most likely served two primary purposes—hunting *and* companionship. But it probably was not until after the rise of the ancient civilizations that some dogs stopped working and turned into full-time pets for the wealthy.

Wild-caught birds were another early pet, kept in cultures as diverse as the Roman and the Aztec Empires. In fact, ancient rulers kept wild animals of all sorts in menageries, which were intended to demonstrate their mastery over nature. Birds, because of their size, could easily be kept indoors in cages. Birds were beautiful, exotic, and could sing—a benefit in the era before recorded music became available. Ornamental fish, another wild but easily kept and controlled wild animal, were also kept as pets—most prominently in Japan and China—where the keeping of pet fish probably originated as early as the seventh century. Cats were another early companion. Similarly to dogs, cats were not domesticated for food—they were enticed by early farmers to live near human fields and granaries to keep down the rodent populations—and thus were good candidates to live indoors with humans. They were famously kept as pets (and sacred animals) by the Egyptians. The Greeks and Romans kept dogs and parrots as pets, and the Romans buried their dead dogs with marble grave markers. Prior to the modern era, pets were kept for a variety of reasons—companionship, for certain, but also because they were beautiful, made lovely sounds, were exotic, or gave the owners some level of status. Because Romans believed pets would accompany them in the afterlife, companion animals were often killed upon their masters' deaths, in order that their fates might be shared.

Dogs were probably the first animal purposely bred *as a pet*—starting about 3,000 years ago—and the first animal to have new breeds developed with no functional purpose. (Purebred dogs did not become popular among commoners until the mid-twentieth century.) Prior to that time, dogs had been bred with guarding, herding, hunting, retrieving, and other traits in mind. This was the first time that dogs were bred for smaller size, different coat colors and textures, and other aesthetic traits. In Europe and Asia, only landowners, royalty, and other wealthy classes could afford to care for animals that did not earn their keep. In Europe, for example, hunting dogs were highly prized by nobility and probably served both as hunters and companions. In China, the Pekinese and the Shih Tzu were bred originally as companion animals, but also had secondary functions. The Shih Tzu, even though small, was used as a watchdog in the Imperial Palace, and the Pekinese were considered to be sacred because they looked so much like the Chinese lions sacred in Buddhism.

During the Middle Ages, even though many nuns and monks kept pets, the Catholic Church saw the practice of pet keeping as a form of heresy, in that it challenged the human-animal boundary. Another reason why pet keeping was seen as trivial (and still is today, to a much lesser extent) is the fact that, historically, women are the primary caretakers of companion animals. Because of women's association with pets, and in particular with lap dogs, pets were seen as feminizing and symbolized women's inferiority. Even worse, a woman in medieval Europe could be accused of being a witch if her relationship with an animal was perceived as being too close. European elites were often satirized in the press for their keeping of pets, although noblemen who kept hunting dogs were exempt from this stigma. The poor, on the other hand, were prohibited from hunting in England (as lands were owned by either the Church or the nobility); consequently, they were prohibited from keeping hunting dogs. In addition, purebred pets were associated with elites, but the poor kept animals that the wealthy saw as dirty and diseased. Even though people from all socioeconomic status keep pets today, high income remains correlated with pet keeping—the higher one's income, the greater the likelihood of being a pet guardian.

The Development of the Modern Pet Industry

According to historian Keith Thomas (1983), it was only starting in the eighteenth century that people began to give pets human names. Pet keeping as we know it today did not really emerge until the nineteenth century, when enough people had the disposable resources to keep animals only for companionship. This period also marked the rise of the commercial pet industry—an industry that began with companies selling food, medicine, and cages for companion animals but which grew to include breeders, dealers, and all of the associated businesses such as groomers and pet spas. In 1860, the first commercial dog food was available for purchase in England, but it did not reach the United States until the early twentieth century; the first commercial cat litter was produced in 1947.

Attitudes toward animals during this period were nowhere near what they are like today. Historian Harriet Ritvo (1987) points out that even though the modern pet industry arose during the nineteenth century, it emerged during a time of incredible cruelty toward animals. Most Europeans and Americans were indifferent to animal pain and suffering. But slowly, during this century, an attitude that we might call the beginnings of the humane impulse began to emerge. Ritvo suggests that one reason for that change was that Europeans and Americans felt that they had "conquered" nature and vanquished much of the threat that it posed to human culture. Only with the conquest of human cities over the natural world could humans begin to incorporate animals into their homes. And with industrialism and the changes in the agriculture industry, which removed farm animals from most communities, animals largely disappeared from many people's lives. This left a gap that was compounded by the fact that middle-class families were having fewer children. This gap was filled by the development of the modern pet industry. Widespread pet keeping was also enabled in part by the rise of a middle class that now had incomes to support what had been an elite, somewhat frivolous hobby. Not only do we see pet keeping rise during this period, we also see the appearance of fancy dog-, cat-, and rabbit-breeding clubs.

Historian Katherine Grier (2006) also points out that the pet movement was fueled by a desire on the part of Victorian families to use animals to teach children middle-class, bourgeois virtues, such as kindness and self-control. Moralists of the period saw having a relationship with a pet as a way to instill positive traits in a child. Grier calls this the "domestic ethic of kindness," which may have played a role in reducing some of the casual violence toward animals prevalent at that time. (Cruelty to animals was seen during this era as a sign of "inward moral collapse.") It was expected that children would learn kindness toward all those that were dependent on others for their care, including pets, the elderly, and even slaves. This domestic ethic led to the expansion of pet keeping in American society, the rise of the animal welfare movement at the end of the nineteenth century, and is still very much with us today. But as Grier points out, although pet keeping changed the way that people thought about and dealt with animals within the home, it did not transform the way that people thought about animals outside of the home, or animals in general.

Even though the nineteenth century marked the rise of the commercial pet industry and the modern pet-keeping movement, it really was not until the mid-twentieth century that many of the practices that we associate with pet breeding emerged. For instance, when a companion animal became sick, it either died on its own or the owners often killed it. (The practice of bringing animals to a veterinarian did not become common until the 1900s.) In the twentieth century, the pet industry also became a truly commodified business. From a handful of pet stores initially specializing in birds to companies producing special food, cages, and equipment to the rise of the big box chains such as Petco and PetSmart, the pet industry has become incredibly profitable and powerful.

Why We Keep Pets

Not everyone loves pets. Many people do not like companion animals—they can be messy, cause great inconvenience, and exacerbate allergies, and many people think all animals should live outdoors. People can get hurt or even killed by their pets. Some exotic pets carry diseases such as salmonella, and every year people die from being attacked by dogs. Many people have pets but do not elevate them to the status of family members; others are concerned about how much money, time, and energy are spent on animals.

It wasn't that long ago that pet keeping was seen as so wasteful and irrational that scholars came up with a number of theories to account for the existence of pets. For instance, Konrad Lorenz (1970) and other animal behaviorists thought that pets are "social parasites." They have evolved with very cute faces and bodies intended to trigger a parental response in humans. A related idea is that we anthropomorphize pets—projecting onto them our own thoughts and desires, and creating in them a form of substitute kin. The idea here is that people who develop attachments to animals are incapable of forming relationships with other humans; we create artificial relationships with substitute people (our pets). There were even two studies conducted in the 1960s and 1970s that purported to show that people who kept pets were psychologically unhealthy, and that pets kept their guardians from forming effective social relationships with other people. (Modern research has demonstrated that these studies were extremely flawed.) Today, however, most scholars agree that people live with animals for a much simpler reason—because it provides concrete benefits to us.

We have discussed a variety of functions that companion animals serve, but the primary reason for keeping pets today is companionship. For example, the American Pet Products Manufacturing Association found that 56 percent of dogs sleep with their caretakers in the bedroom, either in or on the bed. And even though most cats lived outside in the past, today most are kept indoors and are considered part of the family. Even rabbits, among the most recent animals to be domesticated (for food), are now considered by many thousands of households to be companion animals. Rabbits live indoors as house pets, many sleeping in their caretakers' bedrooms. The closeness of these animals is one testament to the major role that they play as companions to us.

Research in the human-animal bond (to be discussed further in chapter 10) shows that living with animals gives people very real emotional, psychological, and even physical benefits. The Victorian motivation for owning pets—to teach children positive skills such as kindness and empathy—is still a factor today. Studies indicate that forming attachments to companion animals may develop nurturing behavior in children, and children with pets may exhibit more empathy than other children.

Why do animals have such an important influence on human well being? Scientists have developed a few theories to try to understand why humans benefit so greatly from living with companion animals. The **biophilia hypothesis**, introduced by biologist Edward Wilson (1984), states that humans and other animals are naturally drawn to each other and that this relationship is mutually beneficial. This explains not only the human-companion animal bond, but also why animals play such a strong role in the literature, art, and games of children. Another explanation suggests that humans are hardwired to pay attention to animals because for much of human evolution we depended on them as a source of food. In other words, the interest in animals as pets is only secondary to our interest in animals as food. A different theory is known as the social support theory and states that anything that provides social support (such as marriage, belonging to a church, or membership in a social club) is beneficial to human health because of our need to have social contact. Whether that contact is with animals or other humans is irrelevant—the point is that we have social contact with another creature. And finally, some scholars suggest that because men tend to have more social support than women do, women may need companion animals more than men.

The Human-Pet Relationship

Companion animals have a "social place" in our family, household, and daily routines. By incorporating them into our breakfast-eating, TV-watching, and holiday-taking routines, they are truly a part of the family. The human-pet relationship is different from most every other human-animal relationship in that it is not based primarily on utility, and in that it is truly a two-sided relationship, in which both parties—human and animal—play a major role. When we interact with a companion animal, we are interacting with an animal that we know as an individual, and whose purpose in our lives is one of companion, friend, and even family member. In the most ideal circumstances, the relationship is structured not only by the human's needs or interests but by the animal's as well. In my home, I live with dogs, cats, rabbits, and a bird. In some of those cases—especially with the dogs and the cats—I can pick up and hug these animals whenever I want, and can usually expect to be rewarded with an affectionate response. I can also expect to be enthusiastically greeted by the dogs when I arrive home, whose behaviors tell me that they miss me and are glad I am back. With respect to the rabbits that share my home, some, such as Maggie—a large mixed-breed black rabbit that lives in my guest room—will also treat me with affection, licking me and nuzzling my hand. Others, such as Igor—a gray dwarf rabbit that inhabits the living room—will only tolerate my attentions. Still other rabbits, such as three-legged Molly—who also resides in the living room—will actively avoid me. Because of her feelings about me, I try to bother her as little as possible. How each partner in the human-pet relationship behaves is structured in part on how the other partner behaves.

We mentioned earlier in this chapter that one of the most important criteria for being a pet is having a name because having a name symbolically and literally incorporates that animal in the human domestic sphere. Having a name also allows for human-animal communication: We can talk to animals. Even though nonhuman animals do not possess human verbal language, we can and do still talk to them, and many companion animals understand much of what we say, based on our tone, inflection, body language, and facial expressions; many animals know the meanings of specific human words including, but not limited to, their names.

Sociologists who have studied human-animal communication have shown that similarly to baby talk, human-pet communication has a clear structure and a distinctive tone, set of bodily gestures, and comportment. Beyond immediate communication, this talk serves as a sort of glue in human-animal relationships and, more broadly, enhances the social lubricative function of companion animals in human-human relationships. Sociologist Clinton Sanders (1999) has studied the communication between humans and dogs, and maintains that language enables human and canine interactants to construct and share a mutually defined reality. Animals, because they lack human language, normally are excluded from social exchange with humans. But in the domestic realm, guardians of pets have made a number of allowances for that lack of language. Ask any caretakers and they will not only admit that they talk to their animals, they will also maintain that their animals understand what they are saying. In addition, we speak "for" our animals—to friends, to family, to the veterinarian. We also speak through them; sometimes people use their pet dog or cat as a sort of mediator to communicate information to another person. Pet owners see cross-species communication as real and possible, and this possibility itself allows for that communication and for the reciprocal relationships that we have developed with our companions.

By opening up the door to cross-species communication, and by including (some) animals in our own worlds, we humanize those animals and give them

Figure 8.2. "Cat Love." (Cartoon by Dan Piraro. Courtesy of http://www.bizarro.com.)

a "person-like" status. Sanders, for example, describes a form of social exchange involving his dog asking (through body language or barking) to be let out, and Sanders acquiescing and letting him out. Both partners played an equal role in that exchange and were able to anticipate and acknowledge the needs and interests of the other.

Not all people relate to companion animals in the same way. Psychologist Michael Fox (1979) has written that there are four categories of pet-owner relationships: The object-oriented relationship, in which the pet is seen as a novelty or a decorative item; the utilitarian relationship, in which the animal is used to provide a specific benefit to humans, such as being a guard dog; and the need-dependency relationship, in which the animal satisfies the human's needs for companionship. The final category is the actualizing relationship, in which the person's relationship with the animal is fully equal and based on mutual respect.

Pet ownership is also gendered. Sociologist Michael Ramirez (2006) shows how pet guardians use gender norms to choose their companion animals and to describe their pets' behaviors. They use their pets to display their own gender identities. For example, the men Ramirez surveyed reported that they consider dogs to be a more "masculine" animal than cats, and both men and women explain their pets' behavior in terms of their sex. Female animals were said to "flirt" and women were more likely to describe their pets in more feminine terms, whereas men were more likely to describe theirs in more masculine terms. In addition, men were more likely to play and roughhouse with their dogs, but women were more likely to kiss and hug them. Gender roles and expectations, then, shape not only how owners see their companion animals but how they relate to them as well.

Love and Grief

One term that is used to refer to pets and not a single other category of animals is "love." One of the major reasons we keep pets is because, for many of us, we value having someone love us unconditionally, regardless of our weight, appearance, income level, or even personality. We find joy in having someone in our lives that displays so much positive emotion to us. In addition, because so many animals are so attuned to our own emotions, when we are feeling pain or sadness, they seem to know it and respond accordingly.

Not only do many animals behave as if they love us, but many Americans feel that their dogs not only await their return home from work, but actually can anticipate their return. Biologist Rupert Sheldrake (1994) tries to explain this phenomenon in his book, *Dogs That Know When Their Owners Are Coming Home* (1994). He attributes it to a form of telepathy between pets and their people. Other scholars feel that there is a more logical explanation for this behavior, such as the dog's "internal clock" alerting him to the end of the workday, his superior sense of smell and hearing, or environmental cues such as a phone call to a spouse or partner. Whatever the explanation, dogs' behavior with their guardians certainly seems like love. Consider the many stories of dogs that waited patiently, sometimes for years, for their dead owners to return home.

One of the problems with living and forming relationships with companion animals is that, almost always, humans will live much longer than their pets. How do people cope with that loss? As with other intimate relationships, the grief that we feel when a pet dies can be overwhelming. The same stages of grief that Elizabeth Kübler-Ross (1969) outlined regarding human grief (denial, anger, bargaining, depression, and acceptance) also apply to those grieving the loss of their animal companions. Sociologists and psychologists have recently addressed the issue of human grief over animal death using tools such as the Grief Experience Inventory and the Censhare Pet Attachment Survey to understand the depth of these feelings, as well as how people cope with them. These studies have shown that grief is higher when people have a greater attachment to the animal, when they have very little support or understanding from other people, and when there are other stressful events in a person's life. Contrary to what some people might think, grief is not greater for people without children than it is for those with children, nor is it necessarily greater for people with a single pet than for those with multiple pets. Finally, although women apparently grieve over pets more than men do, it may be that men mask their feelings because it may be seen as socially unacceptable for men to display feelings in public. Philosopher Yi-Fu Tuan (1984) points out that losing a pet is different from the loss of a person because when a dear friend or family member dies, they are gone forever. When a pet dies, that space may be filled again with a substitute for the dead animal. Thankfully, even though some people today still

Figure 8.3. Gravestone honors Bernie at the Los Angeles Pet Memorial Park. (Photograph courtesy of Elizabeth Terrien.)

do not acknowledge the importance of companion animals, many more do. Pet lovers no longer have to grieve alone. There are also a number of books available on the subject, and people can elect to go to counseling to help them cope with the loss of a companion animal.

Development of Humane Attitudes Through Pets

Pets, it is clear, *do* things. They can influence our behavior, affect our emotions, and even impact our health. What is also clear from recent research is that they influence our attitudes—about and toward other animals, and toward other people.

Research has shown that there are, broadly, five sets of factors related to attitudes toward animals: personality, political and religious affiliation, social status (class, age, gender, education, income, employment, ethnicity), environmental attitudes, and current animal-related experiences and practices (Kellert 1980, 1985, 1994). Living with a companion animal is the most common way in which attitudes toward other animals are shaped (Herzog and Burghardt 1988; Kidd and Kidd 1990; Schenk et al. 1994; Daly and Morton

ANIMALS AND SOCIETY   160

2006). Those people who live with animals as companions may have a more positive attitude toward other animals, but those whose current animal experiences tend to be exploitative will have negative attitudes. In adulthood, attitudes toward pets are correlated positively with having had family pets as children, and having had important pets. When those childhood experiences were good, the adults continue to like and want pets as adults. These studies also show a positive correlation between pet keeping as children and humane attitudes as adults, including vegetarianism, donating to animal charities, and belonging to animal welfare organizations. It seems clear that living with companion animals plays at least some part in our attitudes toward other animal and, perhaps, toward other people.

A number of recent studies points to a correlation between positive attitudes toward companion animals and a more humane attitude toward other animals. Some very preliminary studies are showing there is a link between positive attitudes toward animals and a more compassionate attitude toward people. Anthropologists James Serpell and Elizabeth Paul (1993, 1994) trace the evolution of animal keeping in the West and its association with attitudes toward fellow humans. They point out that, starting in the seventeenth century, many of the most enlightened humanitarians had an affinity for animals; scholars and philosophers dating back to the ancient Greeks thought eliminating violence toward animals would make humans more peaceful. We also know that abolitionists and animal rights activists were often the same people. There appears to be at least a correlation between affinity toward animals and social justice. As we have discussed here, when the commercial pet industry began to develop in the nineteenth century, many saw animal companionship as a way to cultivate virtues such as kindness and self-control in young people. The humane education movement takes as its central premise the idea that childhood pet keeping can be used as a springboard to teach children empathy toward other animals. In 1882, George Angell, founder of the Massachusetts Society for the Prevention of Cruelty to Animals (SPCA), created the Bands of Mercy, which were after-school clubs where children met, learned about animals, prayed, and testified about the animals that they had helped. Currently, nine states have laws mandating that public schools teach some form of humane education, acknowledging the importance of having young people learn about kindness toward animals.

Today, some scholars (Ascione 1992, 1997; Paul 2000; Melson 2003; Taylor and Signal 2005; Henry 2006; Daly and Morton 2006, 2009) think that living with animals may in fact teach empathy and compassion—toward animals and people. However, at least one recent study challenges this notion,

Figure 8.4. Olivia Montgomery plays catch with Sheba. (Photograph courtesy of Robin Montgomery.)

finding evidence that living with animals is not correlated with empathy and that living with cats is, in fact, negatively correlated (Daly and Morton 2003). Another point to consider is that pets occupy a unique position in human society—as cultural theorist Erica Fudge points out, "a pet is a pet first, an animal second" (2002:32). In other words, we can love pets, but still not think too much about other animals because pets are not *really* animals after all.

Contradictory Attitudes Toward Pets

Pets are many things to us. They are beloved family members that are lavished with attention, love, and money. We buy them clothes, toys, grooming items; we take them on vacation with us, and we bury them in pet cemeteries. We have even created a special pet heaven known as the Rainbow Bridge where they go after death. Yet at the same time they are used to demonstrate status and construct identity; they are commodities in an industry focused on profit, often to the exclusion of animal welfare; and they are one of the most egregious examples of America's throwaway culture, abandoned in huge numbers and euthanized in the millions.

How can we simultaneously lavish extraordinary amounts of love, money, and care on our beloved companion animals, yet at the same time allow millions of those same animals to suffer and die? That is the ultimate contradiction of pets in the United States. There is no doubt that the companion animal relationship can be one of the richest, most fulfilling ways that humans and animals can interact, bringing huge benefits to both person and animal companion. Yet the production of those animals, in a world already filled to capacity with domestic animals that either have no home or are living lives of abuse or neglect, is often driven by profit rather than concern for animal welfare. And once they are here, many companion animals experience neglectful and even abusive treatment at the hands of those who are supposed to care for them.

According to the Humane Society of the United States (HSUS), of the six to eight million dogs and cats that enter animal shelters in this country every year, 3 to 5 percent of cats and 2 to 30 percent of dogs are returned to their homes, and about half are euthanized. (Most shelters do not keep statistics on the number of other abandoned or euthanized animals.) In addition to these millions of animals, countless thousands of other loose and stray animals die from starvation, disease, or traffic accidents. The loss of life and the suffering that this entails are enormous. The cost to taxpayers is also substantial. One Washington animal shelter estimates that it costs taxpayers $105 for an animal control officer to pick up a stray dog or cat, transport the animal to the shelter, provide food and water for the animal, euthanize the animal if not adopted or reunited with his or her family, and send the body to the landfill. Animal control programs in this country alone cost $2 billion per year, and this does not count the millions that independent animal organizations spend to rescue and rehome animals.

Pets and Domination

Historian Harriet Ritvo (1987) writes about how the rise of organized and expanded pet keeping in the nineteenth century was linked with a new set of attitudes about animals. In this period, we see the linkage of affection toward pets on the one hand with the notion of control and domination on the other. The early pet fanciers were also pet breeders, and the breeding of pets is one of the most concrete, corporeal ways in which humans exercise control over animals. This is one reason why for so many years mixed-breed animals

BOX 8.1

ORGANIZATIONAL FOCUS: ASPCA

The ASPCA was the first humane organization in North America and is, today, one of the largest in the world. Headquartered in New York City, the ASPCA is one of the largest organizations working to protect companion animals in this country.

The ASPCA was founded in 1866 by Henry Bergh on the belief that animals are entitled to kind and respectful treatment at the hands of humans, and must be protected under the law. Bergh was an animal lover who began a campaign to get better treatment for the carriage horses of New York City that were often mercilessly whipped by their owners. Nine days after the ASPCA was incorporated in New York City, New York State passed the nation's first anticruelty law, banning the cruel treatment of carriage horses, and the ASPCA was given the authority to enforce it.

The idea that animals should be protected from cruelty began to grow, and other humane organizations were founded around the country; most states soon had their first anticruelty laws as well.

Today the ASPCA works in three key areas: community outreach—including partnerships with local shelters and programs and materials to help promote adoption in local communities; animal health services—including grief counseling, an animal hospital, and an animal poison control center; and anticruelty initiatives—including humane law enforcement and lobbying for stronger anticruelty legislation.

were viewed with such distaste (and still today in fancy pet circles). Animals that were allowed to have sex on their own, with their own partners, and create their own "mongrel" offspring were seen as vulgar and uncontrolled.

Today we see this level of control not only in the specialized breeding and genetic manipulation of pets, but in the forms of surgery that pets undergo as well. The reliance on cosmetic surgery for dogs is one result of the breeder's focus on perfection. In order to conform to breed standards, for example, certain dogs require docked tails and/or cropped ears. Some dog behaviorists worry that because dogs use their tails to communicate with other dogs, tail docking puts them at a disadvantage when socializing and may affect physical functions such as stability as well. Scholars see these unnecessary surgeries as symbols of human control over nature and animals.

WHAT'S IN A NAME? GUARDIAN OR OWNER?

In recent years, we have seen a change in terminology referring to pets and the people who live with them. Increasingly, animal welfare advocates and the general public have begun using the term "companion animal" rather than pet when describing the animals with which we share our lives. But another, perhaps more important, change involves switching from the term "owner" to the term "guardian" or "caretaker" when referring to a person who lives with a companion animal.

In Defense of Animals (IDA), an international animal rights organization that has spearheaded the campaign to end the use of the term "owner," argues that this term is linked with the treatment of animals. In particular, using the term owner means that companion animals are considered to be mere commodities or property, not individual beings, and IDA argues that this classification underlies their exploitation. The organization and other advocates argue that it was not so long ago that women, children, and others were seen, in legal terms, as merely property.

In 1999, IDA created the Guardian Campaign as a nationwide platform to redefine society's standards for relating to and treating animals. Since that time, dozens of North American cities and counties, as well as the state of Rhode Island, have changed the language used in their legislation to replace "owner" with "guardian."

In 2006, Carlisle-Frank and Frank conducted a study to test whether using the term "owner" or "guardian" is correlated with attitudes toward pets; researchers found that indeed it was. Those who self-identified as guardians engaged in more responsible behavior toward their companion animals than did those who called themselves owners, and appeared to have more intimate relationships with their animals as well. The study did not demonstrate causation, however, so it is not yet known whether differences in attitude and treatment are shaped by the use of the different terms, or whether different attitudes toward animals shape what term a person chooses to use.

Figure 8.5. "Clipped." (Cartoon by Dan Piraro. Courtesy of http://www.bizarro.com.)

Historian Harriet Ritvo suggests that cats have historically been less popular than dogs as pets in part because they could not be as easily controlled.

Philosopher Yi-Fu Tuan, in his book *Dominance and Affection: The Making of Pets* (1984), deals with the question of dominance with respect to our relationship with pets. Many modern breeds of companion animals are bred for qualities that we find attractive, but that are harmful to the health of the animal. Is it cruelty or playfulness to breed a variety of goldfish with dysfunctional bulging eyes? Even without specific genetic defects associated with certain dog or cat breeds, many modern breeds of dog or cat are unable to survive without close human attention. Although dependency has been bred into domestic animals since the earliest days of domestication, it has accelerated in recent years with the production of animals such as Chihuahuas that are physically and temperamentally unsuited for survival outside of the most sheltered of environments. More disturbing are cats that go by the name of Twisty-Cats, or Kangaroo Cats, all of whom have a genetic abnormality that results in drastically shortened forelegs or sometimes a flipper-like paw rather than a normal front leg, and that are being selectively bred by a handful of breeders. How much do we really love our pets, he asks, if we continue to breed them in a way that makes them less healthy and that shortens their life expectancy?

We can list a whole host of other ways that humans have, throughout history, sought to control animals, including castration, tail docking, declawing, tattooing, microchipping, vocal cord cutting, and ear cropping. But you might argue that, more important than all of these secondary control measures, is selective breeding. By creating animals that are smaller, more docile, more colorful, and more dependent, humans exercise the ultimate form of control because we control the very body of the animal.

The reality is that most people do not really care that we have bred health problems and dependency into our pets because we are not looking to create independent animals anyway. We train our dogs to "sit" and to "lay down," demonstrating that their role is to do what we ask them to do, and not to cause any trouble. We like our pets to be quiet and unobtrusive, like a piece of furniture. For Tuan, a dog's submission to command is their most important attribute.

Suggested Additional Readings

Alger, Janet and Steven Alger. 2003. *Cat Culture: The Social World of a Cat Shelter.* Philadelphia: Temple University Press.

Beck, Alan. 1996. *Between Pets and People: The Importance of Animal Companionship*, rev. ed. New York: Putnam.

Fudge, Erica. 2002. *Perceiving Animals: Humans and Beasts in Early Modern English Culture.* Champaign: University of Illinois Press.

Paul, Elizabeth, A. Podberscek, and J. Serpell, eds. 2000. *Companion Animals and Us: Exploring the Relationships between People & Pets.* Cambridge: Cambridge University Press.

Sanders, Clinton. 1999. *Understanding Dogs: Living and Working with Canine Companions.* Philadelphia: Temple University Press.

Tuan, Yi-Fu. 1984. *Dominance and Affection: The Making of Pets.* New Haven, CT: Yale University Press.

Suggested Films

Dealing Dogs. DVD. Produced by Tom Simon and Sarah Teale. Hastings-on-Hudson, NY: Working Dog Productions, 2006.

Katrina's Animal Rescue. VHS. Written by Kim Woodard. New York: Thirteen/WNET New York, 2005.

Shelter Dogs. DVD. Directed by Cynthia Wade. Burbank, CA: Red Hen Productions, 2004.

Helping People, Helping Pets: Working with VET SOS

CHERYL JOSEPH
Notre Dame de Namur University

"Be sure we have enough of the heartworm meds for large dogs and bring along an extra 50-pound bag of the dry cat food! We almost ran out of both last time," directs veterinarian Ilana Strubel. *"Cheryl, you and your student go with Dr. Durphee. We'll all meet at the entrance to the car camp off Third Avenue. This afternoon, we'll go to Dolores Park."* So begins another morning for the staff and volunteers of Veterinary Street Outreach Services (VET SOS), who will spend the day working with the companion animals of homeless people throughout San Francisco.

Created by Dr. Ilana Strubel in 2002 at the urging of a homeless woman she often encountered in her own neighborhood, Strubel's vision was expansive. She would provide free medical care and health referrals, food, pet supplies, microchips, neuter/spay surgeries, and education about responsible pet ownership to homeless individuals and their pets through the use of a specially equipped mobile outreach van. All the services would be provided by volunteers including other veterinarians, veterinary technicians, and outreach workers.

This is where I got involved and ultimately involved my students from the Sociology: Animals in Human Society undergraduate major at Notre Dame de Namur University (NDNU). While I was grieving the recent death of Ebony, my 16-year-old German shepherd, in 2006, Dr. Strubel suggested I might benefit from participating with VET SOS at one of their events. Having already worked with homeless people and with animals in general for several years, it seemed like an obvious fit.

I still clearly remember that first experience. We were collaborating with Project Homeless Connect that day, a program conceived by San Francisco's mayor, Gavin Newsom, whereby hundreds of nonprofit and for-profit organizations come together under one roof to provide assistance to homeless individuals. There, homeless people can, for example, have lost identification replaced, receive haircuts, obtain legal advice, sign up for housing, and get job counseling all in the same day. VET SOS had parked their mobile van on the curb and secured space outside the building for the intake of the companion animals and other services by the time I arrived. I was immediately assigned to organizing the pet food and pet supplies, then distributing them in an organized and equitable manner. To say it was chaotic and exhausting is an understatement. The people were polite, and the pets were well-behaved, but the need was crushing. An air of desperation converged on all sides of me. Yet the love and devotion I saw people display with their animals were truly uplifting. One woman told me, "My cats are the *only reason I want to wake up in the morning and go on living." A Persian Gulf veteran stated, "This dog is my medicine. He loves me when no one else will. He does for me what no pills can do."*

I learned that VET SOS not only cares for the animals but in doing so, provides an opportunity to make contact with homeless people who would not ordinarily request assistance for themselves but will do so out of concern for their pets. I recall one man who wanted his dog's infected foot treated. While the vets worked on "Rocket," he joked with me about the coincidence of having the same condition on his own foot. I was able to refer him for medical attention for himself as a result. Others shared with me how they had cut back on their drinking or were trying to get in a substance abuse program so they could take better care of their pets. VET SOS volunteers regularly observe similar occurrences.

Impressed with the work and impact I saw, I joined VET SOS again the following month but this time; I was part of the mobile van unit helping with basic exams and procedures at a homeless encampment. People had already been lined up for hours when we arrived at 9 a.m. I was able to comfort nervous owners whose animals had to be transported to clinics for serious ailments and to reassure others that their pets would not be taken from them by law enforcement. I heard numerous stories that conveyed the significant role their animals played in the lives of the homeless. One woman remarked, "It's hard being homeless; you lose everything: the material things, family and friends, security, respect. You find nobody has much compassion. People judge you, censure you. You live with constant fear—they're afraid of you and you're afraid of them. Your animals become everything to you. They're your heart. Without my dogs, I wouldn't want to go on living."

Following these two experiences, I began to bring some of my students with me when I worked with VET SOS. They, too, observed the special bond that homeless people shared with their pets. One student admitted, "I went into this thinking that homeless people shouldn't be allowed to have pets. I mean, they can't even take care of themselves, right? I came away with an entirely different perspective. The dogs and cats I saw were better fed than most of their owners!" As the students continued to return and more students asked to join them, we eventually developed an anecdotal study based on the accounts and sentiments people shared with them along with observations the students made. One student justified the study this way: "It would be a shame to lose such rich data and poignant stories when so many in our society have demonized the homeless. For me, seeing them with their animals has made me personally more humane."

These projects eventually escalated into collaboration with NDNU's sociology department that had, for several years, been hosting a picnic on Thanksgiving Day with the homeless people who live in San Francisco's Golden Gate Park.

Although the university community had already garnered the trust of the park's inhabitants, VET SOS was viewed with suspicion. When the two arrived together on Thanksgiving morning in 2007, individuals came out of their encampment in droves with their dogs in tow. The university community shared sandwiches and pumpkin pie with about 300 homeless people that day and VET SOS treated about fifty of their dogs. Since that time, VET SOS returns bimonthly to treat the rising number of companion animals living in the park while students, faculty, and staff from the university continue to share a picnic lunch each Thanksgiving with the increasing numbers of homeless people.

My involvement with VET SOS and that of my students has resulted in numerous other fulfilling outcomes. I now sit proudly on the Advisory Committee for VET SOS and some of their members speak regularly to my classes on various aspects of the animal-human bond. I've developed a working relationship in Pets Are Wonderful Support – San Francisco (PAWS-SF) that frequently works with VET SOS to better serve some of their mutual clientele. PAWS-SF provides assistance to low-income infirm, elderly, and HIV-infected individuals with pets. These connections have given me the underpinnings to develop classes such as "Teaching, Learning, and Healing through Animals" along with "Animals, People, and their Environments." Finally, organizations such as VET SOS, PAWS-SF, and others provide our students with chances to bring the community into the classroom, and with research opportunities, internships, and jobs upon graduation.

9

Animals and Science

Figure 9.1. "Big Mac." (Cartoon by Dan Piraro. Courtesy of http://www.bizarro.com.)

IN THIS CARTOON BY award-winning cartoonist and painter Dan Piraro, two laboratory rats sit in their cage and wait while a human-size mouse trap, baited with a McDonald's bag, sits on the floor. Rat one says "Quiet, everyone! The test subject is coming!" while a man, presumably an animal researcher, approaches the door. Like many of Piraro's cartoons, this one derives its humor from imagining that the situation of animal research being reversed: The rodents are now the experimenters and the scientists are now the test subjects.

The use of animals in scientific and medical research is one of the most controversial forms of human-animal interaction in modern society. Animal researchers, and most members of the public, feel that we have a right to use animals in research if it will benefit humans. Animal advocates, on the other hand, focus on the ethics of using animals in this manner. In this chapter, we will address these complex issues.

The History of Vivisection

Vivisection, or surgery conducted on living beings, has been practiced on humans and nonhuman animals for thousands of years. The ancient Greeks practiced dissection and other techniques on animals as well as on living and dead human beings (mostly prisoners and slaves). Aristotle, for example, performed experiments on living animals; another Greek scholar, Herophilos, conducted experiments on hundreds of live humans. Dissection of live humans was largely discontinued after the rise of Christianity in the Roman world, but the use of humans as test subjects in drug trials continued; in fact, it was once common for doctors and inventors to test drugs on themselves and their families.

Experimentation on animals became a standard part of medical research starting in the seventeenth century with wild animals, farm animals, and domestic cats and dogs being used as subjects; a number of important discoveries were made based on such research. For instance, some 400 years earlier, Ibn al-Nafis, a thirteenth-century Arab doctor, was able to describe the circulatory system by dissecting live animals. In the eighteenth century, physiologist Stephen Hales described blood pressure by studying a horse and the physician Antoine Lavoisier used guinea pigs to explain how respiration worked. In the nineteenth century, the chemist Louis Pasteur used sheep to demonstrate that infections were caused by bacteria.

Animal experimentation, combined with the dissection of dead humans, allowed scientists to learn about how the human body works and ultimately led to the development of most modern vaccines, medicines, and surgical procedures. The justification for using animals for this kind of work was the philosophical and theological attitude toward animals that we discussed in chapter 2. Descartes, for instance, viewed animals as mechanical bodies with no soul or mind and no ability to feel pain. Anesthesia was not developed until the mid-nineteenth century, so animals were cut apart while fully conscious and, as we know now, were fully able to feel pain.

Experimenting on live humans, even prisoners, was largely replaced by animal experimentation but reappeared, infamously, during World War II when Nazi scientist Josef Mengele experimented on human subjects. These were mostly Jews, Russians, and Gypsies housed at concentration camps. Japanese scientists had experimented on Chinese prisoners of war from the 1930s into the 1940s. Even the United States has its own history of experimentation on humans. During the 1940s, the U.S. Army, in conjunction with the University of Chicago Department of Medicine, infected 400 inmates from a Chicago prison with malaria to test new drugs on them; the U.S.

military and the CIA have conducted numerous experiments on soldiers, prisoners, and other test subjects, many without consent. As early as 1932, doctors at Tuskegee Medical School ran a trial on patients without their consent in the infamous Tuskegee Syphilis Study. During this decades-long study, poor African American men who had syphilis were evaluated by doctors in order to see how syphilis would develop in an untreated patient. The men—hundreds of whom died during the course of the study and whose wives and children were infected—were neither informed that they had syphilis nor offered treatment even though penicillin was widely available as a cure for the disease as of 1947. The study was only called off in 1972 after the press publicized the details, which elicited a huge public outcry. In 2010, it was revealed that American doctors had intentionally infected Guatemalan prisoners and mental patients with syphilis from 1946 to 1948.

Once the Nazi atrocities came to light after World War II ended, the **Nuremberg Code** was developed establishing guidelines for human experiments. These included the mandate for voluntary, **informed consent** from any test subjects, the absence of coercion, a lack of suffering and minimal risk for subjects, and a clear scientific gain for the subject and society in general. The code also mandates that the study should be based on the results of previous animal experimentations; in other words, humans should not be the first study subject. In 1979, a new report was authored by the National Commission for the Protection of Human Subjects in Biomedical and Behavioral Research. Known as the Belmont Report it outlined three basic principles to guide scientists involved in human experimentation: respect for persons, beneficence, and justice. These ethical principals are the foundation of many nations' medical laws today.

For the most part, human experimentation today is limited almost entirely to clinical trials. In other words, after preclinical research has been conducted on animals (or in recent years, **in vitro** or computerized models), drugs, vaccines, or medical devices are further tested in clinical trials, in which humans who have given their consent are exposed to the drug or device in order to further prove that it is safe and effective. Without these clinical trials, no matter how much animal testing has been done, there is no way to realistically gauge either the safety or efficacy of a drug or treatment.

There is no real "Nuremberg Code" for animal research. Animals used in medical studies are covered under the Animal Welfare Act (AWA). The Animal Welfare Act governs the transportation, housing, feeding, and veterinary care of warm-blooded animals in laboratories, as well as animals bred and transported for the pet industry, and for those that are in zoos or circuses. However, the USDA's definition of "warm-blooded animal" excludes mice and rats—the two animals most commonly used in research—as well as birds. Animals raised for food have been excluded since the earliest days of the law.

Laboratories that use covered animals—in other words, animals that are not rodents, birds, or farm animals—fall under the jurisdiction of the USDA, which handles licensing, inspection, and all other compliance and enforcement measures. The AWA also mandates that researchers provide pain medication and anesthesia for covered animals—"if the experiment allows." Unfortunately, because the researcher makes the decision as to whether anesthesia or pain medication is scientifically necessary, there is little mandate

BOX 9.1

IMPORTANT LEGISLATION: THE ANIMAL WELFARE ACT

President Lyndon Johnson signed the Laboratory Animal Welfare Act (LAWA) into law in 1966 to appease public outrage over the theft of pets for animal research. Now named the AWA, this law governs the transportation, housing, feeding, and veterinary care of warm-blooded animals in laboratories, as well as animals bred and transported for the pet industry, and those that are in zoos or circuses.

Laboratories that use covered animals fall under the jurisdiction of the USDA, which handles licensing, inspection, and all other compliance and enforcement measures. Congress has amended the AWA three times. In 1970, it was broadened to include animals used for entertainment and in the pet industry. The 1970 amendment also required the use of anesthetics and painkillers (in instances where they do not interfere with the experiment) as well as a requirement that all dealers have USDA licensing. In 1985, Congress amended the AWA again to require training of animal care staff to provide better care and handling of covered animals, minimal exercise standards for dogs, environmental enrichment for primates, and improvements in housing conditions. Another new provision of the act was the requirement that facilities using covered animals form Institutional Animal Care and Use Committees (or IACUCs) to further oversee the treatment of animals at those institutions. In addition, it mandated that researchers demonstrate proof that they considered alternatives to painful or distressful research and prohibited multiple surgeries on the same animal. Congress also passed an amendment in 1990 that mandated a five-day waiting period before municipal shelters could sell animals to research facilities (to prevent the immediate transfer of pets to laboratories before their owners could claim them).

with respect to pain at all. Sociologist Mary T. Phillips (2008) points out that even today, many scientists make arbitrary decisions regarding whether an animal needs pain medication or not; even though anesthesia is commonly used today (although haphazardly administered), pain medications are still very rarely used, and even less commonly thought about. Because the AWA does not cover mice, birds, and rats, labs that do not receive government funding and that only use those animals are not accountable to any government agency at all. Because of the growing popularity of **transgenic** mice and rats—rodents that have had new genes inserted into their genetic code—the number of laboratories that fall into this category is rapidly increasing.

The Scope of Animal Research and Testing

Today in the West, no drug, surgical technique, or medical device can be used on humans until it has first been tested on animals. In addition, a huge amount of theoretical research is conducted on animals. Animals act as models for human diseases, and also serve as spare parts for organ donations and for tissue and cell use. Animals are also used to teach medical techniques to medical students and veterinary students, and their bodies are dissected by science students. Animals are used in research on subjects ranging from asthma to AIDS, cancer to diabetes, birth defects to biochemical weapons, organ transplants to heart problems, and antibiotics to vaccines.

Another major area in which animals are used is in product testing. Each year, thousands of new or updated household products—everything from shampoo and toothpaste to drugs for AIDS, cancer, and balding to floor cleaners and shoe polish and mascara to new drugs—are sold in the United States. Over the years, the majority of these products or their components have been tested on animals to find out if they can cause harm to humans. The FDA and the EPA mandate that drugs and chemicals be tested on animals; if they are not tested, they have to have a warning label stating that. Drugs cannot be legally sold until they have been tested on animals; this is not the case for household products and personal hygiene items such as cosmetics. Yet these products also are routinely tested on animals before they are marketed. It should be noted that safety testing of drugs or chemicals does not make the products safe—the testing notifies the public when the product is unsafe.

Much of animal research is conducted in the fields of **experimental** and **comparative psychology**, where animals are models not for human anatomy but for the human psyche. Animals are subjects in psychological research projects that cover such topics as depression, obesity, cigarette smoking, anxiety, social isolation, pain, bulimia, and hallucinations. Among the most infamous examples of these studies were psychologist Harry Harlow's "mother love" experiments in the 1950s; for these studies, Harlow took baby rhesus monkeys from their mothers and gave them artificial mother replacements made from wood, wire, and cloth. Sometimes these maternal replacements were benign and provided the lonely babies with someone to hug when distressed, but at other times the "mothers" would shock or otherwise harm the babies when they tried to hug them. The results of this research (and of much other psychological research) demonstrated something that most everyone knew already: Babies, when deprived of the love of a mother, would develop emotional and psychological problems that lasted throughout their lives.

Finally, animals are also used in classroom dissection for middle school and high school education, and in college, medical school, and veterinary school anatomy and surgery classes. Here, they are used to teach surgical and procedural skills to students, to demonstrate physiological functions, and to study and induce disease. Tens of millions of additional animals per year are used in such education.

Every year, the USDA puts out a report detailing the number of facilities the agency inspected (1,088 in 2007); the number of reportable, or covered, animals at each facility; and how many animals were used in experiments involving no pain (approximately 392,000 in 2007), pain in which researchers used pain relief (approximately 557,000 in 2007), and pain for which "no drugs could be used for relief" (77,776 in 2007). Again, because these numbers do not include rats, mice, and birds (let alone amphibians, reptiles, and fish) not protected under the AWA, one can only guess at the numbers of these animals subjected to pain every year.

It is difficult to find out exactly how many animals are in U.S. laboratories today because research and testing facilities do not have to report the number of unprotected animals—mice, rats, birds, reptiles, or amphibians—even though they comprise 85 to 95 percent of the total animals in labs each year. According to the USDA, inspected facilities used more than a million reportable warm-blooded animals in research, testing, or experiments in 2007. By factoring in the estimated number of nonreportable animals, we can estimate that 20 million or more animals altogether are used every year. Scientists who do federally funded research are required to use less sentient animals wherever possible (such as fish rather than monkeys) and because of cost, many scientists choose smaller, less expensive animals (such as mice) rather than larger animals. But many other factors impact the choice of animals, such as the type of research being done (surgical studies for medical or veterinary students typically involve larger animals, for example) or the biological attributes of the animal (zebra fish are often used in embryonic research because the embryo is transparent and develops in an egg outside the mother's body). And nonfederally funded research, such as through pharmaceutical companies, is not bound by many of the requirements that universities face.

In research, rodents are by far the most used group of animals because of their small size, quick reproductive cycles, and economic cheapness—prices start as low as $9 for a rat as opposed to $850 for a macaque monkey. Genetically modified mice and rats are especially popular among scientists. Breeding companies can custom "create" these animals, allowing researchers to manipulate and study various aspects of gene function or expression, model diseases and genetic abnormalities, and more.

Rabbits are the most common reportable animal in laboratories, with approximately 236,000 used in U.S. labs in 2007. They are popular because researchers can purchase them for as little as $30 they are easy to handle and have short gestation periods; and their reproductive cycles can be easily tracked. Also, because their eyes are extremely sensitive and their tears do not easily wash out toxins, researchers are able to conduct tests in which caustic substances can be left on their eyes for days. Rabbits also produce antibodies that some researchers consider far superior to those of any other laboratory animal; their high fertility rate and unique reproductive system also make them popular in reproductive and fertility studies.

After rabbits, guinea pigs (207,000 in 2007) and hamsters (172,000) are the next most common reportable animals in medical research. Guinea pigs have been popular for so long, in fact, that the term "guinea pig" arose years ago when referring to someone or something being used to try something new.

Medical researchers use farm animals (110,000 in 2007) such as sheep and pigs to study heart disease and heart and valve replacement procedures. Pigs are common subjects of **xenotransplantation** experiments, in which researchers attempt to replace unhealthy human organs with animal organs.

Dogs are used next often in research and testing because they are easy to handle—one tragic drawback to being man's best friend. In 2007, U.S. laboratories used about 72,000 dogs for a variety of projects, including toxicity tests, surgical teaching programs for medical students, and dental and heart experiments. They are also research subjects for testing veterinary drugs and pet foods.

In 2007, American research facilities used about 70,000 nonhuman primates. The most commonly used primates are monkeys, specifically

Figure 9.2. The most commonly reported laboratory animal, rabbits like the one shown here may be subjected to dermal abrasion testing. (Photograph courtesy of People for the Ethical Treatment of Animals.)

macaques. Researchers have infected primates with syphilis, hepatitis, and other human viruses; used them in a variety of transplant surgeries, behavioral studies, and neurological studies; and subjected them to crash tests and other experiments in which their humanlike bodies suffer a variety of traumas. We have even sent chimpanzees into space and on space shuttles. The use of chimpanzees is particularly controversial due to their intelligence and emotional and social complexity, their close genetic relationship with humans, and the fact that they are an endangered species. Although countries such as the United Kingdom and New Zealand prohibit the use of apes in experimentation, about 1,300 chimpanzees are currently in U.S. labs. In 2007, citing financial reasons, the National Institutes of Health announced that it would stop breeding chimpanzees for research. (Care for one chimpanzee can reach $500,000 over its 50-year life.)

Once common high school dissection subjects, about 23,000 cats were research subjects in 2007. They were used almost entirely for neurological research and vision studies. Some of the more notable examples include experiments in which researchers severed cats' spinal cords, sutured their eyes, forced them to endure lengthy sleep deprivation, and more.

BOX 9.2

FAMOUS ANIMALS: SILVER SPRING MONKEYS

The Silver Spring monkeys were a group of macaque monkeys that lived at the Institute of Behavioral Research in Silver Spring, Maryland. They are famous because of the legal battle they sparked involving People for the Ethical Treatment of Animals (PETA), the lab, and the National Institutes of Health, and the resulting public attention given to the issue of animal research. The monkeys were being experimented on by psychologist Edward Taub. He disabled part of the monkeys' nervous systems, restrained the animals in stocks, and withheld food from them in order to force them to use their limbs. Alex Pacheco, the cofounder of PETA, found out about the monkeys while doing an undercover investigation at the lab and exposed the conditions that the monkeys were living in to the public via a police investigation and media campaign. (The monkeys lived in wire cages, alone, with no bedding, no food bowl, and no toys or other items whatsoever; many had mutilated themselves and had open sores on their bodies.) Taub was charged with 119 counts of animal cruelty. Ultimately, he was convicted of six, which were later overturned on appeal. The public outrage over the case led to the 1985 amendment to the Animal Welfare Act that required better care and handling of animals and environmental enrichment and that discouraged painful experiments when alternatives could be found. It also led to a public custody battle over the monkeys, which PETA lost, as all the monkeys were ultimately euthanized.

Environmental Enrichment

Many animals in laboratories spend their entire lives in isolation in metal cages, without toys, soft beds, or the barest of comforts. Rabbits, dogs, monkeys, apes, and rats are all highly social animals. In the wild or in other domestic situations where they enjoy companionship, they spend much of their time grooming, communicating, and interacting with each other. These behaviors are impossible when these animals live in cages. More to the point, although it was once assumed that animals used in medical research had only to be physically healthy, it is now understood that forcing an intelligent, feeling social creature to spend his or her life in a cage with no companionship or interaction with others, with little or no opportunity for exercise or play, and with limited and stressful interaction with humans, is psychologically and emotionally stressful. Yet, many millions of animals have endured such living conditions and many millions more continue to do so.

Countless scientific studies that compare animal behavior in conventional laboratory settings to environments modified to better suit them demonstrate that stereotypic behaviors such as rocking, pacing, staring, fur chewing, or self-mutilation are common in the typical barren laboratory cage. These behaviors indicate boredom, unhappiness, and psychological stress. Studies also show that animals' natural behavioral repertoire narrows considerably over time in lab environments, resulting in animals that are listless, tense, despondent, or hostile.

The term "environmental enrichment" refers to ways that animals in laboratories or other environments can have some of their behavioral needs met while living under these artificial conditions. The Animal Welfare Act mandates environmental enrichment for nonhuman primates, including modifying cages and other housing to allow monkeys and apes to pursue "normal" or species-typical behaviors and social interactions, rather than the stereotypical behaviors associated with bored, depressed, or psychologically damaged animals; examples include social housing, "inanimate enrichment items" such as toys, and opportunities to forage for food.

Unfortunately, primates—and to a much lesser extent, dogs—are the only animals the government requires to receive enrichment or exercise. This means that many other dogs, most rabbits, cats, guinea pigs, mice, and rats still spend their entire lives in barren cages with only food and water containers, and possibly a litterbox or other method for containing urine and feces. Toys, blankets, companionship, and an environment that is more interesting than a cage are nonexistent for most of these animals; the one exception is mice as typically they are kept in groups.

Animals as Stand-Ins for Humans

Why are animals so widely used as test subjects for human diseases? The use of animals in biomedical research relies on two related, and contradictory, ideas: Animals are physiologically, mentally, and emotionally similar to humans; the tests will result in meaningful results which can be extrapolated from one species to the next. No credible scientist working on human health issues would bother to test a drug on an animal if he or she did not think the results could be applicable to humans. So we must grant the similarity of humans to not only chimpanzees but also to mice and rats in order for the tests to have validity. Even though researchers do not see animals as "miniature people," they are used as a form of stand-in—substituting for human physiology, anatomy, and even psychological and emotional capacities. Animals can be induced to develop a disease, they can be studied when they naturally develop diseases, they can be studied to understand their resistance to diseases, and they can be studied whether or not the diseases that they manifest are ever present in humans. But in all cases, the animals are chosen because they share certain features with humans—anatomical, physiological, psychological, or even behavioral and emotional.

On the other hand, one of the ways that researchers have historically justified animal research is by distancing animals from humans. Here the argument states that because they are *not* human, animals cannot reason, they cannot remember and anticipate pain, and they certainly do not enjoy the same legal and moral standing that humans do. This is a step up from the Cartesian model of animal consciousness that viewed animals as unconscious, unfeeling machines, but the underlying meaning is the same: Animals are not humans, so they do not feel the same; they cannot register pain, loneliness, or fear, and they do not warrant the same protections.

But many animal advocates have asked how animals can be similar enough to humans to use for product and drug testing and different enough so that they are subjected to treatments that could never be given to humans. And, if we are similar enough that psychologists can study animals in order to understand human depression, why would those same researchers not recognize that intensive confinement, segregation from other animals, and lack of positive stimulation would result in an animal suffering from loneliness, sadness, anxiety, and even depression, just like humans would in similar circumstances? In other words, researchers have tested every psychological drug ever marketed for humans on animals in order to test safety and gauge effectiveness. Scientists claim that their studies show that the drugs work to combat depression in animals, yet at the same time claim that animals do not experience depression. How do scientists justify this disparity between difference and sameness?

The Social Construction of the Lab Animal

Animals must be "de-animalized," just as they are in modern agriculture, in order to justify all the things that are done to them in the lab. This is one reason that lab animal suppliers often do not use the term "animal" at all. The creatures that they sell are research "models." Models are not animals, and they certainly are not *specific* animals. And the distance must increase between researcher and animal to make the behavior acceptable.

"Laboratory animals," as we now know them, did not exist until the twentieth century. Americans have always categorized animals according to the value of their utility, and for most of U.S. history they were classified as farm animals, companion animals, or wild animals. Up until the mid-twentieth century, there was no "laboratory animal" category, and there were no companies that purpose-bred animals for research.

Prior to the 1960s, many of the animals that were used for testing, educational uses, military uses, or research were former pets sold by city or county shelters for research. **Pound seizure** refers to the practice of selling animals that would otherwise be euthanized. Although it is much less common today to find shelters selling animals for research, three states—Minnesota, Oklahoma, and Utah—still *require* that publicly funded shelters do this, and other states allow it. Today, most animals used in laboratory research come from dealers: Some still come from Class B (or random source) dealers, who acquire animals from animal shelters, auctions, other licensed dealers, but most come from companies that breed them specifically for this purpose.

Biologist Lynda Birke (1994, 2003) and psychologist Kenneth Shapiro (1998, 2002) have discussed the animal/not animal dichotomy that is so prevalent in scientific thinking today. Rats, which the public typically thought of as carriers of filth, first appeared in labs "from the wild" at the turn of the twentieth century. Researchers now use thousands of strains of rats for specific scientific purposes, moving them further away from any type of wild animal. Laboratory rats no longer represent disease but the scientific progress against disease, a marvel of symbolic inversion. On the other hand, in order to justify the continuing research on millions of rodents each year, the public still needs to feel some abhorrence or ambivalence toward them. Our feelings for dogs and primates make using them for research more problematic. It is a narrow linguistic and conceptual rope that scientists and the public walk—moving back and forth between rat as animal and rat as scientific tool and not-quite animal—but it is necessary in order to win the war of public sympathy.

Animals are also de-individualized—members of groups of faceless, replaceable, nameless animals "sacrificed" each year. For animals such as inbred mice or rats that look virtually identical and that may not express suffering as visibly as other animals, the process of de-individualizing is even easier. Kenneth Shapiro (1998) points out that animals are also de-specified: They no longer represent their own species, but instead represent ours. Rat no. 29474B is no longer a rat at all, or even an animal, but a human simulacrum. In order to justify using a rat—a species that split off from the human evolutionary line 7–8 million years ago—to mimic human biological functions, the rat can no longer be a rat at all.

Part of the training that animal scientists undergo is learning to suppress emotion and empathy, which would interfere with a scientist's ability to conduct animal research. Biologist Lynda Birke (2003) calls this "objective detachment." One way in which scientists do this is via writing techniques. Scientists, when referring to animal subjects and the procedures they carry out, use a particular kind of language in order to create a distance between animal and researcher, object and subject. In scientific parlance, animals in labs are not killed—they are "sacrificed."

Scientific writing eliminates or reduces ambiguity and unease through the use of distancing words such as "hemorrhage" instead of bleed, and passive voice. Another technique is the use of "inscription devices:" graphs, charts, and so on that mediate between scientist and the objects of science, which, again, are often living creatures. The individual living, feeling animal is what is known as the absent referent in scientific writing; it is absent from the text, yet is the very animal being referred to. The animal is even absent in the photos in such articles. In photos included in scientific journals, only a specific body part, rather than the whole animal, is included.

Animals in labs do not have names. They have numbers. They do not act, choose, or play a role in what happens to them. They are acted *upon*. In short, in the language of science, animals are objects—and never subjects—of their own lives. If they were granted subjectivity, it would be much more difficult to experiment on them—not because scientists are animal haters or lack compassion. After all, many scientists and laboratory workers have companion animals at home. But *those* animals are seen as individuals and are treated as such; it is only the animals in the lab that must be objectified. Animals in the lab have moved from being biological creatures to objects or tools for human use.

Interestingly, the development of the modern scientific language used to describe animal research is a relatively recent phenomenon. In 1847, Canadian physiologist Horace Nelson dissected live dogs to demonstrate the efficacy of using ether as an anesthetic. After removing a dog's ear, amputating his leg, and slicing him open from leg to neck, the dog began to awaken, and Nelson wrote that the dog's "violent efforts and cries giving everyone present to understand, that he was no more sleeping" (Connor 1997). The dog was then strangled. Nelson himself noted that these experiments, in which the dogs were always strangled at the end, were cruel. Today, scientific accounts of experiments involving animals would never use such graphic language or frank assessments.

Figure 9.3 Petunia, shown here with Kate Turlington, was rescued from a university animal laboratory. (Photograph courtesy of Ed Turlington.)

The History of the Anti-Vivisection Movement

Although many people think of the animal rights movement as being very modern, it actually originated in nineteenth-century England, with groups who were opposed to vivisection. The anti-vivisection movement was made up of feminists who were involved in the suffragist movement in England (and later the United States), religious leaders who were opposed to vivisection on moral grounds, and humanists who saw vivisection as a crime against God's creatures.

Of all the religious groups voicing their opposition to animal experimentation, the Society of Friends (or Quakers) were the most vociferous. Quakers were unusual among Christian groups in that they believed in an afterlife and a present day when humans and other species could live together in peace. Furthermore, they believed that women and men were spiritually equal; in fact, women were able to preach alongside men. Quakers such as Anna Sewell denounced the cruelty inherent in vivisection. In 1877, Sewell wrote *Black Beauty*, a story about a horse that experiences a great deal of cruelty in his life. *Black Beauty*, considered by some to be the *Uncle Tom's Cabin* of the animal protection movement, was extremely influential in the growing anticruelty movement in England. And because it was nominally a children's book, it served to instill in many young readers an empathetic understanding of animals.

Suffragists too saw the cruelty of vivisection, and many saw women as being victimized by men in the same ways that animals were by humans. Neither women nor animals had rights at that time, and many feminists could not help but see the parallels between the treatment of women, who were in those days strapped down during childbirth and forced to have hysterectomies, and animals. In 1875, the National Anti-Vivisection Society, the world's first such organization, was founded by a woman, Frances Power Cobbe. In 1898, she founded a second group, the British Union for the Abolition of Vivisection. Because of the activities of Cobbe and other anti-vivisectionists, England passed the world's first animal protection law, the Cruelty to Animals Act of 1876, which governed the use of animals in vivisection. The law mandated that experiments involving the infliction of pain only be conducted "when the proposed experiments are absolutely necessary . . . to save or prolong human life" and that animals must be anesthetized, could only be used in one experiment, and must be killed when the experiment was concluded.

Working-class men, too, for a time took a stance against vivisection. Because the bodies of poor people and criminals were still being used for dissection, many in the working class feared that they would be next. In 1907, a number of different groups coalesced together in the fight against vivisection in a series of events known today as the Brown Dog Riots. They were inspired by the death of a dog that two female medical students claimed had been experimented on multiple times, contrary to the conditions of the Cruelty to Animals Act. The women later installed a memorial to the dog in a park in Battersea, England (the home of an anti-vivisection hospital), that became the focal point of the battle between pro-vivisectionists—mostly medical students—and anti-vivisectionists—made up of feminists, trade unionists, and socialists. The labor groups saw the medical establishment, largely made up of wealthy elites, as oppressive and thus aligned themselves with the anti-vivisectionists: Both thought of themselves as underdogs. In her book on the riots, Coral Lansbury writes, "The issue of women's rights and anti-vivisection had blended [in the late nineteenth century] at a level which was beyond conscious awareness, and continually animals were seen as surrogates for women who read their own misery into the vivisector's victims" (1985:128).

The anti-vivisection movement arrived in the United States with the opening of the first animal laboratories in the 1860s and 1870s, and the

Figure 9.4. Original brown dog statue erected at Battersea Park. (Photograph courtesy of the National Anti Vivisection Society, Wikimedia Commons.)

subsequent formation of the American Anti-Vivisection Society (AAVS) in Philadelphia in 1883. Originally, the AAVS was founded to regulate the use of animals in scientific research, but it eventually adopted its current mission of abolishing such research. Similar to the National Anti-Vivisection Society, the AAVS was begun by women who were also involved in other types of social reform such as the struggles for women's suffrage, child protection, and temperance. Many of these women had also been active in the antislavery movement in the mid-nineteenth century. Although England's anti-vivisection reform resulted in the Cruelty to Animals Act in 1876, it would not be until almost a century later that the United States passed its own piece of parallel legislation: the Animal Welfare Act. Unfortunately for animal welfare advocates, neither act goes far enough in protecting animals used in research.

BOX 9.3

ORGANIZATIONAL FOCUS: AMERICAN ANTI-VIVISECTION SOCIETY

Founded in 1883, the American Anti-Vivisection Society (AAVS) is the first organization in the United States dedicated to ending experimentation on animals. One of the Society's main programs is Animalearn, which focuses on ending vivisection and dissection in the classroom. Animalearn provides a lending library, The Science Bank that offers alternatives to using animals in schools. The Alternatives Research & Development Foundation (ARDF), an affiliate of AAVS, awards grants to scientists and educators working to develop nonanimal methods.

The anti-vivisection movement lost much of its steam in the early part of the twentieth century, and the bulk of those working to protect animals on both sides of the Atlantic Ocean focused on the enactment of anticruelty laws for the protection of companion animals. With the rise of the modern animal rights movement in the 1960s, the modern anti-vivisection movement also emerged, as well as the first steps toward finding nonanimal alternatives to animal research.

Alternatives to Animal Research and Testing

Animals are just one form of research model used to study medical issues relating to humans. Others include computer simulations, in vitro tests, and epidemiological studies, all of which have been useful in studying genetic function, drug development, nutrition, psychology, disease, anatomy, and more. Could these and other not-yet-developed methods completely replace the use of animals in medical research and product testing?

One goal of the Animal Welfare Act is the minimization of animal pain and distress via the use of alternatives to animal research and testing; many organizations and researchers are indeed working on this. In addition, universities and other federally funded research facilities are expected to work harder to search for alternatives to animal models. The Alternatives Research and Development Foundation, for example, provides grants to scientists developing alternatives for product testing. The Johns Hopkins Center for Alternatives to Animal Testing also works to develop alternatives in research and testing through collaborating with scientists, animal welfare groups, and the biomedical industry. The Institute for In Vitro Sciences is a nonprofit organization providing in vitro research and testing services, as well as training for other scientists in the use of alternative methods. Finally, the National Institute of Environmental Health Sciences established the Interagency Coordinating Committee on the Validation of Alternative Methods to develop and validate nonanimal testing methods. Today, thanks to the work of organizations such as these, hundreds of companies no longer use animals in testing their products or ingredients.

The AWA also requires that biomedical researchers consider alternatives to animal use when formulating their proposals and that they investigate all alternatives to the use of live animals before settling on the use of animals for their research projects. Even with this mandate, millions of animals are still used as research subjects in this country; the development of or investigation into alternative methods, although encouraging, seems to be a second thought for many researchers. Why are more alternatives not being developed, and why are more researchers not utilizing nonanimal methods in their research? Sometimes it is just a case of inertia: Changing methods, even when it would not be difficult or expensive, involves doing something new or different and for scientists, like the rest of us, change is often hard.

Developing alternatives for animal testing and research is one part of what is known in the research community as the **three Rs**: reduction, refinement, and replacement. Reduction means reducing the use of animals to get more results from fewer animals, refinement means minimizing animals' pain and distress and/or enhancing their well being, and replacement means replacing animals with nonanimal methods.

Alternatives to animal use are numerous. They include clinical research and observation—in other words, testing drugs and therapies on consenting human populations in a controlled experiment that must take place after animal testing and before a drug is released to the public. Microdosing involves giving human volunteers very small doses of drugs in order to test safety and efficacy. In epidemiological studies, entire populations are examined in order to study health trends; these studies can, for example, link diseases such as lung cancer to smoking. Genetic research can reveal which genes can lead to hereditary health problems. In vitro research uses cell and tissue cultures in a test tube or a petri dish, and one of its uses is drug development. Postmarket surveillance involves tracking drug side effects after the drug has been cleared by the FDA and made available to the general public. Human autopsies and noninvasive imaging technologies such as CAT scans and MRIs allow the human body to be explored, and the use of human stem cells and tissues is useful in developing vaccines. Artificial tissues are good tools for determining toxicity. And computer and mathematical models can simulate physiological processes and provide a way to chart the action of toxins in cells and its effects on the whole body.

In vitro research is particularly valuable for testing diseases at the microscopic level because that is the most fundamental level at which diseases manifest. A number of important in vitro products such as EpiDerm (an in vitro human skin model) and EpiOcular (an in vitro human corneal model) are being used to replace animal toxicity research such as the **Draize tests**. Artificial tissues such as Corrositex are another excellent and inexpensive way of conducting toxicity tests. Mattek Corporation, the makers of EpiDerm and EpiOcular, has also released other models derived from human cells that mimic the human trachea, the inner cheek, and even the vagina.

Unfortunately, the FDA does not require postmarketing drug surveillance, which would force scientists and the government to further study the side effects of drugs after their release. Sometimes this happens when patients suffer adverse effects or even die after taking the drugs. But it is a powerful incentive for companies to remain ignorant of the side effects resulting from their drugs because a popular medication can make millions of dollars each day it is on the shelves. For example, Merck, the maker of the anti-inflammatory drug Vioxx, tested the medication on mice and rats in several experiments. Yet, in 2004, the company withdrew Vioxx from the market because as many as 55,000 people died from heart attacks and strokes after taking the drug. Since then, thousands of lawsuits have been filed against Merck for patient deaths and other damages, which demonstrates the dangers of relying on animal testing rather than clinical research.

A serious obstacle to the use of many already-developed alternatives is the fact that the appropriate government agency has not yet validated most of them. In 1997, the National Institute of Environmental Health Sciences created the Interagency Coordinating Committee on the Validation of Alternative Methods (ICCVAM) as well as the National Toxicology Program Interagency Center for the Evaluation of Alternative Toxicological Methods (NICEATM). These were formed in order to coordinate the development, validation, and acceptance of alternative product testing methods. In 2000, the ICCVAM recommended that federal agencies replace the Draize skin tests for measuring skin corrosion with Corrositex. In 2001, the United States also signed onto the international Organisation for Economic Cooperation and Development's (OECD) decision to phase out the use of the **Lethal Dose 50 test**, which could impact animal testing in the OECD's thirty member countries. Many more methods still await validation, however.

There are also a number of models available now to substitute for the use of live animals in biology and anatomy classes, and even in surgical courses in veterinary and medical schools. They include mannequins and simulators, multimedia software and virtual reality technology, in vitro technology using animal tissue and cells, and the use of "ethically sourced" animal cadavers—that is, animals that have died naturally or in accidents, and whose bodies have been donated for such use. In addition, clinical work with animal patients would, like clinical work with human patients, give veterinary students opportunities to gain hands-on experience; a good example is when veterinary students are able to perform spay/neuter surgeries on shelter animals in conjunction with trained veterinarians.

The Battle Over Animal Research Today

As a measure of the importance of animal research, advocates for and against it make competing claims about how much animal research has or has not played a part in the awarding of Nobel Prizes. According to an analysis conducted by The Humane Society of the United States, two-thirds of the Nobel Prizes awarded in the fields of physiology and medicine went to researchers who primarily or entirely used nonanimal-based methods for their research. But the Foundation for Biomedical Research shows that the winners of seven of the last ten Nobel Prizes in medicine have relied at least in part on animal research. So who is right? Is animal research good science or bad?

Animal researchers have made important discoveries, although anti-vivisectionists point out that many of these same discoveries could have been made without the use of animals. There are also numerous examples in which animals were very poor models for human health or anatomy, lead-ing in some cases to tragic results where drugs (such as thalidomide and, more recently, Vioxx) caused severe health problems in humans even after extensive animal testing. Sometimes, the FDA approves drugs even when they cause problems in animals but clinical tests do not indicate human health issues. Conversely, drugs that show promise in human clinical trials are sometimes pulled back by the FDA when animals that are administered the drug develop side effects. And often the tests that are done on animals are so far removed from the way that humans use drugs that the results are nonsensical. For example, for years, U.S. antidrug agencies have claimed that marijuana kills brain cells. It turns out that they were citing a study that was done in 1974 in which a group of rhesus monkeys were forced to inhale the smoke from over five dozen joints through a gas mask in a very short period of time. After less than three months of this, the monkeys suffocated to death, which resulted in a loss of brain cells.

As close as chimpanzees are to humans genetically (they share more than 98 percent of our genes), even they are not even perfect human models. Even though most HIV researchers now think that HIV originated in primates as simian immunodeficiency virus (SIV), HIV in chimpanzees operates very differently than it does in humans. Chimpanzees infected with HIV for research purposes respond asymptomatically and show no signs of infection; many chimpanzees even reject the HIV virus through their own natural immunity. For that reason, chimpanzees are probably not good subjects to test a potential HIV vaccine because their natural immunity is not shared with most humans. Chimpanzees also do not get sick when infected with hepatitis B, although the hepatitis B vaccine was originally tested on chimpanzees. They also respond differently to hepatitis C than humans.

Xenotransplantation research—transplanting nonhuman animal organs into humans—is another area in which researchers consider the use of animals to be critical for saving human lives. Because of the worldwide shortage of human organs available for transplant—more than 92,000 Americans are currently awaiting an organ transplant and more than half will die while waiting. The result is a growing black market in organ sales as well as illegal harvesting of organs and tissues for transplant; some of these black market organs and tissues make recipients sick. Even though most organ donation advocacy groups focus on recruiting new donors or improving procurement practices in order to meet the growing demand (some are even proposing incentives for organ donation), there is support in the biomedical community to focus instead on xenotransplantation.

Scientists have been studying xenotransplantation for decades. Aside from the ethical concerns about creating "body part farms" where animals are raised in order to provide organs for humans, xenotransplantation has, from a medical perspective, been a total failure. Since 1964, when doctors transplanted a chimpanzee heart into a man, to 1984, when a baboon heart was transplanted into an infant, to 1994, when a pig's liver was transplanted into a woman, the result always has been the same: In all cases, the patients died shortly after their surgeries. This failure demonstrates that even the relatively small genetic differences between, say, chimpanzees and humans are actually quite large when it comes down to the ability of one species' organs to survive in another, even in a closely related species' body. Today, scientists are genetically modifying pigs to make them more "human-like" so that eventually these pig-human organs can be transplanted into humans successfully. Activists opposed to xenotransplantation argue that rather than spending years to make pig organs more human, we should make human organs more easily accessible. In the United States, one must choose to be an organ donor. If the law made all citizens default organ donors, but gave us the option of opting out (as do some European countries), we would immediately add millions of organ donors to the system.

Animal researchers claim that the use of animals in medical science and product testing is critical to human health. Yet opponents of animal experimentation, including a number of scientists and doctors, counter that it is wasteful, that animals are poor models for human disease, that clinical testing of drugs and vaccines often reveals side effects and problems not seen in the original animal testing, that the animals suffer needlessly, and that the use of animals in biomedical research diverts money away from nonanimal alternatives. Clinical research, for example, focuses directly on human patients but receives far less funding than animal research. The other major reason why animal advocates oppose the use of animals in research is a moral issue—even if animal research provides concrete health benefits to humans (or to other animals, as in the case of veterinary research), anti-vivisectionists argue that it is wrong to inflict pain and suffering on sentient creatures in order to benefit others.

Part of why vivisection is still a life-and-death issue for so many is the way the biomedical industry has typically framed the debate. Those who promote the use of animals in science pose the issue as a question of sacrificing animals to save humans or, as industry groups put it, "Your child or the rat." If the choice really were between a child or the rat, everyone would want to see their children saved; they would sacrifice a few animals to achieve that end.

But is that really the most accurate way to see the debate, and is that really the question that best sums it up? Or are there other ways of envisioning the issue that do not sacrifice either the child or the animal, but that allow for both to survive and thrive? It might be better to ask: How can we save the child without sacrificing the rat?

Animal researchers are engaged in an intense public relations battle with anti-vivisectionists—and oftentimes the public—on the morality and efficacy of animal research. Although the animal researchers appear to be winning the battle based on public opinion polls supporting the use of animals in research when respondents are given no alternatives to the use of animals, when respondents are asked how they feel about the use of animal alternatives, the response is overwhelmingly supportive. In addition, these polls show that the majority of the public only supports animal research when it definitively can lead to cures for life-threatening human illnesses, when there is no pain involved for the animal, and/or when no alternatives exist.

Overwhelmingly, polls show that Americans and Europeans want to see more being done to increase the use of nonanimal alternatives in research and testing. The public is uncomfortable with the idea that animals often suffer and die for human benefit, especially when it is becoming clear that effective alternatives exist and are becoming more prevalent all the time. The public is also clear on pain and suffering: One poll indicated that support for animal research plummeted more than 40 percent when animal suffering goes from mild to severe.

Animal protection advocates are often accused of picking the most egregious of animal experiments to attract public sympathy—those in which animals are poisoned, electrocuted, intentionally crippled, and worse. But the

Figure 9.5. "Insensitivity." (Cartoon by Dan Piraro. Courtesy of http://www.bizarro.com.)

medical research lobby engages in the same practices, highlighting the lifesaving drugs and procedures that have alleviated suffering, particularly for children, and downplaying all of the animals and money wasted on useless research.

Why do so many people feel that if we do *not* test on animals, we are letting people suffer and die from preventable or curable diseases? Philosopher Katherine Perlo writes, "You would not be accused of 'letting your child die' because you had refrained from killing another child to obtain an organ transplant" (2003:54). In other words, *not* supporting future animal experimentation does not equal letting children die, and it certainly does not equal killing them, as animal protection advocates often accuse. Perlo also points out that supporting animal testing and experimentation is the default action, whereas opposing them is seen as the aggressive anomaly, as dangerous as withholding a lifesaving drug from a child. Furthermore, just because most people would choose to save their own child or companion animal over a neighbor's child or animal, this commonsense favoritism should not serve as a basis for social policy.

Most animal researchers, government labs, and private corporations that use animals maintain that they are committed to the three Rs: refinement, reduction, and replacement. But this has not translated into many concrete changes. Animal advocates feel that changes still need to be made if animals are going to continue to be used. For instance, housing and environmental enrichment is still lacking in many facilities for many species, leaving animals bored, stressed, lonely, and depressed. Even given all of the oversight and many of the legal changes discussed in this chapter, as well as the many alternatives now available, the use of animals (including nonreported animals) is actually on the rise. And because so many animals are exempt from any government regulation, millions of animals per year have no protection from pain or suffering at all. Ultimately, the welfare of these animals rests on the scientists and students who oversee them, at least until a point when the public demands a different level of protection for them.

Suggested Additional Readings

Birke, Lynda. 1994. *Feminism, Animals, and Science: The Naming of the Shrew*. Buckingham, UK: Open University Press.

Groves, Julian McAllister. 1997. *Hearts and Minds: The Controversy over Laboratory Animals*. Philadelphia: Temple University Press.

Lansbury, Coral. 1985. *The Old Brown Dog: Women, Workers, and Vivisection in Edwardian England*. Madison: University of Wisconsin Press.

Rader, Karen. 2004. *Making Mice: Standardizing Animals for American Biomedical Research, 1900–1955*. Princeton, NJ: Princeton University Press.

Rollin, Bernard. 1989. *The Unheeded Cry: Animal Consciousness, Animal Pain, and Science*. Oxford: University of Oxford Press.

Rowan, Andrew. 1984. *Of Mice, Models and Men: A Critical Evaluation of Animal Research*. Albany: State University of New York Press.

Suggested Films

Chimpanzees: An Unnatural History. VHS. Directed by Allison Argo. New York: Thirteen/WNET New York, 2006.

The Laboratory Rat: A Natural History. DVD. Directed by Manuel Berdoy and Paul Stewart. Oxford: Oxford University, 2002.

One Rat Short. DVD. Directed by Alex Weil. New York: Charlex, 2006.

IO

Animal-Assisted Activities

IN ONE OF MY animals and society classes in the fall of 2010, one of my students was a dog. Actually, one of my students was a veteran of the Iraq war with **post-traumatic stress disorder** (PTSD) who brought his psychiatric **service dog**, Rock, with him to class. Service dogs that assist veterans are able to help men and women cope with problems such as depression, anger, social isolation, nightmares, and panic attacks. Service dogs protect veterans from crowds and situations that might make them anxious. They provide a loving, calming presence to these people, can act as a facilitative lubricant in social situations, assist them in reentering society, and turn on lights and check rooms to help their person feel safe. As we continue to fight wars, and our soldiers continue to come back from these wars with emotional and psychological traumas, dogs such as Rock will increasingly become a part of the fabric of our society.

Animals as Human Assistants

The first animal to be domesticated, at least 15,000 years ago, was an assistant to humans. The dog was man's first nonhuman partner, and was initially brought into human culture in order to provide assistance with hunting in exchange for a share of the kill. It is easy to imagine that the dog's hunting skills were not all that were valued by Mesolithic-era humans. The dog's ability to protect, playful nature, and social qualities all probably came into play very early, and early hunter-gatherers most likely benefited from all of them. Other animals that were used as hunting partners were **raptors**, popular in the Middle East, and mongooses in Egypt.

Dogs are still used as hunting aids today, and there are dozens of breeds that have been developed with specific hunting skills in mind—such as sight hounds that sight prey from a distance and then stalk and kill it; scent hounds that hunt by scent; and terriers that are bred and trained to locate small mammals, often rooting them out of their dens. After the development of guns, new breeds of dogs were developed, known generally as gun dogs. These include retrievers that find and return animals shot by the hunter, setters and pointers that can locate and point toward game shot by hunters, and spaniels that can flush animals out of brush. Today, because hunting is less commonly practiced, hunting with dogs has become controversial. In the United Kingdom, where mounted hunters sent packs of dogs to chase and then kill a domestically raised fox, the practice was finally banned in 2005 after years of public protests. In the United States, the controversy over hunting dogs stems from the practice of using dogs to chase animals such as bears into trees, where hunters then kill them. Fox hunting, too, is growing in popularity in the United States and with it a rising tide of protest.

Dogs have also been used for transportation by people with no access to large domesticated animals such as horses; dogs were harnessed to sleds loaded with goods, and helped Asian nomads who crossed the Bering Strait and settled on the North American continent.

With the domestication of plants in the Middle East's Fertile Crescent beginning about 10,000 years ago, human reliance on animals increased. Crops, and especially stored grain, would have attracted mice and rats, so human communities encouraged cats to take up residence in and around villages in order to control the rodent population. The cow, domesticated about 8,000 years ago, was the first creature to be used as a draft animal; its ability to pull a plough would have a monumental impact on the development of human culture. Thanks to the plough, agricultural communities could increase their crop yield significantly, feeding far more people and producing more surplus foods, which in turn were used to support nonworking classes, to provide goods for trade, and to allow cultures to expand.

New beasts of burden followed soon after, with donkeys, water buffalo, camels, and horses all being domesticated within a few thousand years in Asia and the Middle East. These animals could all be used to carry loads on their backs, pull ploughs and carts, and, importantly, carry riders. (Llamas and alpacas, domesticated in the New World, were never ridden nor did they work in agriculture.) Interestingly, horses were initially too small to ride and were originally domesticated as a meat and dairy source. Eventually, the largest and strongest horses were bred to each other to create the modern horse. With the ability to carry riders and goods, horses and other beasts of burden enabled the creation of long distance trade routes, communication among far-flung cultures, territorial expansion, and, ultimately, warfare.

In fact, some archaeologists feel that the horse has contributed more toward human civilization than any other animal. Using horses to deliver mail may have first occurred as early as 2,500 years ago in the Persian Empire and was popular in ancient Rome as well. At least 5,500 years ago, Eurasian cultures were using horses, initially while harnessed to chariots or carts and later as part of mounted cavalry units, to attack other cultures. With new inventions such as shoes, saddles, and stirrups, the use of horses in battle intensified. The Romans, nomadic cultures from Central Asia, Muslim warriors, and the knights of the Middle Ages all used horses to great advantage, as did the Spanish conquistadors. The latter used horses to efficiently conquer the great civilizations of the New World, none of whom had domesticated animals that could be ridden. (The Carthaginian general Hannibal famously used elephants in battle in the third century BCE, and their use in war probably goes back to the first millennium BCE in India.) After conquest, Spanish and later English colonists used horses to do everything from deliver mail to control cattle to settle new frontiers, allowing for the westward expansion of the North American continent.

Dogs' roles expanded after the domestication of animals for food. Dogs began to be selectively bred and trained to guard and herd animals such as goats and sheep, a task for which they are still used today. Guard dogs, for example, are trained to protect a herd of animals, while other dogs, known as stock dogs, are trained to work them—sorting them, herding them, and driving them. The ancient Greeks and Romans used dogs to guard their communities and military outposts. Most dogs can alert humans that there is a danger present (even tiny dogs can serve as watchdogs), and larger dogs can also defend humans and animals from attack. Livestock dogs do not live with people, but live full-time with the herds that they are assigned to guard. This illustrates one way in which animals whose ancestors would have had a predator-prey relationship develop brand-new relationships over time. Other animals used to guard animals include llamas and donkeys.

Even animals that we do not consider working animals had tasks—pigs and sheep were pulled across fields to push seeds into the tilled earth, and sheep used their feet to thresh grain after harvest. Dogs also assisted horses in some of their activities by running alongside carriages in Renaissance Europe

Figure 10.1. Sunshine packs in supplies for backpackers David and Wayne on a trail maintenance project in the Gila Mountains, New Mexico. (Photograph courtesy of Kerrie Bushway.)

to guide the horses through busy city streets. Dogs, like horses, also assisted in warfare; for example, they were trained to attack Native Americans after the Europeans settled North America. When he arrived on the island of Jamaica in 1494, Christopher Columbus set a dog upon a party of natives instantly killing six of them. Hernán Cortés and Francisco Pizarro used dogs to conquer native peoples in Central and South America. And long before this period, Alexander the Great brought his own greyhound, Peritas, into battle with him; according to legend, Peritas died while saving Alexander's life. It would not be an overstatement to say that the outcomes of many of histories wars might have been very different if it were not for the role that animals such as horses and dogs played in them.

With the Industrial Revolution in the nineteenth century, dogs, cattle, and horses—the most important working animals—saw some of their traditional activities supplanted by machines. Cars and trains replaced horse-drawn carriages, tanks replaced horses in battle, and tractors replaced the plough. Today, dogs have been replaced by fences and electronics to guard property, and even cats have been partially supplanted by chemical or electronic products that kill or repel rodents. Many of the activities once performed by these animals, however, are now conducted in sports. Horse racing, dog racing, herding competitions, falconry, and virtually all of the competitions in the modern rodeo are derived from working animals' roles and responsibilities.

Working Animals Today

Today, even after industrialism, many countries still rely heavily on animal labor. Undeveloped nations and even many developing nations still use animals for transport, to plough fields, haul goods, thresh grains, and herd livestock. But animals have also been trained to develop new skills, or to put old skills to new use, in the modern world.

In Europe, for instances, dogs and pigs have been used for hundreds of years to sniff out truffles, small fungi that grow underground. Because of their superior sense of smell, dogs are involved in a whole host of new jobs today. Dogs are used to detect mold, destructive insects, or the chemical residue associated with bombs. Dogs are used in fire departments to sniff out accelerants that may have been used to start fires. Dogs are employed by agencies as diverse as Immigration and Naturalization Services, the Drug Enforcement Agency, and the U.S. Department of Agriculture to sniff vehicles at international borders and travelers at airports for illegal drugs or weapons, banned plants or animals, and even illegal immigrants. A more recent discovery is that some dogs may have the ability to smell cancer cells in human patients; in the last ten years, scientists have been conducting studies to determine whether or not this is possible, and if so, how dogs can be used to prevent and treat disease.

Dogs are heavily used in law enforcement today, but this history goes back to nineteenth-century England when police officers brought their own dogs with them on patrol. By the turn of the century, the practice of training dogs specifically for police work had begun in Europe and the United States. Today police dogs work as part of canine units paired with human handlers who not only work with them but take their canine partners home at night. These dogs are highly trained animals that do not just sniff out contraband but play a major role in chasing and apprehending suspects. Because the work is dangerous and many dogs have died while on duty, a number of organizations work to raise funds to buy police dogs bulletproof vests.

Search and rescue is another type of work for which dogs, with their excellent sense of smell and highly evolved sense of loyalty, are especially well suited. These dogs are trained to do either search and rescue, where dogs and their human partners search for survivors of disasters, or body recovery work, where dogs are trained to sniff out the scent of human remains. Both types of work take place in the aftermath of natural or human-caused disasters such as earthquakes, avalanches, and bomb attacks. Other dogs are used by law enforcement to search for missing persons. There are hundreds of such dogs in the United States, some associated with law enforcement agencies, and others that live with private persons who volunteer at disaster sites around the world. As is the case with dogs working in other areas of law enforcement, search and rescue dogs can face dangers on the job, physical and psychological. In the aftermath of the September 11 attacks in New York City, hundreds of dogs worked to find survivors and bodies, and many suffered from upper respiratory conditions (because of the poor air quality), torn and burnt paws, and psychological distress. When dogs that are trained to find human survivors locate very few living people, they often experience stress and anxiety, perhaps in part because of the reactions of the humans around them.

Even though most police officers now drive cars, many law enforcement agencies still have officers mounted on horseback who are used to patrol parks, rural and backwoods areas, and even city streets. Mounted units are often used for crowd control during large public protests, and these situations can cause injury to the horses and the public.

If the military use of animals was one of the earliest and most far-reaching in human culture, it remains today one of the most controversial. Horses and elephants no longer participate in modern warfare, having been supplanted by tanks and other modern weapons. World War I was the last conflict that used horses in great numbers. During that war, millions were killed as well as tens of thousands of dogs and other animals such as pigeons that were used to deliver messages across battle lines. Dogs, however, have become increasingly important in twentieth- and twenty-first-century wars. In past wars, dogs had been used as sentries, pack animals, and messengers, and occasionally for attack purposes. They also provided companionship to soldiers and acted as mascots, trackers, and guides. Today, they not only provide watchdog services but are often used to detect mines. Dogs have also been used to pull injured soldiers to safety and to act as scouts, moving ahead of soldiers to alert them to dangers.

As one might guess, the life of a military dog is very dangerous. There are no estimates for how many dogs have been killed or injured during wartime, but thousands died in Vietnam alone. War is not only dangerous for dogs but, just as for people, it is stressful. It is only in recent years that scientists have recognized that dogs can be afflicted with PTSD. Another very sad reality for military dogs is that, since World War II, when surviving military dogs were able to return home after the end of the war, it has been U.S. military policy not to allow war dogs to return to the United States. Instead, the dogs that served their countries, and saved the lives of U.S. soldiers, have either been left on the battlefield or killed outright. At the end of the Vietnam War, most of the U.S. military dogs were either killed or left for the South Vietnamese Army; no doubt many of those dogs were eaten. In 2000, after years of public outrage, legislation was passed to allow retired military dogs to return home, where they either live with the soldiers with whom they served or can be adopted into new families. Even with the new law, it is difficult for many soldiers to bring the dogs that they served with home with them. Many remain at war to serve with a new soldier after their previous handler has been discharged. For those dogs that are allowed to retire, they are carefully evaluated for aggression or other temperament issues that would make the dogs dangerous at home. They are only placed with families (or law enforcement agencies) who understand the responsibilities of owning such unusual dogs.

Many other soldiers in Iraq and Afghanistan have befriended local dogs and cats, and upon discharge from the military have sought to bring them home. Historically, this would have been impossible as the U.S. military did not allow soldiers to bring home animals adopted from other countries. In fact, the military often killed animals found wandering around U.S. bases overseas. In recent years, though, animal protection organizations have rallied support and resources to allow soldiers to bring home animals that they have befriended. The ASPCA International has a project called Operation Baghdad Pups that helps soldiers bring animals home from Iraq; a British-based organization called Nowzad, founded by a former soldier, brings back dogs from Afghanistan and Iraq. These animals have provided comfort to soldiers during war, and most people agree that they should be allowed to come to the United States with their adopted soldiers. In addition, these dogs often risk death or starvation if left behind, as neither Afghanistan nor Iraq currently has a culture that is very animal-friendly—especially when it comes to stray dogs or cats. Unfortunately, even when the U.S. government permits the animals to come home, soldiers must fill out mountains of paperwork and must arrange for the dogs to be vaccinated and transported out of a war zone—which can cost thousands of dollars.

A more recent animal doing military work is the giant African pouched rat. These rats are being trained by the U.S. military to detect buried landmines. Thanks to their highly developed sense of smell and relatively small size (they weigh between 10 and 15 pounds at maturity), they can smell explosives and do not set them off by standing on them. These highly intelligent rodents are trained as infants with clicker reward training to be led on a leash by a handler through areas thought to contain buried land mines. Pouched rats are also beginning to be trained and used to find victims of natural disaster, and to detect tuberculosis in human patients.

Assistance Animals

A very special type of working animal is known as an **assistance animal** or service animal. Assistance animals are animals that have been trained to provide physical assistance to people with disabilities or other impairments. Some of these animals are rescued from shelters, but it is more common today that they are specially bred and trained by assistance animal organizations and then are given to people who need them. Training is extensive. Typically, a foster family does the initial training, and then follows a final period of advanced training with the new owner, so that the animal learns to work specifically with that person. The animal also has to undergo extensive training to be comfortable in the public, and to not get distracted by the attentions of strangers. Training can cost as much as $60,000 for a single animal.

In the United States as in many nations, there are federal laws (for example, the Americans with Disabilities Act) giving people with disabilities rights to access public and private facilities, prohibit discrimination, and allow for disabled people to lead normal lives. These laws allow people with documented disabilities to live and travel with service animals.

The oldest form of service animal is the guide dog for the blind. These dogs are trained to help those with visual impairments get around, cross streets safely, and otherwise negotiate daily life. In Europe, guide dogs have been used for hundreds of years, and were introduced into the United States after World War I through an organization called The Seeing Eye—this is why guide dogs are often called "seeing eye dogs." Helen Keller, for example, was given a guide dog in the 1920s. But the idea of dogs guiding humans is much older than this—in the mythologies of many cultures, dogs serve as the guide for the dead to enter the afterlife. The first guide dog was a German shepherd; although there are other dogs that work as service animals today, such as golden retrievers and Labradors, German shepherds still remain very popular.

Other types of service dogs include those that assist people with physical impairments or conditions such as Parkinson's disease. These dogs can push or pull wheelchairs, open doors, pick up dropped objects, and turn lights on and off. Some can even help do the laundry, assist their owners with dressing, and help with grocery shopping. Importantly, service dogs—especially guide dogs—must be able to ignore their handlers' commands when it is apparent that following those commands would lead their owners into danger.

Dogs can be trained to assist the deaf or hard of hearing by alerting their handlers to sounds such as car alarms, doorbells, smoke alarms, or cries. Some dogs can even alert epileptic owners that a seizure is imminent, so the owner can sit down and take their medications before the seizure strikes. However, thus far it has not been possible to train specific dogs to do this. Seizure response dogs, on the other hand, are trained to provide help to a person after a seizure has occurred. Finally, some dogs are trained to work as psychiatric service dogs. These dogs live with patients who have psychiatric conditions, providing a calming influence for these patients, relieving their anxiety, and grounding them. Patients who experience hallucinations, for example, use their dogs to help them understand what is real and what is not, and, as mentioned, those suffering from PTSD benefit from having their dogs turn on lights and check rooms to make sure that they are safe.

If the owner's condition changes, assistance animals can adapt to those changes; for instance, patients with conditions such as Parkinson's and multiple sclerosis will often experience a decrease in abilities to which the animal is able to adapt. Service animals live with their owner and accompany them outside of the home. Only about 1 percent of all people in the United States who have disabilities live with service animals. One reason for this low number is that, at this point, the demand for these highly trained animals is greater than the supply.

Service animals do not "work" for their entire lives. Typical service dogs only work eight to ten years on average and are then "retired" to live as a pet either with the family that they have served or with a new family. When working, service animals wear brightly colored vests that identify them as service animals. These markers serve two purposes. They notify the public that the dog (or other animal) is a service animal, which means that in the United States, for example, they can be admitted into buildings not normally open to animals. But they also notify the public that the dog is working. This is important because often in public when people see a dog (or a more unusual animal such as a miniature horse), their first impulse is to pet the dog, which can be very distracting to a working animal. When these dogs get home after a long day, their owner takes off the animal's vest, and the dog is allowed to relax, play, and be a dog again. Service animals, then, occupy an interesting position in our culture—they are tools, in the

Figure 10.2. Haruna Matsumoto is a student of architecture at UC Berkeley. Miles is her hearing dog. (Photograph courtesy of Karen Diane Knowles.)

sense that they are used to provide a very specific function for humans and are only truly valuable as long as they provide that function. Yet they are also animals, and as such form close bonds with their human guardians, in many cases becoming part of the family.

Over the last twenty years, other animals have entered the assistance animal world, including miniature horses and capuchin monkeys. In all of these cases, the animals not only perform important functions for their owners, they also provide companionship. This function could be considered just as important given the social isolation that many disabled people experience in society. Studies show that people with disabilities who live with service animals have greater self-esteem and less anxiety and are less socially isolated than those who do not live with service dogs. In 2010, the Americans with Disabilities Act was amended by the Department of Justice to narrowly define "service animals" as dogs and, in some cases, miniature horses that assist people with physical disabilities. That means that all other service animals, including psychiatric service animals, do not have ADA protection, and public and private facilities no longer need to admit them.

BOX 10.1

ORGANIZATIONAL FOCUS: PUPPIES BEHIND BARS

Puppies Behind Bars is an organization that trains puppies to act as psychiatric service dogs for Iraq and Afghan war veterans. This organization pairs Labrador retriever puppies with prisoners who are taught to raise and train the dogs from eight weeks to fifteen months of age. The dogs are trained by the prisoners to walk alongside wheelchairs, open doors, push lights on, bring telephones, and accompany the veteran anywhere that they go. After they are trained by the prisoners, the puppies are given to the veterans. They are useful not only for physical assistance but also for assisting veterans affected by PTSD by keeping them calm and making them feel safe, whether at home or in public. They help veterans reintegrate into society. Puppies Behind Bars does not just help the veterans. Studies have found that prison animal programs such as Puppies Behind Bars reduce recidivism rates in prisoners. Because the animals treat prisoners the same as they do other people—without judgment—they make prisoners feel like they are human again, putting them in touch with their emotions, reducing aggression, and allowing them to develop empathy. Studies also show that prisoners assign the dogs they work with a social identity, which positively impacts their concepts of themselves. These prisoners have a much better chance of reintegrating into society once they are released.

Animal-Assisted Therapy

One of the most recent forms of animal assistance is the field known as **animal-assisted therapy** (AAT). AAT uses specially trained animals to help with the mental, physical, and emotional care of patients who suffer from a variety of complaints. The animals used in AAT are known as therapy animals, and are chosen because of their gentle temperaments and their ability to help patients to heal. They include dogs, cats, rabbits, horses, birds, guinea pigs, and dolphins. Unlike service dogs that live with the person with the disability or condition, therapy animals typically live with handlers (sometimes at special facilities) who train them and bring them to meet patients. Very often, these animals are companion animals as well as therapy animals.

AAT officially began during World War II when a Yorkshire terrier named Smoky that had already served in combat missions alongside his human

BOX 10.2

ORGANIZATIONAL FOCUS: THE DELTA SOCIETY

The Delta Society was founded in 1977 by psychiatrist Michael McCulloch, physician Bill McCulloch, and veterinarian Leo Bustad, who had noted the beneficial impact that animals could have on human health. In 1981, Bill McCulloch helped initiate the American Veterinary Medical Association's Human-Animal Bond Task Force to review the profession's role in recognizing and promoting the human-animal bond. The Delta Society's programs now include Pet Partners, which trains and screens volunteers with their pets so they can visit patients in hospitals, nursing homes, hospice, physical therapy centers, schools, libraries, and many other facilities. They also provide resources for healthcare, educational, and other professionals so they can learn how to incorporate therapy animals into their practices. The Society has a National Service Animal Resource Center that provides information and resources for people with disabilities who are considering getting, or are currently partnered with, a service animal.

partner Corporal William Wynne, visited injured soldiers in a military hospital. Smoky's presence made the wounded soldiers feel better. It is most likely that this sort of occurrence had been going on for as long as humans lived with animals. For instance, dogs were used in some Greek temples to help heal the sick and wounded and, in ninth-century Belgium, animals were sometimes used to help the disabled cope with their conditions. But it was not until WWII that animals were formally recognized as playing a role in human healing. The idea that animals could play a role in psychiatric treatment did not develop until the 1960s when a New York psychotherapist named Boris Levinson brought his dog Jingles to work. He found that patients would communicate with Jingles present, where they would not in his absence.

AAT has exploded in the last thirty years, with new organizations and college programs emerging to study the role animals play in human healing and to promote the use of therapy animals.

Today therapy animals are used in hospitals, nursing homes, rehabilitation facilities, orphanages, and hospices. For instance, Zosia, an Australian shepherd, recently won the American Kennel Club Award for Canine Excellence for the thousands of visits she has paid to veterans recovering from injuries in Veterans Administration hospitals, as well as to cancer patients, amputees, and people in psychiatric hospitals. She also takes part in "reading to dogs" programs at her local library; these programs help children gain confidence and reading skills as they read aloud to dogs.

Therapy animals are also used in a rehabilitation setting to improve physical health by improving motor skills and balance for the disabled or injured and those with neurological disorders. **Hippotherapy**, or therapeutic horse riding, uses horse riding as a form of therapy for physically and emotionally disabled people, or for people recovering from conditions such as strokes. Learning to ride a horse in a safe setting teaches balance and flexibility to the disabled, and gives a sense of accomplishment and companionship to those involved (Burgon 2003). Some programs are related to the care of horses as well, teaching responsibility and horsemanship skills.

Therapy animals are also used in areas where the healing is entirely emotional or psychological, rather than physical. For example, therapy animals are used at prisons and juvenile detention centers, at schools and libraries, and with people suffering from autism, speech disorders, Parkinson's, dementia, schizophrenia, and a whole host of other conditions. Because of the documentation demonstrating that the presence of animals can lower stress in people (see Hansen, Messinger, and Baun 1999), animals are now starting to be used in courtrooms when children are asked to testify in stressful cases. Cooper, a golden retriever from Hobbs, New Mexico, is that state's first courthouse dog. As of October 2010, he is available at the Lea County courthouse to visit with children who are either witnesses to or victims of a crime. Cooper's presence makes children feel safe; when he accompanies a counselor or attorney to meet with a child, he helps the child to trust the adult as well. Counseling and animal-assisted therapy professor Cynthia Chandler notes (2005) that one of the reasons for the success of therapeutic animals is that dogs act as a surrogate for therapeutic touch because it is often inappropriate for human therapists to touch their patients during treatment. **Equine-facilitated psychotherapy** refers to the use of horses in psychotherapy for patients with psychiatric problems. For people with anger control issues, for example, learning to care for a horse without expressing anger is one benefit. Other patients develop self-esteem, self-acceptance, trust, empathy, and communication skills.

Finally, the use of therapy animals has been shown to increase social health by facilitating verbal interaction, increasing attention skills, decreasing loneliness, and increasing self-esteem. Using animals in an educational setting can help children increase vocabulary skills, improve reading, and assist with memorization (Hergovich et al. 2002; Gee, Crist, and Carr 2010). Therapy animals provide comfort, someone safe to talk to, safety and security; they can also bring withdrawn children or adults suffering from PTSD out of their shells. It is not uncommon for patients who are uncommunicative and withdrawn to open up and begin talking after receiving a visit from a therapy animal. Some facilities, such as nursing homes, are adopting their own cat, rabbit, or dog to live in the facility so that the patients can have access to an animal all the time. Even substance abuse treatment can benefit from the participation of animals in therapy sessions (Wesley, Minatrea, and Watson 2009).

The Human-Animal Bond: Benefits to Humans

Although the history of working animals outlined in this chapter is primarily a utilitarian one—animals recruited, generally without their input and often against their wills, to provide services to humans—the more recent examples of the ways in which animals work for humans are a better indication of the human-animal bond. In particular, the rise of animal-assisted therapy could not have occurred if it were not clear that humans benefit from animals in ways that are not strictly utilitarian. Assistance animals, for example, do not just push wheelchairs or help guide the blind across streets; they make people feel less lonely, and more connected. Military dogs, too, do not just detect bombs or guard installations; they also give soldiers comfort during extremely stressful times. The wealth of research on the power of the human-animal bond tells us that the benefits that animals can provide to humans go far deeper than we ever thought before.

And not only that, the healing power of animals is not anecdotal anymore but scientific. For example, animals encourage more exercise, which results in better physical health. Seniors who own dogs go to the doctor less than those who do not (Siegel 1990), and cope better with stressful life events without entering the healthcare system. Pet owners have lower blood pressure, triglyceride, and cholesterol levels than nonowners (Anderson, Reid, and Jennings 1992), fewer minor health problems (Friedmann et al. 2000), decreased rates of heart attack mortality (Friedmann, Katcher, Lynch, and Thomas 1980), and higher one-year survival rates following coronary heart disease (Friedmann et al. 1980). Children exposed to pets during the first year of life have a lower frequency of allergic rhinitis and asthma.

There are also measurable emotional benefits of living with companion animals. The companionship of animals decreases loneliness (Sable 1995), helps children in families adjust better to the serious illness and death of a parent, and increases psychological well being and self-esteem. Pets provide

Figure 10.3. Thomas Cole poses with Mari. (Photograph courtesy of Kate Turlington.)

emotional comfort and support for children whose parents are going through a divorce, and to seniors who have faced the loss of family (Hart 1995). Pet owners feel less afraid of being a victim of crime when walking with a dog or sharing a residence with a dog. The presence of a dog at a child's doctor's appointment or during a jury trial or other court appointment decreases the child's stress and provides comfort (Hansen et al. 1999). For children, animals can help focus attention, enhance cognitive development, and have a calming effect. Their presence may improve attendance at school, compliance with authority, and learning and retention of information. A few studies show that contact with pets develops nurturing behavior in children who then may grow to be more nurturing adults. And people with HIV/AIDS who have pets have less depression and reduced stress (Allen, Kellegrew, and Jaffe 2000).

The health and emotional benefits of living with companion animals are so strong that, in the last two decades, organizations have emerged to help those with physical disabilities or serious health problems keep companion animals in their lives. Known as **human-animal support services**, these agencies provide financial or practical support to help people keep their pets. San Francisco's Pets Are Wonderful Support (PAWS) is one such group, giving aid to people with HIV and AIDS by providing veterinary care for their animals, pet food, or transportation to a veterinary clinic. PAWS volunteers visit with HIV/AIDS patients in their homes to make sure that their companion animals have what they need in order to keep the patient happy and healthy.

Finally, living with pets has documented social benefits as well. Animals are known as "social lubricants." They increase our ability to affiliate with others around us by stimulating conversation, and aiding in the reduction of anxious feelings. They stimulate social interaction, conversation, and rapport in a variety of situations (McNicholas and Collis 2000; Wells 2004). For instance, pets in nursing homes increase social and verbal interactions among residents (Bernstein, Friedman, and Malaspina 2000), and children with autism who have pets have more prosocial behaviors and less autistic behaviors (such as self-absorption). Children who live with pets are more involved in activities

Figure 10.4. "Valentine." (Cartoon by Dan Piraro. Courtesy of http://www.bizarro.com.)

such as sports, hobbies, clubs, or chores, and score significantly higher on empathy and prosocial orientation scales than children who do not.

This research is now so well-known that the National Institutes of Health encourages adding pet-related questions to future national health surveys in order to better understand the relationship between human health and the presence or absence of animals (NIH 1987). In any case, it is clear that, even though animals' presence in human lives may have started out primarily as a utilitarian issue, it has clearly evolved into something that goes to the very heart of how we function and how we feel. One final question arises, however. Do these animal-assisted activities benefit animals at all?

What About Benefits to Animals?

Domesticated animals certainly benefit from living in a situation in which the human-animal bond is facilitated. Companion animals that live in close contact with humans where their primary function is to provide companionship, benefit from having their physical needs met and from the emotional bond that they also experience. Animals that live in a close relationship with humans will likely have better physical, emotional, and mental health, and more social interaction than those who live in isolation, or who have different functions such as racing, entertainment, food, etc. But there is very little research on the benefits to animals of living with humans. The studies are, overwhelmingly, focused on human benefits. There are a handful of studies that have attempted to test whether human-animal interaction is good for animals. One study (Gantt, Newton, Royer, and Stephens 1996), for instance, showed that dogs' stress levels (as measured by their heart rates) decreased when being petted, and another study (Hama, Yogo, and Matsuyama 1996) showed the same thing for horses.

Those assistance animal programs that rescue and rehabilitate animals (such as shelter animals) and place them as assistance dogs do certainly give something back to the dogs themselves. Sadly, those programs are becoming increasingly focused on purpose-bred dogs because the success rates after training are higher with dogs raised and trained from puppyhood, than with adult rescued dogs.

Suggested Additional Readings

Abdill, Margaret N. and Denise Juppe, eds. 1997. *Pets in Therapy.* Ravensdale, WA: Idyll Arbor.

Beck, Alan and Aaron Katcher. 1996. *Between Pets and People: The Importance of Animal Companionship.* West Lafayette, IN: Purdue University Press.

Becker, Marty. 2002. *The Healing Power of Pets.* New York: Hyperion.

Fine, Aubrey, ed. 2000. *Handbook on Animal-Assisted Therapy: Theoretical Foundations and Guidelines for Practice.* San Diego: Academic.

Jalongo, Mary Renck, ed. 2004. *The World's Children and Their Companion Animals: Developmental and Educational Significance of the Child/Pet Bond.* Olney, MD: Association for Childhood Education International.

Levinson, Boris and Gerald P. Mallon. 1997. *Pet-Oriented Child Psychotherapy.* Springfield, IL: Charles C. Thomas.

McCormick, Adele and Marlena McCormick. 1997. *Horse Sense and the Human Heart: What Horses Can Teach Us about Trust, Bonding, Creativity and Spirituality.* Deerfield Beach, FL: Health Communications.

Wilson, Cindy and Dennis Turner, eds. 1998. *Companion Animals in Human Health.* Thousand Oaks, CA: Sage.

Suggested Films

Kids and Animals: A Healing Partnership. VHS. Directed by Michael Tobias. Fur, Fins, and Feathers M.D., 2000.

War Dogs: America's Forgotten Heroes. VHS. Directed by William Bison. San Fernando, CA: GRB Entertainment, 1999.

Websites

Delta Society: http://www.deltasociety.org

Equine Assisted Growth and Learning Association (EAGALA): http://www.eagala.org

Equine Facilitated Mental Health Association: http://www.narha.org/SecEFMHA/WhatIsEFMHA.asp

Guide Dogs for the Blind: http://www.guidedogs.com

Latham Foundation: http://www.latham.org

National Center for Equine Facilitated Therapy: http://www.nceft.org

North American Riding for the Handicapped Association: http://www.narha.org

Nowzad Dogs: http://www.nowzad.com

Operation Baghdad Pups: http://www.spcai.org/baghdad-pups.html

Pets Are Wonderful Support: http://www.pawssf.org

Therapy Animals: http://www.therapyanimals.org

The Healing Gifts of Animals: Animal-Assisted Therapy

CYNTHIA KAY CHANDLER
University of North Texas

There is no doubt how comforting it can be to hug or hold a pet, especially when you are hurt or feeling sad or lonely. The comfort and joy that animals provide to humans through their love and companionship are highly valued. The benefits of interacting with animals are so valuable that many therapists from different occupations bring their pets to work to provide service as pet practitioners and this type of practice is referred to as animal-assisted therapy or AAT. AAT is most commonly practiced by occupational therapists, recreational therapists, physical therapists, speech therapists, and mental health therapists, such as counselors, psychologists, and social workers. To practice animal-assisted therapy professionally one must obtain the proper training and credentials in a professional health or mental health field and additionally obtain some training in AAT interventions. The most common pet practitioners are dogs, followed closely by horses. Other pets that frequently work in a therapeutic role include cats, rabbits, birds, Guinea pigs, gerbils, hamsters, llamas, and various farm animals, including pot-bellied pigs, cows, and chickens. To work as a therapy animal a pet must desire social contact with humans, be friendly and well behaved, and have a high tolerance for noise and stress. It is recommended that the animal handler and the animal as a therapy team pass a standard evaluation, usually about 30 minutes in length, to determine if the handler and the pet have the appropriate temperament and aptitude to perform animal-assisted therapy. So what kinds of benefits do therapy animals provide in their service as therapy animals? For one, clients who enjoy animals are often more motivated to attend and participate in therapy if they can interact with a therapy animal. And, participating in therapy that involves petting and playing with an animal can be entertaining and fun, which often distracts a client from the pain that may accompany the performance of therapy. Holding or petting an animal can soothe anxiety and calm a client. Researchers have demonstrated that when a person pets or holds a friendly animal, such as a dog, human hormones associated with stress decrease significantly, and human hormones associated with health and well being increase significantly. This effect is so profound that other researchers were able to demonstrate that patients in a hospital recovering from heart problems who received weekly visits from a therapy dog actually recovered more quickly and required less pain medication during the recovery process than patients in a control group, who did not visit with a therapy dog. The benefits of interacting with a therapy animal are social, emotional, and physical in nature. This is why so many health and mental health therapists choose to integrate interaction with their pet as part of a client's therapeutic regiment. Interaction with a therapy animal facilitates a client's healing and recovery process.

I have worked with dogs, a cat, and horses providing animal-assisted therapy in my capacity as a licensed professional counselor in north Texas over the past ten years. I began the first animal-assisted therapy training program at a university accredited counselor preparation program in the United States. When I began teaching my university course on AAT, there was no adequate textbook available to teach from at the time so I had to pull my educational material from many different resources, including what I knew from my own experiences providing animal-assisted therapy to clients. Thus, I decided to write a book that would integrate my experiences and existing AAT material, thereby making the material more convenient to teach and more accessible to the public. The book was published in 2005 and titled Animal Assisted Therapy in Counseling. *It was the first comprehensive text dedicated to teaching mental health professionals how to practice AAT. One of the great joys of being a teacher is the opportunity to inspire others to achieve. I have had the pleasure of mentoring graduate students who shared my interest in AAT. I have directed the research of graduate students and been honored to coauthor research articles with them on AAT.*

When you have a love for a field such as I have for AAT, you spend a great deal of time contemplating ideas that can lead to greater possibilities. I am currently working on the second edition of my AAT book that will include detailed discussions resulting from the evolution of my thoughts since the book was initially published. For instance, an important contribution to AAT in the mental health field is my explanation of how AAT can be integrated into counseling practice regardless of which primary counseling theory a therapist follows. The practice of mental health is based on theories developed by predecessors in the field, people you have probably heard about or would hear about in a basic psychology course: Sigmund Freud, Carl Jung, Alfred Adler, and so forth. Several different counseling guiding theories exist, though some of the theories share or have an overlap of ideas. To promote the practice of AAT in mental health, I felt it important to help practitioners understand how AAT interventions are consistent with various counseling guiding theories. And I have dedicated much time to this topic in the upcoming edition of my book, as well as in an article published recently (October 2010) in the Journal of Mental Health Counseling. *I also felt it important for the upcoming second edition of my book to explain, in depth, the basic psychodynamics of AAT, as I currently understand them; something that has not yet been adequately addressed in the field. Understanding the psychodynamics of a counseling session means to understand the therapeutic meaningfulness of interactions, such as between client and therapist. Understanding the psychodynamics of an AAT session is to additionally comprehend the meaningfulness of interactions involving a therapy animal. This not only includes a client's interactions with and reactions to a therapy animal, but, of equal importance, a therapy animal's interactions with and reactions to a client. Also of great value is appreciating the meaningfulness to a client of the client's observation of interactions between a therapist and a therapy animal; these interactions are useful for modeling healthy and appropriate human-animal interactions and human-animal relationships.*

Animal-assisted therapy is a professional modality that is rising in popularity. Not just because it is enjoyable to take your pet to work with you, but because interaction with an animal can significantly contribute to the healing process for a client. I recall how comforting it was for survivors of Hurricane Katrina, who had been displaced to shelters in north Texas, to hold and pet my therapy dogs Rusty and Dolly, red and white American cocker spaniels, while the survivors shared their emotional distress from the impact of the disaster. Both young and old thanked me and my dogs for the opportunity to be comforted by the dogs' loving affection. Performing animal-assisted therapy not only allows more time for you and your pet to spend together, but also provides opportunity to share your pet's healing social gifts with persons in need. Animal-assisted therapy is a most joyous profession.

III

ATTITUDES
TOWARD ANIMALS

11

Working with Animals

SINCE THE FIRST WOLVES began partnering with humans, at least 15,000 years ago, humans have worked with nonhuman animals. As we discussed in chapter 10, humans have raised animals for food and have used them as draft animals, for protection, as hunting partners, and for tasks ranging from search and rescue to guiding the blind and disabled. In all of these cases, humans and nonhumans are engaged in a relationship, and even though many of those relationships are not reciprocal, and many are coerced, these relationships do form the basis for many types of human-animal interaction. In this chapter, we will deal more specifically with many of the relationships that exist between humans and the animals with whom, or on whom, they work. We will cover animal rescue volunteers, who save animals because of a deep love for them; shelter workers and veterinarians— people who are often drawn to working with animals because of a love or affinity toward them, but who often have to kill them; laboratory workers, who either experiment on animals or care for the animals experimented on by others; ranchers, who raise animals for slaughter and yet often have complicated relationships with them; and slaughterhouse workers, whose job it is to kill animals for human consumption. In all of these cases, we will find that the relationships that exist between human and animal are often complex, and often form (and are informed by) the different attitudes that people have toward animals.

Ethnographic Fieldwork

What we know of the relationships that people who work with animals have with these animals comes primarily from sociology, and especially from the work of sociologists Arnold Arluke and Clinton Sanders. In their ground-breaking book, *Regarding Animals* (1996), these men utilize an **ethnographic approach**, studying in countless settings in which humans and animals intersect, such as veterinary clinics, research laboratories, medical schools, animal shelters, circuses, dog training schools, and pet stores.

Arluke and Sanders cleaned cages, helped with surgeries, participated in medical experiments on animal subjects, and even euthanized animals, all to gain perspective on the conflicting attitudes of the people who work in these industries as well as the coping mechanisms they use to carry out their day-to-day responsibilities. Doing ethnography involves trying to understand the "native's point of view." In this case, that means the people who work with animals. How do they feel about what they do? What do animals mean to them? (It doesn't mean understanding the animal's point of view, however. Students may want to consider what the ethical issues may be in euthanizing animals as part of ethnographic research.) Participant observation is another term to describe the type of work carried out by anthropologists and sociologists in settings such as this; it combines subjective (*participant*) knowledge gained through personal involvement and objective (*observer*) knowledge acquired by disciplined recording of what one has seen. In participant observation, the researcher participates in a research setting while observing what is happening there. Observing the interactions between human and animal is one way to understand how people feel about and relate to animals. This type of observation may show, for example, that **informants** may say one thing, but their interaction (or lack thereof) with animals often indicates an entirely different sentiment.

Engaging in ethnographic research in environments that are generally closed to the public—such as slaughterhouses or animal research labs—is often difficult to do. The people who work in these facilities are often suspicious of outsiders because the work they do is either not well understood by the public, or it is heavily stigmatized. In addition, these facilities are often targets of undercover operations by animal rights activists who may even break in, document activities, and vandalize the facilities or free the animals. In order to gain access to these types of facilities, scholars must often find a gatekeeper—someone who is in a position of power and can help open doors to access—and negotiate a mountain of paperwork to obtain permission to study there. Anthropologists and sociologists protect the anonymity of their sources by changing the names not only of the individual people but the facilities as well.

Ethnographic fieldwork is never totally neutral; whether studying lab animal workers or prostitutes, the researcher's personal characteristics—such as gender, age, ethnicity, personality, and background—play a role in how he or she relates to the informants, and how they relate to the researcher. Especially in human-animal studies work, where the researcher may have a bias about the subject because of his or her feelings toward animals, maintaining neutrality and objectivity is often difficult—and may not even be desirable.

People Who Work with Animals

There are some obvious points that we can make about people who work with animals. These people develop different relationships with animals, and have different understandings of them, than those who live with companion animals. People who work with animals, for example, often feel that they *know* animals better than those who do not work with them; ranchers and farmers are heavily critical of animal lovers who oppose their husbandry practices—for instance, the branding or tail docking of animals. Dog trainers, too, feel that their specialized knowledge separates them from people who lack this training; they use this knowledge to justify their methods, even when those methods are criticized by outsiders.

People who work with animals also *see* animals differently than those who do not work with them. Because our relationship with and understanding of animals are shaped by what they mean to us and what their value is to us, animals will mean something very different to different people. For example, the social constructions of a pet dog and a racing dog are very different, and these differences derive from the conditions in which the animals live, and what the animal's function is. A person who lives with a companion greyhound sees this dog as a loving friend, with a history and interests and desires. A person who breeds or trains racing greyhounds, however, sees these dogs as having an economic value that results from their ability to win races; they are not part of a family, they have no real history (outside of the track), and their interests do not matter. Because of the different ways that these dogs are socially constructed, they are treated differently—one lives in the house and sleeps on the bed with his or her people, whereas the other lives in a kennel, races other dogs, and is killed or given away when he or she can no longer perform.

These interactions and relationships do not occur accidentally, or incidentally. They are the product of a very specific working environment in which animals often are seen as products or machines or units of value. People working in animal shelters or research laboratories construct boundaries between themselves and animals in order to protect themselves from the emotional connections that otherwise may emerge. Some, such as shelter workers or veterinarians, must care for and about the animals, but must also be able to detach themselves from them. Others, such as slaughterhouse workers, must learn immediately to detach themselves from the work that they do, and the animals that they do it to.

Clinton Sanders (1996) calls much of the work that people do with animals "dirty work," in that it often involves dealing with disgusting substances—blood, pus, feces, and urine—and can be degrading to one's identity. In addition, it is emotionally dirty work—it often invokes messy emotions, such as grief, anger, and depression. On the other hand, some workers, such as animal rescue workers and others, often embrace this kind of dirty work as a badge of honor, bragging about how easy it is for them to open up and clean abscesses, for example.

Many people who work with animals do so because they say they love animals. Others do so because it is a job. Others do so because animals provide a profit, or are a useful tool. And for many people, the reasons overlap in complicated ways. For instance, pet breeders often say they do what they do because they love animals. But many make a profit off of animals. In addition, although they say that they love, say, dogs and this is why they breed them, their very activity—dog breeding—directly and indirectly results in the death of dogs via culling (in which they directly kill unwanted or imperfect babies) and euthanasia because of overpopulation. As is the case with dogfighters and cockfighters, many of these men say that they "love" their dogs or their birds. Yet again, this love is full of contradictions because when the dog or cock performs poorly, they must be killed; and even when the animal performs well, he will most likely die during one of his fights.

As we discussed in chapter 8, studies show that living with animals affects our attitudes toward them, and even toward other people. It should not surprise us, then, to hear that working with animals shapes our attitudes toward them as well. Ecologist Stephen Kellert (1989) created a nine-category typology of basic attitudes toward animals that traces one's attitudes about the use of animals. Kellert's research has shown that there are three primary sets of factors that shape animal attitudes, including social status (class, ethnicity, gender, etc.), environmental attitudes, and animal-related experiences and practices. In terms of status, white, female, urban, middle-class, and young people are more associated with positive attitudes toward animals than all other groups. Those with positive attitudes toward the environment are more likely to have positive attitudes toward animals.

In terms of working with animals, people whose livelihoods do not depend on animal use have more positive attitudes toward animals than those whose livelihoods do. For example, people with farm experience tend to be highly utilitarian (that is, they see animals in terms of their use value), believe that laws concerning the treatment of farmed animals are adequate, and state that farmers treat animals well. This correlation is not surprising, but it is important nevertheless. People who work with animals, and especially those who do so in a utilitarian or even harmful fashion (such as ranchers or slaughterhouse workers), have statistically more negative attitudes toward animals; these attitudes will shape their treatment of those animals. It is unclear from the research, however, just what is the causative factor. In other words, does a negative attitude toward animals cause one to work in a profession that mistreats animals or uses them for human gain? Or does working in such a profession cause one's attitudes to shift?

Animal Rescue Volunteers

Animal rescue groups are typically privately funded groups, often made up of volunteers that rescue domesticated animals and place them for adoption. The animals can either be surplus animals from public or private animal shelters, unwanted pets from the general public, or stray former pets. Today, in most locations in the developed world, animal rescue groups operate alongside of city- and county-run shelters. Many groups are species- or breed-specific, rescuing only rabbits, Great Danes, or Chihuahuas.

Animal rescuers who have a relationship with their local shelter are generally contacted by staff at the shelter when an animal meeting the breed or species requirement is brought in and the rescuer will then pick the animal up. Rescuers aid their local shelters by cutting down on the volume of animals the shelter must deal with, and are often better able to find a suitable adoptive home for animals that because of their breed, species, or temperament may be difficult for the shelter to place. Many rescuers operate foster homes that provide permanent sanctuary care to animals that by virtue of their age, health, or temperament are deemed unadoptable. Others specialize in certain kinds of animals, such as disabled animals or seniors, often keeping them as sanctuary animals, but also offering them for adoption.

Animal rescue volunteers who work on the front lines rescuing abandoned and unwanted animals collectively spend millions of dollars per year on everything from spaying/neutering and other medical costs, food and other animal maintenance costs, and the other expenses involved in rescuing animals. Not surprisingly, rescuers see themselves as fighting to save animals against a never-ending tide of breeders who produce too many animals, pet stores that sell animals to the public with no screening or education, and members of the general public that abandon animals. No matter how hard rescuers and animal advocates work, the animals just seem to keep coming, with no end in sight.

For these volunteers, sometimes the fatigue of knowing how many animals continue to be abandoned and euthanized feels overwhelming. Known as **compassion fatigue**, animal rescuers are at risk for being overwhelmed and traumatized by the constant animal suffering, and the knowledge that what they do is never enough. Many rescuers are depressed, and deal with that depression in unhealthy ways. Many, for example, use food, alcohol, or drugs to self-medicate.

Another difficulty faced by animal rescue volunteers is that they often have an antagonistic relationship with the workers at the animal shelters they assist. Rescue volunteers often feel that they care more about animals than shelter workers and that they are more knowledgeable about the particular breed or species of animal that they rescue; they sometimes look critically at

---

BOX 11.1

COMPASSION FATIGUE

Compassion fatigue affects people who are exposed to the traumatic suffering of others, such as doctors, nurses, emergency service personnel, counselors, social workers, clergy members, and animal shelter workers. Left untreated, symptoms can worsen and the condition can evolve into burnout, which can cause employees to quit their jobs.

Compassion fatigue may result in poor job performance and plummeting self-esteem, and it can even drive some people who experience it out of their professions entirely. Those who suffer from it can also experience tension in their home lives, or even fall into clinical depression or other mental-health problems.

The signs of compassion fatigue can mimic those of post-traumatic stress disorder, which can afflict people who have survived a traumatic event such as combat, rape, or assault. Sleeplessness, irritability, anxiety, emotional withdrawal, avoidance of certain tasks, isolation from coworkers, feelings of helplessness and inadequacy, and even flashbacks are among the symptoms.

Figure 11.1. Ed Urbanski and Dick Westphal flush prairie dogs out of a burrow. (Photograph courtesy of Yvonne Boudreaux, Prairie Dog Pals.)

shelter workers whose jobs require that they euthanize animals. In return, shelter workers often see rescue volunteers as untrained, unprofessional, and unaware of the realities of working in a shelter and the hard realities of, and need for, euthanasia.

Shelter Workers and Veterinarians

A number of recent studies have examined the stress experienced by workers who have to kill animals as part of their jobs, including shelter employees and veterinarians. These studies show that significant numbers of workers experience **perpetration-induced traumatic stress** related to the killing of animals in their care. Sociologist Arnold Arluke (1996) calls this the "**caring-killing paradox**" in which many people who are drawn to work involving animals do so because of an attachment toward them, which paradoxically results in their participation in the animals' deaths. As with animal rescue volunteers, depression, substance abuse, and high blood pressure are a few of the health issues from which these workers suffer.

Veterinarians have not always treated companion animals. The veterinary profession began in the field of livestock medicine, and veterinarians' work was focused on keeping economically valuable animals healthy for work or for eventual slaughter. Only with the rise of pet keeping in the twentieth century and the invention of the internal combustion engine (and the resulting decline in the use of horses as transport) did "small animal medicine," or companion animal medicine, become a viable specialty.

Today, far more veterinarians practice companion animal medicine than large-animal or livestock medicine. For these veterinarians, their job is to help animals—and they do, which is tremendously satisfying. On the other hand, veterinarians (and veterinary technicians and other veterinary staff) must also deal with pet owners who are often irresponsible and whose actions can sometimes cause animals to suffer. They must deal with a great deal of ambiguity; for instance, how do they respond when clients ask them to perform procedures on their animals that they may disagree with, such as declawing cats (which is really amputating the first knuckle of the cat's paw) or removing the tails or cutting the vocal cords from dogs? Worse, veterinarians must regularly deal with the inevitable deaths of their animal patients, and the suffering of their human clients. Veterinary schools, in fact, now routinely offer courses that deal with "end-of-life" issues, demonstrating how important this subject is to veterinarians.

For shelter workers, many are attracted to the job because of wanting to help animals, but for others, the job is just a job. For those who do love animals, it can be tremendously satisfying because they are able to save lives. But like veterinarians, they are also faced every day with irresponsible owners who abandon their animals, and some workers must themselves be tasked with the job of killing animals that were brought to their shelters. For this reason, shelter workers score very high on compassion fatigue surveys, and shelter workers and veterinarians can experience perpetration-induced traumatic stress.

Sociologist Leslie Irvine (2004) outlines the top three reasons why people give up their companion animals to shelters: moving, allergies, and behavior problems. Interviews with those surrendering animals often reveal multiple reasons rather than the primary stated reason. Irvine also found that many owners were ignorant of basic facts about companion animals, and this lack of knowledge was another contributing factor in their decision to abandon their animals. This is especially unfortunate given the amount of resources that shelters usually offer to adopters—from behavior training classes to literature to help lines and consultants who are willing to help new adopters make successful adoptions. The fact that so many people who surrender their animals are ignorant of these resources may well illustrate their lack of interest in trying to solve the problem that brought them to the shelter.

Other research (Frommer and Arluke 1999) shows that many people who surrender their animals to a shelter feel guilt and shame for what they did. Those who had pets as companions, rather than for utilitarian reasons, feel more guilt when they surrender an animal than do people with guard dogs, for example. As with shelter workers, these animal owners must find ways of coping with their guilt. They often do so by displacing the blame onto others, such as their landlord or partner who "made them" give up their pet. Many owners also pretend that no other alternatives existed other than bringing their animal to the shelter; that way they can't accept the blame for their animal being killed. Many blame society for the situation that they are in (saying, for example, that "nobody wants dogs"), all to keep themselves from accepting the responsibility of abandoning their pet. Many people will blame the animal themselves, assuming that there is something wrong with the animal that made him or her so difficult to care for. Finally, and most unfortunately for shelter workers, many owners blame the shelter workers or other rescue workers themselves for not finding a home for their pet.

Some owners will describe the animal that they are abandoning in glowing terms, so that if they are euthanized, they can assure themselves that it is the fault of the shelter for not trying harder to find them a home. Also, by talking up the good points of the animal, the owner can feel better about the animal's chance at adoption; they can feel less guilt and can talk themselves into thinking the animal will get a good home. Owners also pick shelters where they believe their animals are more likely to be adopted rather than euthanized. Other coping strategies include directly blaming animals for their surrender (because they are chewers, biters, or just not friendly enough), and some justify their actions by saying that the animal would be better off dead than in another situation.

How do the shelter workers cope with seeing animals abandoned daily at their facility, with the guilt that is often directed at them by owners, and with the fact that they actually have to interact with the perpetrators of animal suffering? Shelter workers feel stress combined with guilt due to their role in euthanizing animals. These feelings come from the obvious conflict between caring for and killing animals—both are the contradictory, but necessary, halves of a job which demands that people—often animal lovers—must kill animals because other people have failed them.

One way that shelter workers deal with this is by displacing their feelings onto others, just as the pet owners do. Many workers blame shelter management, but most blame the person surrendering the animal. They not only blame the owners but also want the owners to accept responsibility for their actions. Many shelter workers also blame society in general, and also pet breeders for breeding too many animals. Another coping mechanism employed by shelter workers is to take the moral high ground, seeing themselves as morally superior to everyone else in society. Although society has created this problem, it is only a select group of people—the shelter workers—who have the fortitude to help solve it.

Animal shelter workers often also borrow coping mechanisms from owners, maintaining that euthanasia is often the "best alternative" for an animal, rather than getting one's hopes up for an animal to be adopted. In general, many coping strategies employed by shelter workers involve emotional distancing. Although new employees often get emotionally attached to individual animals, and do all that they can do to prevent an animal from being euthanized, more experienced staff know to keep one's distance in order to maintain one's emotional health. For example, many shelter staff know to evaluate animals based on their adoptability or marketability, in terms of making the decision about who lives and who dies, rather than on any emotional connection they may have with an animal. Others don't talk about the work that they do when at home in order to create an emotional separation between the stress of work and the sanctity of home.

To make matters worse for shelter staff, in recent years the **no-kill movement** has emerged within the animal humane movement, bringing with it new conflicts. This movement aims at ending the euthanasia of healthy animals in shelters in the United States, and often pits so-called and often private no-kill shelters against public, open-door facilities that still euthanize animals. Sociologist Arnold Arluke (2008) has studied the ways in which workers in no-kill shelters and those in open-admission or "kill shelters" shift the blame onto each other for the problems facing them. Workers in open-door facilities blame no-kill shelters that often pick and choose the animals they take in, and thus end up with more adoptable animals than those in the open-door, "kill shelters." Many shelter workers also charge that

Figure 11.2. Dr. Gerald Givan examines an iguana. (Photograph courtesy of Mary Cotter.)

no-kill shelters warehouse animals, keeping them in large numbers for lengthy amounts of time, and contend that this is a form of cruelty as well. No-kill workers, on the other hand, pride themselves on being able to give a chance to animals that would otherwise be euthanized at the kill shelters: the biters, the scared animals, the unsocialized animals. Their own identities hinge upon the idea that they are not engaged in animal cruelty, and they are fighting to save all animals; this identity also rests, in part, on making kill shelter workers the bad guys. Finally, no-kill workers also suggest that kill shelters are resistant to change and are invested in maintaining the status quo, even if it means animals must die. No-kill workers, then, construct themselves as the outlaws of the animal protection movement.

Ranchers

Cattle ranchers have very complicated relationships with animals. On the one hand, the animals that they work with are raised for one ultimate purpose—to produce milk (for dairy cattle) or to produce beef (for beef cattle). They are ultimately a product with a clear economic value. On the other hand, scholars working with ranchers have shown that ranchers recognize cattle as beings with minds, and that many even have affection toward them. This creates a complex set of interactions between human and cow and, as with animal shelter workers, a number of coping strategies. Sociologist Colter Ellis (2009) points out that it is only through the labor of cattle ranchers and others like them that the rest of society can enact loving relationships with one set of animals—pets—while eating other animals—livestock—that were raised and slaughtered by others.

Ellis's work with cattle ranchers in Colorado has shown that many do form emotional attachments to their cattle; women ranchers, who are often relegated to the task of bottle raising babies, are generally more open about their feelings. Some ranchers feel conflicted when bringing their cattle to slaughter, for example, even when doing so brings profit to the rancher. Most ranchers also report that they take a lot of pleasure in calving season, when ranchers often sleep with or near the cattle that are close to giving birth, assisting in their labor. They often have affectionate relationships with young cows or those that are deemed "special," but as these animals get older, these relationships often shift. Legal scholar Gary Francione (2000) has coined the term **moral schizophrenia** to discuss the cognitive confusion that occurs when people who exploit and cause suffering to animals may also love and care for those same animals (or different ones). Like shelter workers, who must distance themselves from animals that are going to die, many ranchers will utilize distancing strategies; for instance, by minimizing the discomfort that they feel when a cow dies. Ellis maintains that without these strategies, these ranchers would be unable to do their jobs.

Ellis and Irvine (2010) have written about the ways in which children, who appear to have a natural affinity for animals, must be socialized to learn utilitarian attitudes toward them. Children who become ranchers learn, through programs such as 4H and FFA, to develop new attitudes toward animals; for many, this is an emotionally trying process. Children who started out as animal lovers must learn to say goodbye to the animals they've raised—animals that will be slaughtered for food. Ellis and Irvine call these children "emotional apprentices" who must learn to manage their emotions and especially learn not to get attached to their animals. One strategy that older kids learn, for example, is to no longer name their animals; when they have a name, it's too easy to get attached to them.

Laboratory Workers

The men and women who work in medical research laboratories that use animals also have conflicted and contradictory relationships with the animals under their care. This includes the trained scientists whose research protocols are being followed, and the animal care workers who spend much more time with the animals than the scientists do.

A number of recent studies have examined the stress experienced by workers who have to kill animals as part of their jobs, including those who work in laboratories. In his research, Arnold Arluke (1996) found that although many laboratory technicians ended up in their occupations to make money or are using it as a stepping-stone to another job, many others were attracted to the work because of their love of animals. Not surprisingly, those who saw their work as just a job also saw animals as just part of their work, and in many cases, they viewed animals quite negatively. These workers, for example, hated the way that the monkeys displayed their antipathy toward their treatment and conditions—by screaming, pulling, grabbing, fighting, and biting. It shouldn't surprise us to find out that these workers were unmoved by the death or suffering of the animals, and they did little to improve the well being of their animals. Workers who took their jobs because of their affinity for animals, on the other hand, developed relationships with animals, spent their free time with them, advocated on behalf of them, and because of their strong attachments to them, suffered greatly when the animals under their care suffered or were killed.

Many researchers and workers cope with the unsettling aspects of their work by compartmentalizing, or separating their scientific and common-sense responses to animals, which allows them to go home to their dog without feeling bad about what they just did to the dogs at the lab.

How do those who experiment on animals justify what they do? One way is by denying animals the capacity to feel pain. Some laboratory workers and researchers use terms such as "discomfort" rather than "pain" to describe what the animals are feeling. Today, even after the passage of the Animal Welfare Act (AWA), scientists can inflict pain on animals without giving pain medication when the researcher proves that it is "scientifically necessary." Unfortunately, because it is the researcher who makes this decision, there is no limit as to the number of studies that can be done on animals that inflict pain. In addition, because mice and rats are not classified under the AWA as animals, they are exempt from even this regulation. Painkillers after surgery are almost never given in laboratory research, either because the researchers never even think about it, or because it would introduce another variable into the data.

We discussed in chapter 9 the ways in which scientists "de-animalize" the animal when engaging in scientific writing. These methods—using the passive voice, emphasizing graphs and charts, and using terms such as "sacrifice" rather than "kill"—also serve to distance the researcher from the animal, and from what the researcher is doing to the animal. Biologist Lynda Birke (1994) calls this "objective detachment," which involves acquiring the skills of appearing not to be affected by one's emotions. She also points out that it's easier for men to learn this than for women, because of the way that women have been socialized to feel empathy in our culture. For laboratory workers, suppressing emotion and empathy is necessary for the worker's emotional well being and also for the work itself.

Some researchers and research technicians definitely feel something for the animals used in research. In 1993, the University of Guelph in Canada held a unique memorial service to commemorate and honor the animals used in research, and the school has held similar events in subsequent years. And while other universities and even private animal-testing or research facilities have created their own memorials to the animals used at those facilities, most animal research institutes do not openly or otherwise acknowledge the lives or deaths of the millions of animals used annually. There is no doubt that these types of tributes provide some satisfaction and an alleviation of some guilt for the workers who participate in them. Yet, ultimately, they provide a justification for biomedical research by asserting to the workers, students, and even the public that animals have made a great contribution to science—a contribution that, the argument goes, humans could surely not live without.

Slaughterhouse Workers

Slaughterhouse work has always been stigmatized. In many cultures, the work of slaughtering animals was done by slaves. In other cultures, it is performed by the underclasses, or in India, the untouchables. Slaughterhouses themselves are typically located on the outskirts of town, so that normal citizens do not have to hear the screams of the animals and smell the blood. French anthropologist Noelie Vialles in her book on slaughterhouse workers, *Animal to Edible* (1994), notes that although society craves meat, it has no desire to either see animals being transformed into meat, or invite the slaughterhouse worker to dinner.

At the time of Upton Sinclair's 1906 novel *The Jungle*—based on the life of a fictional Chicago meatpacker named Jurgis Rudkus—most Americans had no idea what conditions in U.S. slaughterhouses were like. Sinclair's novel exposed readers to the horrors that animals faced: being slaughtered while fully conscious, kicking and screaming in pain, and the dangers faced by the workers, many of whom lost their limbs and even their lives. *The Jungle* so upset Americans that Congress passed the Meat Inspection Act and the Pure Food and Drug Act of 1906. In the 1930s, with the rise of the labor movement in the United States, slaughterhouse work became unionized and workers began to enjoy better pay and better working conditions. The work was still brutal and dangerous, but federal agencies such as the Occupational Safety and Health Administration provided oversight into workplace conditions, making the work safer. In the 1970s, many slaughterhouse workers made $18 per hour, and could put their children through college.

These recent improvements have largely been lost as the large meatpackers have deunionized, consolidated, and targeted immigrants for their labor force. Today, hundreds of thousands of immigrant workers—most from Central America, and many here illegally—work in horrific conditions in the slaughterhouses. The pay today averages $7.70 per hour. Because they don't speak English, and many are undocumented, they can neither organize for better pay or treatment nor request government help to fight the conditions in which they work. Journalist Charlie LeDuff (2000) investigated a North Carolina slaughterhouse and found that not only did the slaughter and meatpacking industry cut its costs by recruiting Mexican laborers, but the jobs themselves are also segregated by race, with white managers, mechanics, and boxers; Indians who do warehouse work; African Americans on the kill floor; and Mexicans doing the butchering.

As we discussed in chapter 7, slaughterhouse workers spend long days doing repetitive work at rapid speeds using dangerous equipment and sharp tools. How do these workers cope with a job that is stressful, hard, dangerous, and which dehumanizes both animals and humans? It should not surprise anyone to learn that animals are treated as machines in this environment, and the workers learn to shut out any connection to suffering. When discussing how many animals will be killed in a day, Noelie Vialles points out that workers say "5300" rather than, say, "5300 cattle," thus pushing the live (soon to be dead) animals a bit more into the background. Eventually, most workers get used to the killing, although not all do. Workers suffer not only physical problems from the hard work and unsanitary conditions, but many suffer psychological trauma as well.

Attorney Jennifer Dillard investigated slaughterhouse work and argues in a 2008 article that prolonged work on a kill floor exposes workers to the risk of psychological damage, including post-traumatic stress disorder. And **sociologist** Amy Fitzgerald and her colleagues (2009) have documented a spill-over effect from the violent work of the slaughterhouse into the surrounding community. This research shows that U.S. counties that have slaughterhouses consistently have higher rates of violent crime than demographically similar counties that don't. A number of studies now document the negative effects—primarily higher crime—of slaughterhouses moving into rural areas in the United States. Some scholars have suggested that the increases in crime can be traced to the demographic characteristics of the workers, the social disorganization in the largely immigrant communities, and increased unemployment rates, but Fitzgerald theorizes that there is a clear link between the increased crime rates and the violent work conducted in slaughterhouses, and that that link can be explained by the

Figure 11.3. "Masculinity." (Cartoon by Dan Piraro. Courtesy of http://www.bizarro.com.)

loss of empathy experienced by the workers. As workers become desensitized to suffering, they can more easily cause suffering in humans as well.

Gail Eisnitz, then working for the Humane Farming Association, interviewed slaughterhouse workers (2007) who told her that they have participated in extreme types of violence, even for a slaughterhouse. Many reported that they have, due to the line speeds and quotas that the workers must meet, beaten, strangled, boiled, and dismembered animals alive. (Legally, animals are supposed to be stunned before death in the United States.) These workers told Eisnitz about the effects this violence has had on their lives; the results included self-medicating with alcohol or drugs, and domestic abuse.

Suggested Additional Readings

Arluke, Arnold, and Clinton R. Sanders. 1996. *Regarding Animals.* Philadelphia: Temple University Press.

Irvine, Leslie. 2004. *If You Tame Me: Understanding Our Connection with Animals.* Philadelphia: Temple University Press.

Sanders, Clinton R. 1999. *Understanding Dogs: Living and Working with Canine Companions.* Philadelphia: Temple University Press.

Vialles, Noelie. 1994. *Animal to Edible.* Cambridge: Cambridge University Press.

Young Lee, Paula. 2008. *Meat, Modernity and the Rise of the Slaughterhouse.* Durham: University of New Hampshire Press.

Suggested Films

Katrina's Animal Rescue. VHS. Written by Kim Woodard. New York: Thirteen/WNET New York: PBS Nature, 2005.

Shelter Dogs. DVD. Directed by Cynthia Wade. Burbank, CA: Red Hen Productions, 2004.

War Dogs: America's Forgotten Heroes. VHS. Directed by William Bison. San Fernando, CA: GRB Entertainment, 1999.

Working with People Who Work with Animals
CLINTON SANDERS
University of Connecticut

I am an ethnographer. My research doesn't involve asking a large number of people simple questions to find out about their attitudes as do survey researchers. I also don't put people in small sterile rooms and have them engage in relatively meaningless experimental tasks. Instead, I believe the richest, most meaningful, and true-to-life (i.e., "valid") information (i.e., "data") comes from directly participating within those groups of people whose lives and experiences I am interested in understanding. This research activity (sometimes also called fieldwork or participant observation) is time-consuming, confusing, messy, exciting, scary, and, in a variety of other ways, just like real life.

My personal and professional interest in how people do things together has frequently led me to investigate occupational activities. Earlier in my career I studied and wrote about professional musicians, narcotics police, tattoo artists, drug dealers, and other (often marginal) workers. Recently, however, I have been interested in an issue that is somewhat more ordinary but also the focus of a wide variety of occupations. I investigate the relationships of people with nonhuman animals. My special interest is in interactions between people and companion dogs. An important principle in ethnographic research is "start where you are." It is typically easier and more productive for the investigator to focus on issues and groups with whom he or she is already familiar. This basic guideline led me to the study of human-animal interactions. Some years ago, following the death of my first dog, I brought two Newfoundland puppies into my household. As a practicing social scientist I watched with fascination as the puppies interacted with each other, noted how they responded to me, and observed the impact they had on my encounters with other people when we were all together in public. I had enrolled both dogs in a "puppy kindergarten" program that met weekly in my local veterinary clinic. I began to write down what I encountered there and soon was adding to these "field notes" with observations of the dogs playing and what happened while we were out walking in the woods and parks around my house. Eventually, I used this information to write about how dog caretakers make excuses for their dogs when they misbehave in public and to explore how companion animals engage in the ultimate social activity of behaving in ways that are intended to shape the actions of those with whom they interact (central to this is what sociologists refer to as the ability to "take the role of the other").

As I became more involved with human-animal issues, I decided to go back to the veterinary clinic to see if they would allow me to "hang out" there and watch (and eventually participate in) what went on there. Initially, I was most interested in continuing to look at animal "owners" but I soon began to focus more on the occupational lives of the veterinarians and other workers in the clinic. I spent over two years there and eventually wrote about such key issues as the characteristics of problematic animal patients and human clients and how veterinarians handled these problems. I also explored the often painful encounter surrounding the euthanasia of a companion animal.

Calling on some of the relationships I had made while working at the clinic, I then moved on to examine a particularly intense human-animal relationship—that between those with visual disabilities and their assistance dogs. I was fortunate that a "guide dog" training program had been established some years before in a town not far from where I lived and, after gaining permission from the various directors, I began to investigate this unique form of interspecies interaction. Of course, at the guide dog program I also spent a lot of time with the trainers themselves (as it turned out I spent more time with trainers than guide dog users) and came to be very interested in how they saw their jobs and, especially, how they came to understand each dog/trainee as an individual and shaped the training process on that basis. In essence, I was interested in the process through which they defined the minds and unique personalities of beings who could not use conventional language. This became one of the central issues I have researched, thought about, and written about for the past two decades.

The experiences with guide dogs sparked my interest in highly emotional and intensely interdependent relationships between dogs and people. I decided that the shared lives of K-9 police officers and their dogs might be a fruitful interactional world to explore. Anyone who has studied police has come to realize that access to their occupational world is difficult for anyone other than someone with a police background. While my way was eased somewhat by my now fairly broad experience with dogs, being allowed to spend time with police and their dogs (I was seeking to observe a few classes of novice officers training their new dogs and more experienced officers going through mandatory recertification as handlers) proved to be quite a difficult project. Rather than wait for approval, I spent this interim time back at the veterinary clinic interviewing veterinary technicians about their job. What struck me here as most interesting was how dirty and emotionally trying the job was and, despite this, how much the vet techs loved what they did. This led me to write about veterinary technicians' involvement in physical and emotional dirty work and the things that made them feel their occupational lives were meaningful. As was clear from the interviews, working with, helping to heal and comfort, saving some companion animals from death, and easing the passing of seriously ill animals helped technicians to see themselves and their work as being of special value. The animal-related rewards of their occupation made the dirt and sorrow they endured worthwhile.

Eventually, I was allowed to spend time with the K-9 training program. This was an amusing, instructive, and sometimes scary experience (on various occasions I was attacked by patrol dogs in training, though I was never injured). As I have already suggested, ethnographic research is an "inductive" process. One does not start with a hypothesis or specific "research question." Instead, the investigator "lets the data talk" and comes to focus his or her attention on issues that emerge as the research proceeds. As I watched and participated in the training of patrol dogs, I noticed something interesting that spoke to an issue that one found in much of the literature on human-animal relationships. The ambivalence that characterizes people's orientation to animals—as food or family members, as experimental subjects or valued patients—has been referred to as the "constant paradox." Nonhuman animals are defined and treated as objects, on the one hand, and sentient individuals on the other. This ambiguous definition of and interaction with animals was glaringly apparent to me in the relationships that developed between K-9 police and their patrol dog partners. As coworkers, the dogs were occupational resources and weapons. They were trained to track people and find certain objects and were used to threaten or apprehend unruly citizens. In these roles they were required to be disciplined, attentive, and occasionally violent. At the same time patrol dogs were also part of the officer's household and were frequently taken to public situations such as schools and town fairs where they were expected to be docile, nonthreatening, and reliably obedient. I found this ambivalence built into the officer-dog relationship to be of considerable interest. These dual and conflicting expectations often resulted in a significant level of tension in the officer's relationship with his animal partner (who, despite the popular cultural image, is frequently unreliable). As one policeman put it when we were talking about the fact that he spent more time with his dog than he did with any other member of his family, "It is like coming to work every day with a three year old . . . a three year old that weighs a hundred pounds and carries a chain saw."

12

Violence to Animals

The killing of animals is a structural feature of all human-animal relations. It reflects human power over animals at its most extreme yet also at its most commonplace.
—ANIMAL STUDIES GROUP (2006:4)

ONE WEEK IN THE summer of 2010, two animal-related stories from far-flung locations hit the news, and both quickly went viral on the Internet. The first involved a street-cam video recording of a woman in England who stuffed a cat that she had just finished putting into a trash can. (The cat was rescued 14 hours later by a passerby who heard the cat crying.) The second involved a video of a girl in Bosnia throwing a litter of mewling newborn puppies into a river to their deaths. Both videos were posted on YouTube and both became media sensations, especially after the participants in popular online message and image board, 4Chan, became involved. (In both cases 4Chan users were able to help identify the perpetrators, resulting in both individuals being arrested and charged in their own countries with animal cruelty.)

I was teaching my Animals and Society course when both of these stories broke, and they offered an interesting teaching moment for my class when one of my students questioned (with some anger) why stories like this generate huge amounts of public outrage, when millions of people, including children, are starving to death or suffering from war, disease, or poverty. Why the disproportionate focus on these two incidents involving animals, and so little interest in human suffering? The answer to this question is complicated, and served as a teaching moment for me. I took the opportunity to point out to my students the difference between public outrage at an isolated incidence of animal cruelty, as in the two events posted on YouTube, and our concerns about and social policy regarding large-scale suffering, whether to animals or humans.

Institutionalized Violence to Animals

When we think about violence toward animals, we tend to think about instances in which an individual has harmed, often publically and without justification, a single animal or perhaps a group of animals. But what we rarely think about is how prevalent violence to animals is in our own society, and how much a part of our social fabric it really is. In fact, violence to animals is truly universal—it is found in every culture and in every time period.

Most animals raised for food or clothing—almost ten billion animals per year in the United States alone—are born, reared, and die in the most extreme forms of confinement, never experiencing even the slightest bit of kindness or mercy. Animals used for medical experiments and product testing live—from birth until death—often painful, lonely lives in small cages. Other animals such as those used in circuses or rodeos live only to entertain us, receiving little if anything in return for the amusements they provide. And wild animals suffer in other ways—losing their lives as their habitats disappear, being removed from their land for the exotic animal trade, and being hunted for trophies. Even those animals that we welcome into our families suffer through the pet industry that produces them. We buy millions of animals as if they were furniture or clothes, keep many of them in inadequate conditions, and discard them when we grow tired of them. But most of us never stop to think about these occurrences, and most would never consider them to be an example of "violence." They are certainly not illegal.

In fact, the suffering and death of animals—in the meat, clothing, and vivisection industries, through hunting and wildlife control, or because of overproduction and euthanasia in the pet industry—are not only *not* considered to be a form of violence, they are also considered socially acceptable. We can call this type of brutality *institutional violence*. It might be useful to look at the sociological understanding of racism to help us understand what this means.

Racism is an ideology of superiority based on the concept of race (a concept which social scientists now show to be illusory) that includes prejudice and discrimination. Individual racism is expressed through personal attitudes and behavior that is directed toward certain racial or ethnic groups defined as inferior by individuals who define themselves as superior. Individual racism, such as performing hate crimes, using racial epithets, or denying someone work, is manifested in actions that are typically deplored by "respectable" members of society. Institutional racism, on the other hand, occurs when racist beliefs become blended into the institutions of a society. Discrimination is then built into the laws and policies of the system. In particular, institutional racism refers to established customs, laws, and practices that systematically reflect and produce racial inequalities whether or not the individuals maintaining these practices have racist intentions. Institutional racism persists because of the economic and political power that accrues within dominant groups because of their position in social institutions. It can be seen in persistent economic inequality, racial profiling in the criminal justice system, segregation within the educational system, and lower tracking of minority students with the same test scores as higher tracked whites.

Institutionalized violence toward animals refers to the "regular" forms of violence toward animals that are part and parcel of the biomedical industry, the agricultural industry, the entertainment industry, and the pet industry. The agricultural industry which inflicts violence on billions of animals per year and the biomedical industry (which inflicts violence on hundreds of thousands of reported animals, and millions more nonreported animals) are the best examples of settings in which wide-scale violence occurs. This violence is not seen or thought about by most people. Another example is state wildlife services that kill millions of wild "pests" such as coyotes, wolves, or prairie dogs that are seen to cause problems to ranchers or homeowners. This is an example of *human-centered destruction*—we disturb habitat, causing wild animals to come into contact with humans—combined with an *animal-centered response*: Our solution is to kill those animals.

What is less obvious is that even the pet industry participates in and profits from institutional violence. Here, the culling of kittens, puppies, and bunnies that do not conform to breed standards is seen as an acceptable part of the breeding process. Killing millions of once-cherished companion animals that are taken to our nation's municipal shelters because they are too old, too much work, or too unruly is just another "inevitable" part of the system. The question is, if we can acknowledge that these activities are by definition violent, why do we not care about them?

Another form of institutionalized violence is sport hunting, in which trophies—the antlers, horns, tusks, heads, or bodies of animals—are sought. Animals are not killed in order to be consumed, but because hunters enjoy tracking and killing the animal and then showing off their kill at home. The Safari Club International, the biggest sport-hunting advocacy organization in the country, offers hundreds of hunting achievement awards. Some of

Figure 12.1. Cockfight at a wedding in Mexico. (Photograph courtesy of Jo Simon, Wikimedia Commons.)

the awards are for a single animal taken (for example, antler size), but many awards are awarded for the killing of a combination of animals. The Africa Big Five Grand Slam means that a hunter has killed an African lion, a leopard, an elephant, a rhino, and a Cape buffalo.

We live in a time in which people seem to care more about animals than ever before. Pet keeping has never been more popular, and millions of Americans and people worldwide could not imagine their lives without the joy that companion animals bring. At the same time, contributions to animal welfare organizations have never been higher; millions of people enjoy watching animal videos on YouTube and animal shows and documentaries on the National Geographic Channel, Animal Planet, and at movie theaters. Yet, as we have discussed, billions of animals suffer and die every year and most of us are either ignorant of this reality or choose to ignore it. We not only tolerate violence in a context that is invisible and distant, we essentially commission it. On the other hand, what we do condemn is violence to an individual animal. And finally, we very definitely notice the ways in which animals are treated in cultures that are different from ours.

Culture-Specific Violence

In addition to the institutionalized forms of violence previously mentioned, every society has a number of practices that may not be fully institutionalized but which are culturally acceptable within a particular setting. For instance, bullfighting in Spain, foxhunting in England, and cockfighting in rural America are examples of violent activities toward animals that may not be built into our social institutions, but that are seen as culturally acceptable—at least by a certain segment of the population. Even though cockfighting was criminalized in the last two states in the nation, Louisiana in 2007 and New Mexico in 2008, and cockfighting is now seen as deviant behavior by most Americans, it remains culturally acceptable in some areas. On a more extreme level, the eating of dogs and the sacrifice of animals for religious purposes are both heavily stigmatized in the United States, and yet are culturally acceptable elsewhere. How do we explain differences like this? How do cultures develop such different attitudes toward animals?

One way to try to understand these differences is to look at people's livelihoods. In chapters 4 and 11 we have discussed the ways in which people have historically used animals as resources, and how that relates in part to their attitudes toward them. What is animal cruelty to one person is simply a way of making a living to another. Sometimes, however, it is not so obvious that economics lies at the heart of the ways in which we use and view animals. Often, cultural traditions demand certain types of uses for animals that make very little sense to those outside of that culture. In the United States, this difference comes into stark perspective when looking at new immigrants and how they deal with animals.

For instance, there have been a number of newsworthy instances in which immigrants from Southeast Asia have been arrested for killing dogs—practices that are considered cruel and deviant in the United States, but that are commonplace in the countries in which these immigrants were brought up. In 1995, a Hmong immigrant from Laos was arrested for beating a German shepherd puppy to death in Fresno, California. The public was outraged and demanded that the man be brought to justice. His own community, on the other hand, defended the actions of the man (a shaman) who killed the dog in order to appease a spirit that was tormenting his wife. By sacrificing the dog, the man believed that the dog would track down the spirit and get his wife's soul back. The Hmong perform ceremonies to release the souls of sacrificed animals so these animals can be reborn. To the Hmong, these practices are definitely not considered cruel.

Another Southeast Asian practice relating to dogs, spurred the creation of a California law a few years prior to the 1995 case. In 1989, after two Cambodian men were arrested for killing and eating a puppy, the public realized that there was no law against eating dogs in the United States. After the men were released from jail, California's legislature passed the first law in the country specifically banning the killing and consumption of animals "traditionally or commonly kept as a pet or companion." Even though many animals, such as rabbits, are traditionally or commonly kept as pets, the law has only been used to prosecute those who have eaten cats or dogs. This demonstrates the capriciousness of our laws—we can legally kill some animals for food, but killing other animals for food (even if some people do keep them as pets) is illegal. Even more perplexing, the law protects dogs and cats from being eaten, but no such law protects dogs and cats from being euthanized in animal shelters. Although very few dogs and cats were probably eaten in California prior to the passage of this law, hundreds of thousands of these animals are in fact legally killed every year in the state. Unfortunately, the wide variety of cultural practices involving animals and, in particular, those used as food has resulted in ethnic tensions and claims of racism in states where immigrants such as Cambodians, Vietnamese, or Hmong live. For instance, only Southeast Asians are prohibited from eating the animals that they want to; the rest of us are free to eat cows, pigs, chickens, and rabbits to our hearts' content.

Animal sacrifice is not only practiced by the Hmong. African immigrants and members of Afro-Caribbean religions such as **Santeria** also sacrifice animals as part of their worship. These practices incite outrage on the part of mainstream Americans, who see animal sacrifice as barbaric. It does not help that those who engage in it are not white and do not seem "American." For instance, Santeria was in the news when the Church of Lukumi Babalu wanted to open a church in Hialeah, Florida, in the mid-1980s; in response, the city council passed a law extending the Hialeah's anticruelty laws to include animal sacrifice, and other cities soon followed. Ultimately, the church took their case to the U.S. Supreme Court, which ruled in 1993 that the city's ban against animal sacrifice was a violation of the First Amendment to the Constitution.

Geographers Glen Elder, Jennifer Wolch, and Jody Emel (1998) argue that cases such as this outrage Americans not because animals are killed. After all, billions of animals, including goats, chickens and sheep—just like those killed in Santeria rituals—are slaughtered in the United States annually. Instead, Americans find animal sacrifice problematic for a number of reasons: These killings are done in public (rather than hidden away

Figure 12.2. This tattoo mallet is used to tattoo sows at a pig factory where workers hit the pigs to drive the spikes into their bodies. (Photograph courtesy of Mercy for Animals.)

in a slaughterhouse), the reasons for the killings are suspicious (many do not consider Santeria to be a legitimate religion), and the killings seem barbaric—even though most Americans do not realize how "barbaric" the killings are in U.S. slaughterhouses. Ultimately, it may be that it is not the animals themselves, or even how or where they are killed, that is problematic for most Americans. Rather, it is the fact that the people engaged in these acts are not seen as American—they are Brazilian, or Haitian, or Cuban, or Cambodian, or Vietnamese. In addition, Elder et al. point out that shamans and Santeros (priests) are not seen as having the proper "credentials" to slaughter animals (as if American slaughterhouse workers are so credentialed). It is these issues that make certain kinds of killing legitimate and other kinds illegitimate, or deviant.

Deviant Violence

Deviant violence refers to forms of violence toward animals that are unacceptable in modern society and that, typically, are criminalized. This would include the killing of companion animals, or even the killing of individual farm animals when they were killed in a method not sanctioned by society or the law. Psychologists Frank Ascione and Randall Lockwood define cruelty as "socially unacceptable behavior that intentionally causes unnecessary pain, suffering, or distress to and/or death of an animal" (1998:85). It is important to note that it is not the behavior itself that is judged to be cruel; it is whether it is "socially unacceptable" or "unnecessary." As we have discussed, if it is deemed socially acceptable or necessary, then it is not considered to be cruelty. Like institutionalized violence and culturally acceptable violence, this category is also socially constructed—what is unacceptable and thus deviant in one context is certainly acceptable in another. Raising, killing, and eating dogs are absolutely unacceptable in North America and Europe but are acceptable and legal in China or Korea.

For most of us in the United States, companion animals are part of the family, and we would do anything to protect them. However, millions of other animals are not as fortunate, and cases of cruelty and neglect appear regularly in the news. Although many of these high-profile cases attract media attention to a serious problem, the overwhelming majority of suffering animals receive no attention. Tragically, most do not survive, either facing death at the hands of their abusers or euthanasia at a veterinarian's office or at an overcrowded shelter.

Of all animals, the majority of reported abuse cases involve companion animals, yet there are no solid numbers documenting the occurrence of animal cruelty and neglect. A recent Humane Society of the United States (HSUS) study of high-profile cases (The Humane Society of the United States First Strike Campaign 2003 Report of Animal Cruelty Cases) indicated that 57 percent involved intentional cruelty, and 43 percent involved extreme neglect. The study revealed that the most commonly reported cruelty offenses involved shooting, animal fighting, torturing, and beating. Of neglected animals, 70 percent were malnourished, and 30 percent suffered from starvation. Other animals suffered drowning, stabbing, or even being burnt alive. Animal cruelty is prevalent across the country in rural and urban areas, and cuts across socioeconomic boundaries.

The HSUS also reports that dogs are by far the most common victims of animal cruelty, comprising more than 60 percent of all cruelty cases, with cats ranking second. Pit bulls, in particular, constitute an increasing percentage of victimized animals. Livestock abuse cases only come to light during undercover investigations, so there is no real way to know how many livestock suffer beyond the normal abuse of the stockyard and slaughterhouse. In fact, many states specifically exclude livestock or any "common" agricultural practices from their cruelty laws. Even when good laws exist, it can sometimes be difficult to convince law enforcement to make an arrest or to seize farm animals that are being neglected or abused. Still, arrests sometimes do happen.

In May 2010, the Chicago animal rights group Mercy for Animals released an undercover video showing workers at Conklin Dairy Farms in Plain City, Ohio, kicking, stomping, stabbing, and beating dairy cows and their calves. A question that emerged after the release of this video (and countless other undercover videos) was: Is this "deviant" animal abuse, or is this kind of

BOX 12.1

CRUSH VIDEOS

One extremely deviant form of violence to animals is known as a **crush video.** Crush videos are film in which a woman is shown crushing small animals to death, usually while wearing stiletto heels.

Those who watch crush videos (also known as "squish videos") are said to have a crush fetish, in which they derive sexual pleasure from watching the crushing of such animals. Some crush fetishists enjoy seeing nonanimate objects such as grapes or toothpaste tubes being crushed (generally by women with very large feet), but the term is generally reserved for those who watch animals being killed in this fashion. This type of fetish is known as hard crush. Hard crush videos range from those involving large and small insects and worms to those involving the deaths of mammals such as guinea pigs, rabbits, kittens, and puppies. Many videos showcase the torture of these animals, where it can take up to thirty minutes or longer for the animal to die.

In the United States, creating, distributing, or possessing crush videos was prohibited in 1999, and many other countries also have banned them. Nevertheless, they are widely available on the Internet. In the last couple of years, Chinese videos have circulated featuring the deaths of cats, dogs, rabbits, and toads. Known there as GTS (which stands for "great women, small men"), the videos may not be illegal in China, although public sentiment in China is widely against them.

In April 2010, however, the Supreme Court overruled the 1999 law banning crush videos, saying that the law was too broad and could outlaw hunting videos and other videos showing legal violence to animals. Just one day after the Supreme Court's ruling, two members of Congress introduced a new bill that would more narrowly focus on crush videos and thus not antagonize the hunting industry who sided with the Supreme Court's ruling. The bill was passed in November 2010.

abuse endemic in the industry? Conklin Dairy responded swiftly by saying that the video only demonstrated that there are a few "bad apples" in every industry or workplace, and that the actions of these few workers do not reflect on the industry as a whole. In Conklin's case, the employee featured in the video was fired by the company, and remaining employees were to be retrained, according to a company statement.

The Link Between Violence to Animals and Violence to Humans

One of the reasons that the public has been paying more attention to animal cruelty in recent years is that psychologists, social workers, and law enforcement professionals have recently begun to document a disturbing fact. There is now wide consensus that there is a strong causal connection, known as **"the link,"** between violence toward animals and violence toward people. Cases of serial killers who started out their careers torturing animals are well documented. We see this especially among young killers, such as school shooters Eric Harris and Dylan Klebold of Littleton, Colorado, and other teenaged killers such as Andrew Golden, Luke Woodham, and Michael Carneal. All of these boys tortured and mutilated animals, including cats, squirrels, cows, and dogs, before they turned their guns on parents, classmates, and teachers. Adult killers, too, such as Jeffrey Dahmer, the Boston Strangler, and Ted Bundy, started their careers by killing animals as children. Violence toward animals has been connected to a host of criminal behaviors, violent and nonviolent alike.

Even though this link has only been discussed in recent years, it has been known to some observers for hundreds of years. Eighteenth-century English artist William Hogarth, for example, created a set of four engravings in 1751 called "The Four Stages of Cruelty," which depicted a fictional person named Tom Nero as he progressed from a young boy into a man, engaging in ever more escalating kinds of cruelty. In the first stage, he is shown as a boy torturing a dog, while other boys torture other kinds of animals. In the second stage, Nero (now a man) beats a crippled horse while another man beats a lamb to death and an overloaded donkey struggles to move. The third print, "Cruelty in Perfection," shows Nero progressing to violence toward humans, and depicts him having brutally murdered his lover. The fourth and final print, the "Reward of Cruelty," has Nero being hanged for his many sins, his body being dissected, and his heart being eaten by a dog. Hogarth was not an animal rights advocate but recognized that cruelty begat cruelty; he intended his prints to teach this lesson to the poor of England.

One way to understand how this connection begins is to look at who the perpetrators of animal violence are. What are their demographic factors, and are there links that we can make between the violent behavior that they engage in and their childhood experiences? Overwhelmingly, men and boys are more likely than women and girls to commit direct, physical acts of animal cruelty and to subject animals to fighting competitions; in fact, about 95 percent of all defendants in animal cruelty cases are male. We noted in chapter 11 that women are more likely than men to have positive attitudes toward animals, and these statistics show that there is a very real, and very distressing, result of those differences. However, women have a higher percentage of involvement than men in hoarding cases, and neglect cases are more evenly split among the genders, although men are still more likely to be culprits than women. Whether male or female, in most cases, the abuser is the owner of the animals.

We also know that the typical animal abuser is a child—and, overwhelmingly, a boy. This finding is important for a number of reasons. Children who are cruel to animals exhibit more severe psychological, emotional, and behavioral problems than other children. Some recent studies on children and animal cruelty have shown that half of all school shooters had a history of animal cruelty (Verlinden et al. 2000). A 1985 study (Kellert and Felthous) found that 25 percent of violent criminals had been violent to animals, compared with 6 percent of nonviolent criminals. A Canadian study found that 70 percent of people arrested for animal cruelty had past records of violence toward humans, and that most animal abusers commit another offense within ten years of their arrest. Another study (Ressler 1988) found that 36 percent of sexual murderers had abused animals—usually sexually—during childhood. Today, "cruelty to animals" is considered by the American Psychiatric Association to be a symptom of conduct disorder, which refers to a pattern of antisocial behavior that can persist into adulthood.

Why do children, primarily boys, abuse animals? Quite often, the children themselves have been the victim of abuse, and often sexual abuse. They may have been victims of bullying, and some were bullies themselves (Henry and Sanders 2007). Children who harm animals generally come from families that are dysfunctional and where there are other kinds of abuse and neglect present; even harsh parenting styles are linked to animal abuse. Up to 88 percent of families with a history of child abuse also exhibited some form of animal abuse (DeVinney et al. 1983); one study (Duncan et al. 2005) finds that children who abused animals are twice as likely to have been abused themselves. Alternatively, they may have witnessed animal abuse, often committed by their father or other adult figure. In one study (Flynn 1999), those who had witnessed animal abuse were more likely to hurt a companion animal, and 8 percent said they had killed one. If sexual abuse is found in the home, the child may also sexually abuse animals. Witnessing a parent's violence toward another parent is also a causative factor; watching a parent get battered leads a child to harm animals. In two studies (Currie 2003; Baldry 2005), children who witnessed domestic violence in the home were three times more likely than other children to harm animals. Children who witnessed abuse to animals or family members may be acting out their own

Figure 12.3. The Four Stages of Cruelty, by William Hogarth, demonstrates the link between violence to animals and eventual violence to humans. Pictured here is the fourth stage of cruelty. (Courtesy of Wikimedia Commons.)

traumas, or regaining a sense of power, when abusing an animal. Perhaps it is just that violence has become the norm for children such as this, or perhaps their sense of empathy has been disrupted. Even children who have not been themselves abused or witnessed abuse can act out emotional problems associated with parental conflict on animals.

On the other hand, children often intervene to protect their mothers and pets from being battered, and some children can even allow themselves to be victimized to save a pet from being harmed or killed. The links go both ways—not only are children in households with violence at more risk of committing violence themselves, they are also at greater risk of being harmed if they live in a household in which an adult harms either another adult or an animal. Unfortunately, although the research is clear that abuse in the home is correlated with abuse toward animals, there are no laws mandating that animal abuse be reported to officials, so early warning signs are often ignored.

Domestic Violence and Animal Abuse

Another well-documented connection is the link between violence toward animals and domestic violence. Studies (Ascione et al. 1997; Ascione 1998; Flynn 2000; Faver and Strand 2003) have found that between 50 percent and 85 percent of women who escaped their homes due to domestic violence reported that their partner had abused the family animals as well. Every year about 1.5 million women are victims of rape or physical assault by an intimate partner, and domestic violence is the most prevalent form of abuse for women in the United States today. The HSUS estimates that nearly one million animals a year are abused or killed in connection with domestic violence, based on the number of women assaulted by their partners and the number of women who reported that their abuser also targeted their companion animals.

This is a very real threat to women, children, and their animals. Abusers often use violence, or threats of violence, against animal companions in order to exercise control over their partner. Killing or threatening to kill an animal reinforces a woman's isolation and lack of control over her own life, keeping her, in many cases, bound to her abusive partner. This threat of violence is used to punish her, to frighten her, and to keep her from leaving him. (Although many perpetrators of domestic violence are women, the vast majority of people who cause serious harm to their partners are men, and the vast majority of partners who harm animals are men as well.) Companion animals have been beaten, nailed to doors, drowned, suffocated, and thrown out of cars. In one survey of domestic violence perpetrators (Carlisle et al. 2004), men considered the animals to be property and used them as scapegoats for their own anger; as children half of these men had watched their own pets be abused.

In some households with domestic violence, the abuser will often sexually abuse the animal, or force the woman to have sex with the animal. In addition, bestiality is often practiced by sexual predators, either prior to having sex with adults or children, or when no person is available. Studies show that people who rape animals are likely to rape people (Quinn 2000). Anywhere from 20 to 37 percent of children who sexually abuse other children have histories of sexually abusing animals. An abuser who rapes a woman or rapes an animal, or forces an animal to rape a woman, is treating the woman and the animal as objects. Forced sex with an animal is the ultimate form of humiliation for a woman. Interestingly, bestiality is still legal in fifteen states, but because of increased knowledge about the connection between the sexual abuse of an animal and domestic abuse, other states are considering criminalizing bestiality. And again, domestic violence also affects children. In one study (Carlisle 2004), children in domestic violence shelters were twenty times more likely than children outside of this context to have witnessed not just domestic violence but also animal abuse.

Because most domestic violence shelters do not accept companion animals, even on a temporary basis, women who escape their abusers often must leave animals at home to be further victimized, or they must remain in their homes with their animals. This is another factor in why some women do not leave their abusive partners. In one survey of women escaping domestic violence (Faver and Strand 2003), one in four women reported that they had delayed entering the shelter because of concerns about their pets' safety. Luckily, there are a few **safe haven** programs available to help women in these circumstances, such as the Companion Animal Rescue Effort (CARE) in New Mexico, provide temporary foster care for the companion animals of abused women, or the Sheltering Animals of Abuse Victims program in Wisconsin, which runs a network of temporary homes, shelters, and farms to confidentially care for animals from violent homes. In addition, more domestic violence shelters are working with local animal shelters to provide refuge for animal victims of domestic violence. Still, most domestic violence shelters still neglect to ask about abuse toward pets in their intake interviews.

It seems clear that more needs to be done to prevent these kinds of violence and to help women and children and their pets. Domestic violence shelters should ask women about violence toward animals in the home, in order to assess whether or not the partner's violence is escalating, and should work with animal shelters and other agencies to provide safe haven for pets that live in abusive homes—to protect the animal and to encourage the woman to leave the abuse. Police, too, could ask women who are leaving their partners about their animals. These women should then be provided with information about animal shelters and rescue groups who could help. And finally, more animal protection organizations should reach out to domestic violence care providers in providing foster care for animals of abusive homes. These issues are complicated, however, by the American system of pet "ownership." If the animal is jointly owned by both partners, as the law presently sees companion animals, this complicates matters. How should an animal shelter respond when the man wants to reclaim his pet if the pet was placed into shelter to protect him/her from the abuser? And finally, how should adult abusers and children who are witnesses to, victims of, or even perpetrators themselves of animal violence be treated, so that this kind of violence can be stopped?

Treatment and Prevention

Thanks to the efforts of social workers, sociologists, psychologists, and members of the law enforcement and animal welfare communities, we now recognize that the link between violence to animals and violence to humans is real. The questions then become: How do we prevent this kind of violence, and how do we treat it when it occurs?

The good news is that a variety of programs exist to counsel and rehabilitate young abusers before their violence escalates. One national program is the National Cruelty Investigation School, a program of the Law Enforcement Training Institute at the University of Missouri, which trains law enforcement officials in how to conduct cruelty investigations, focusing on the link between animal cruelty and human violence. Many local agencies are following suit with their own special units. Another program is the HSUS First Strike program that raises awareness about the connection between animal cruelty and other violence. First Strike works with local animal protection, law enforcement, and social services agencies to reduce animal, family, and community violence. First Strike, through its HSUS connections, is also working to pass felony-level anticruelty laws around the country.

As of this writing, twenty-eight states currently recommend or mandate that judges require counseling for persons convicted of animal cruelty. Unfortunately, not all therapists are trained in how to counsel people who have committed such acts of cruelty. This is another area in which animal protection organizations can participate. For example, the Animals and Society Institute (ASI) also works with health, education, and criminal justice professionals to recognize animal cruelty as a law enforcement issue and as an indicator of violence toward others. ASI provides treatment programs as well: The AniCare Model of Treatment for Animal Abuse is a psychological intervention program for animal abusers over the age of 17, and AniCare

Child is used to treat offenders under age 17 (see sidebar 12.1). The organization also has a program called Rapid Response which complements AniCare by providing outreach to judges, prosecutors, and others in court systems around the country, as well as to the media and the general public.

For instance, we mentioned earlier in this chapter the case of a dairy farm employee who was arrested, thanks to undercover video captured by an animal rights organization, for beating dairy cattle. He pleaded guilty to six misdemeanor counts of animal cruelty and was sentenced to eight months in jail (four of which were suspended) and a $1,000 fine, which animal advocates found to be exceedingly light given the amount of suffering the man inflicted on the cows and calves. As part of his sentencing, however, the judge ordered that he undergo counseling for animal abusers such as is offered through ASI's AniCare program.

Legislation

The United States presents an odd picture of animal protection legislation. The country has three main federal laws—the Animal Welfare Act, the Humane Methods of Slaughter Act, and the 28-Hour Law—that range from very broad (the Animal Welfare Act covers animals in laboratories, during transport, in entertainment venues, and at breeding locations) to very specific (the Twenty-Eight-Hour Law deals only with the transportation of animals, but the Humane Slaughter Act covers their slaughter). On top of that, we have a patchwork of state and local laws that protect different animals from a variety of practices. Currently, forty-six states and the District of Columbia have laws making certain types of animal cruelty felony offenses. That means that in four states (Idaho, Mississippi, North Dakota, and South Dakota), no matter how many animals a person tortures or kills, the greatest penalty that they can face is a misdemeanor; misdemeanors typically only warrant a sentence of probation with a maximum six-month jail sentence in rare cases or a fine of up to a thousand dollars. Within the forty-six first-offense felony states, several have a first-offense provision for aggravated cruelty, torture, companion animal cruelty, etc., in addition to a second offense provision for cruelty to animals. Even for felony-level offenses, which can warrant a fine of up to $100,000 or a sentence of up to five years, perpetrators are very rarely prosecuted or sentenced to the fullest extent of the law—but at least these are steps in the right direction. In addition, twenty-three states prohibit the ownership of an animal for some period of time after being sentenced for abuse, and twenty-eight states require psychological counseling for certain cruelty offenses.

There is no national system which compiles animal abuse statistics, which makes it difficult for lawmakers to get a real handle on the problem. What we do know about animal cruelty comes primarily from media reports. In 2009, there was a bill before the Senate Judiciary Committee that would require the National Incident Based Reporting System, the Uniform Crime Reporting Program, and the Law Enforcement National Data Exchange Program to list cruelty to animals as a separate offense category. Assigning the crime of animal cruelty to its own reporting classification would enable law enforcement, social service agencies, researchers, and others to track trends at the state and national level and to determine the demographic characteristics and other factors associated with animal abuse. As of this writing, the bill has not left committee.

California was recently ranked number one in a Humane Society of the United States survey of how animal friendly each of the fifty states is in terms of its animal-related legislation. California won thanks to its laws protecting companion animals, horses, farmed animals, and wild animals from a variety of abuses. California's animals may have even more protection, if a proposed state law creating a criminal registry for animal abusers passes the state legislature. This law (SB 1277), which would be the first of its kind in the nation, would require anyone convicted of felony animal cruelty to register with the police, as sex offenders are required to do under the Jacob Wetterling Crimes Against Children and Sexually Violent Offender Act. In addition, the law would also mirror Megan's Law, which requires states to notify the public of sex offenders in their communities. SB 1277 would require not only that California maintain a database of animal abusers but also that names, addresses, and photos would be posted online. (Currently, there are a handful of private websites that list such abusers' names, but none are comprehensive.) The bill would be funded by a tax on pet food.

What would be the effects of such a law? According to a 2009 study, Megan's Law has failed to deter sex crimes or reduce the number of sex crime victims. However, the public still supports the sex offender registry as well as the public notification requirement, and proponents claim that it allows parents to better protect their children because parents can find out whether sex offenders live in their communities. SB 1277 could be expected to function in a similar manner. Although it would likely not discourage people from abusing their animals, it would give the public—in particular those who either sell or adopt out animals to the public—a way to find out whether potential adopters are convicted animal abusers. Currently, animal rescue organizations have no way to find out the background of potential adopters, and this could be one more tool to help them to evaluate their qualifications.

Finally, one reason why sex offenders—and not bank robbers, drunk drivers, or even murderers—are the target of legislation such as Megan's Law is that they are especially prone to recidivism. Animal abusers are also. Hoarders, for example, are especially likely to offend again, and a law such as this one would provide the public with enough information to make it at least more difficult for them to acquire animals again.

Suggested Additional Readings

Ascione, Frank. 2008. *The International Handbook of Animal Abuse and Cruelty: Theory, Research and Application*. West Lafayette, IN: Purdue University Press.

Ascione, Frank and Phil Arkow, eds. 1999. *Child Abuse, Domestic Violence, and Animal Abuse: Linking the Circles of Compassion for Prevention and Intervention*. West Lafayette, IN: Purdue University Press.

Carlisle-Frank, Pamela and Tom Flanagan. 2006. *Silent Victims: Recognizing and Stopping Abuse of the Family Pet*. Lanham, MD: University Press of America.

Jory, Brian and Mary-Lou Randour. 1999. *The AniCare Model of Treatment for Animal Abuse*. Ann Arbor, MI: Animals and Society Institute.

Lockwood, Randy and Frank Ascione. 1998. *Cruelty to Animals and Interpersonal Violence: Readings in Research and Application*. West Lafayette, IN: Purdue University Press.

Randour, Mary Lou and Howard Davidson. 2008. *A Common Bond: Maltreated Children and Animals in the Home: Guidelines for Practice and Policy*. Englewood, CO: American Humane.

Randour, Mary Lou, Susan Krinsk, and Joanne L. Wolf. 2002. *AniCare Child: An Assessment and Treatment Approach for Childhood Animal Abuse*. Ann Arbor, MI: Animals and Society Institute.

Suggested Films

Beyond Violence: The Human–Animal Connection. VHS. Ann Arbor, MI: Animals and Society Institute, n.d.

Breaking the Cycles of Violence. DVD. Alameda, CA: The Latham Foundation, 2004.

Websites

Animals and Society Institute Anicare Program: http://www.animalsandsociety.org/anicare

HSUS First Strike Campaign: http://www.hsus.org/firststrike

Latham Foundation: http://www.latham.org

The Humane LINK: http://www.thehumanelink.com

AniCare: Treating Animal Abuse

KENNETH SHAPIRO

Animals and Society Institute

One of the policy implications of the co-occurrence of animal abuse and violence toward humans is the importance of helping individuals who abuse animals. To do that, we must identify juveniles and adults who are prone to this behavior and then provide them with an appropriate intervention. Depending on the severity of the problem, this might involve education, parent guidance, individual or group counseling, or residential treatment.

But first, how do we identify these individuals? If animal abuse is not taken seriously, if for example it is viewed as a normal rite of passage ("boys will be boys"), then taking notice of incidents will not be on the agenda of parents, teachers, and other social agents. For the sake of other animals and humans, taking animal abuse seriously is an important policy change if we are to break the cycle of violence in this country.

Studies document the prevalence of animal abuse—a significant minority of male college students (30 to 40 percent) report abusing animals as children (Miller and Knutson 1997). Yet, relatively few of those reported incidents are prosecuted and an even smaller proportion leads to conviction (Arluke and Luke 1997). Fortunately, in the past two decades, 46 states have passed legislation that makes their anticruelty statutes felonies. With animal abuse now recognized as a more serious crime, more prosecutions and convictions are occurring.

Another policy development that recognizes the importance of animal abuse as a social policy issue is the passage, beginning with California in 1998, by 27 states of legislation allowing or mandating judges to include psychological treatment in the sentences of convicted animal abusers.

These legislative efforts have produced a demand for a range of programs dealing with the problem of animal abuse. Primary prevention involves efforts to educate the general public about the seriousness and importance of animal abuse and its relation to other forms of violence. Humane education is part of the curriculum in many schools, usually beginning at the elementary level, and increasingly is becoming a recognized specialization in graduate education programs. Secondary prevention programs identify people "at risk"—those more likely to become abusers. Children at risk might be those who do not have adequate supervision in their home. The Forget Me Not Farm in California takes children such as these and provides them with gardening and husbandry of animals on a farm, teaching them responsible care giving.

Finally, for those who already "rely" on animal abuse as a way of expressing or dealing with their emotional problems, therapists and clinical researchers have developed interventions to assess and remediate that poor adjustment. The AniCare approach treats children and adults whose problems require more than parent guidance or education but less than residential treatment.

AniCare Child (Randour, Krinsk and Wolf 2002) includes training in empathy (taking an animal's point of view) and self-management techniques (better problem-solving skills). Many children who abuse animals have attachment problems: They may have failed to develop a secure bond with a parent and may be overreliant on a relationship with a companion animal; or they may be taking out the frustration of their own unmet needs on a companion animal.

The adult version of AniCare (Jory and Randour 1998) emphasizes helping an individual be accountable for his or her behavior. Often animal abusers do not admit to themselves or others that what they did is wrong. They are not willing to accept that their behavior is a problem. They develop "stories" that deny the presence or importance of the abuse, that distort their role in the abuse, or that somehow justify it. The counselor must work to help the client see that there is a problem and that he or she must accept responsibility for it.

For children and adults, their justificatory story often is heavily influenced by the subculture in which they are being or were socialized—"they are just animals." Part of the work of the therapist is to help the client come to terms with this cultural layer, reframing it in ways that do not involve abuse.

Educating criminal justice personnel about available treatment programs provides the demand; training therapists to work with this population provides the supply to meet that demand. Together these developments can reduce animal abuse and co-occurring violence against humans. Other policy innovations can complement this positive cycle: cross-training and cross-reporting so that human service personnel recognize and report animal abuse and humane agents do the same for child abuse and domestic violence; safe havens for victims of domestic violence with a place for their companion animals; protective orders that include companion animals; and registration of convicted animal abusers. We do not have to look far for policy suggestions; parallels among child, women, and elder abuse provide models for the needed remedies.

13

Human Oppression and Animal Suffering

Let me say it openly: we are surrounded by an enterprise of degradation, cruelty, and killing which rivals anything the Third Reich
was capable of, indeed dwarfs it, in that ours is an enterprise without end, self-regenerating, bringing rabbits, rats, poultry, livestock
ceaselessly into the world for the purpose of killing them.
—FROM J. M. COETZEE'S *The Lives of Animals* (1999)

Interlinked Systems of Exploitation

As we have discussed throughout this text, nonhuman animals experience
an enormous amount of exploitation by humans. But it is also true that
many animals—primarily those animals defined as companion animals—
experience a great deal of love, care, and humane treatment. So we can say
that not all animals are treated the same. Likewise, not all people are treated
the same by other people. Great numbers of people suffer from poverty, disease, warfare, and crime, and about half of all humans on the planet live on
less than $2.50 per day. At the same time, a small number of people control
the vast majority of the world's wealth and resources, and many theorists see
this vast wealth coming at the expense of everyone else.

According to many scholars, these two situations—animal suffering and
exploitation, and human suffering and exploitation—are linked. In other
words, the same systems of oppression that keep humans from reaching
their full potential, such as the class system, the **caste system**, racism, or
slavery, also work to oppress animals. The reverse may also be true: The systems of animal exploitation found in, for example, the meat industry or the
biomedical industry, can also be said to exploit some humans, while giving
other humans profit.

Feminist theorists and scholars who study inequality have shown us that systems of exploitation are linked—racism, for example, does not work in a vacuum. It is linked to sexism, classism, and other systems of oppression such as homophobia. A Native American woman, for example, may be subjugated by her position as a woman and as a Native American because racism and sexism operate together and even reinforce each other. The same may well be true of animal exploitation: It is woven into the larger systems of oppression in our society and, similar to those other systems, it is well hidden, which allows it to be perpetuated generation after generation. This chapter discusses these linked systems of oppression.

The Roots of Oppression

We located the roots of speciesism and the dominance of humans over other animals in chapter 2, with the domestication of animals. Even though one can argue that hunting is certainly a demonstration of human power over animals, true control did not really develop until the first "food animals" were domesticated beginning about 10,000 years ago. With the rise of agriculture, especially in Europe, the Middle East, and Asia, a new concept of animals and humans emerged, with humans transcending and controlling animals and nature.

Not coincidentally, the rise of agriculture also coincided with the rise of human oppression over other humans. With the rise of intensive agriculture (marked by the use of irrigation, continuous planting, fertilization using animal feces, and plowing using animal labor), human population sizes grew from villages to cities and, ultimately, to states in a handful of areas around the planet. The Nile Valley, Mesopotamia, China, India, and, in the New World, the highlands of Peru and the Yucatan Peninsula were among the locations at which the first state-level civilizations flourished. These civilizations had trade relations with other cultures, used warfare to expand their territories, and were marked by extreme forms of inequality. Peasants supported nonworking elites with their labor, as did slaves. It may well be that the decision to use some humans as slaves derived from the use of animals as food and labor sources. Certainly, the ways that human slaves and animal property were bought, sold, branded, and confined were very similar.

Women, too, saw their status in society drop with the rise of state level civilizations. Women's status in foraging cultures was relatively equal to men; only men's control over hunting increased their status. But for the most part, women in such cultures contributed a great deal to the economy and held relatively equal status to men. But after the domestication of plants and animals, women saw their status drop as their role in productive labor declined and their role in domestic and reproductive labor increased. Men began to control not only production but also the fruits of production—the crops and the animals—and, with that, gained higher status.

The ancient states, with their built-in systems of inequality, led to other forms of **social stratification**. Social stratification can be defined as an institutionalized form of inequality, in which categories of people are ranked in a graded hierarchy on the basis of arbitrary criteria, and resources and opportunities available are distributed according to an individual's placement in the social hierarchy. The result of these systems is that some individuals and groups are considered more worthy, are ranked higher, and receive more of the society's resources than others. Although the earliest (and one of the most enduring) forms of social stratification is slavery, other forms soon developed. For instance, in India, the system that developed there is known as the caste system. It is a religious and occupational system in which people are born into a certain caste, or *jati*, and then are bound by social norms to marry within it, work within it, and die within it. There is no social mobility within the system, and those at the lower end of the system are deemed impure, even as those at the top are pure. In medieval Europe, on the other hand, the estate system was practiced, which was also based on birth, and which restricted land ownership to those of noble birth.

With the rise of colonialism in the fifteenth century, and the conquest of non-Western cultures by the European superpowers, a new system of inequality arose that was based on race. Like the caste system, this structure too is based on birth, and it is rigid—one's physical characteristics serve as the basis for one's treatment in society. In racialized systems such as that of the United States, the physical differences among the groups known as races are thought to correlate with important psychological, intellectual, emotional, and cultural differences; those supposed differences are then used as the justification for segregating out the "inferior" groups and denying them access to resources and power.

Othering and Essentializing

Othering refers to the practice of making people—or animals—different in order to justify treating them differently. Racism, for example, depends on othering: By claiming that people are racially different—that there is something inherent or essential within us that makes us not only look different but also think, feel, and act differently, and that these differences can be summed up by the term "race," we can more easily subjugate others and exclude them from power. The further we can distance those that we do not like or do not want to share resources with, the more we can mistreat them. It can be argued that the Holocaust would not have happened had the Nazis not conducted a brilliantly successful public relations campaign in which they portrayed Jews as alien, foreign, dangerous, and nonhuman. Once the Jews had been successfully "othered," Germans were more easily able to participate in one of the worst genocides in history. Clinton Sanders and Arnold Arluke (1996) call this "boundary work"—creating boundaries between some people and other people, between animals and people, and between some animals and some people. In the case of the Nazis, they created boundaries between Aryans and people such as Gypsies and Jews by animalizing the latter, calling them vermin, for example, and at the same time morally elevating some animals such as purebred dogs. (Americans participated in othering Jews as well, calling them ferrets, weasels, and even vampires in the press.)

Of course, animals are "othered." In chapter 2, we discussed the rise of the human-animal divide, in which a border was erected that separated out all of the animal species from a single animal species—human—and gave that species power over all the others. Without the border, human domination over animals could not occur. But as we have seen, the border is a tenuous structure, having been built with arbitrary characteristics such as the soul, language, rationality, or the mind. As we will discuss in chapter 17, many of these properties can no longer be denied to animals and some, such as the soul, can never be proven for either humans or animals. But most humans are convinced that humans and (other) animals are qualitatively and quantitatively distinct from each other; it is only because we have that assurance that we can deny them anything close to equal treatment.

Another aspect that links the treatment of animals with the treatment of some humans is **essentializing**. Essentializing refers to the treatment of individuals as if they were the same as others in their race/sex/species. All women, for example, are the same because all women share the same hormones and reproductive organs, which means that they all act the same. Women are nurturing and men are aggressive—these are characteristics that are thought to be essential. Racist thought is by definition essentialist thought—if all blacks were not the same intellectually, emotionally, or culturally, how could whites have created laws that ensured that all blacks, without exception, were to be deprived of the rights given to whites? And certainly all animals of a species (or perhaps even all animals) are the same. All mice are the same, and none have anything close to individual personalities, needs, or wants.

One critical way in which women, the poor, and racial minorities in particular have been othered historically is by using animal terms to refer to them. In 1790, British political philosopher Edmund Burke used the term "swinish multitude" to refer to the supporters of the French Revolution (Burke 1790/2001), whom he saw as ignorant, dangerous, and "swinish." Burke, like other elites of his time, saw the poor as being poor because of their inherent, essential qualities, or more properly, their lack of the qualities that were shared by the elites, such as intelligence, rationality, and a moral sensibility. Only the upper classes shared these qualities, leaving the lower classes in a semi-human state, and deserving of ill treatment. Dehumanization and essentializing take place because they serve to justify oppression.

Burke's choice of the term "swinish" is just one example of how people who are considered inferior are associated with animals. Seventeenth-century English Pastor Robert Gray wrote in 1609 that "the greater part of earth was possessed and wrongly usurped by wild beasts . . . or by brutish savages, which by reason of their godless ignorance, and blasphemous idolatry, are worse than those of beasts" (Thomas 1983:42). Gray was a prominent promoter of colonization and maintained that the English deserved to take the lands of the "Indians" because of their beast-like natures. Indeed, the association of "natives" with animals was a major factor in the colonial practices of the period.

The reality is that the line is a fuzzy one between those supposedly deserving good treatment and those who supposedly do not deserve it. The line must constantly shift to accommodate the desires of those in power. Just as not all humans are perceived as deserving fair treatment—slaves, the poor, minorities, immigrants, untouchables, and gays are just a few examples of categories some people have deemed different and thus inferior—not all animals deserve poor treatment. Some animals are considered to be more worthy than others, either because of the pleasure that they give us or the economic value that they possess. Just as humans on one side of the line have more rights than those on the other side, animals on one side of the line have more rights than those on the other side. And the line itself may be shifted in such a way that some humans are lumped together with some animals below the line, and other humans remain separate. The danger lies in the existence of the line itself—as long as there exists in society a line separating

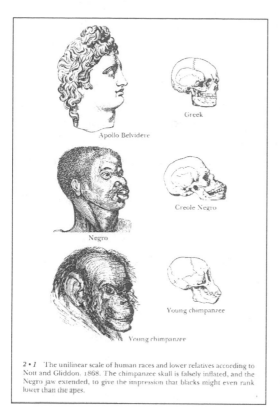

2 • 1   The unilinear scale of human races and lower relatives according to Nott and Gliddon. 1868. The chimpanzee skull is falsely inflated, and the Negro jaw extended, to give the impression that blacks might even rank lower than the apes.

Figure 13.1. This image, a comparison of African Americans to apes, is from Josiah Clark Nott and George Robert Gliddon's *Indigenous Races of the Earth* (first published in 1857) and was reprinted in Stephen Jay Gould's *The Mismeasure of Man* (1981). (Image courtesy of Wikipedia Commons.)

some from others, then no group is truly safe from being on the losing side of it.

Interestingly, there is virtually no research within social sciences (outside of human-animal studies) that addresses prejudice toward nonhuman animals, even though the psychology and sociology of prejudice are well documented otherwise. One recent study (Hyers 2006), however, has attempted to use **social dominance theory** to try to explain feelings of superiority toward animals. According to this approach, subjects who score high in what is called *social dominance orientation* will prefer a more hierarchical view of society in which they sit on top. Psychologist Lauri Hyers tested this model to see whether it predicts superior attitudes toward animals and found that it does, demonstrating that superiority toward some groups tends to be generalized to include other groups as well. She also found that those who viewed themselves as superior to animals create or utilize "legitimating myths" that validate their use of animals, just as those who see themselves as superior to others based on race or sexuality do the same.

Sexism and Speciesism

Feminist scholars have often discussed the ways in which women are defined and even constrained by their bodies. Although social scientists agree that much—if not all—of gender is socially constructed, biology still plays a major role in the cultural views of women's bodies. Because women menstruate, give birth, and lactate, most cultures have usually treated women as fundamentally different from men, and have created rituals and taboos associated with those female activities. Women's bodies are often the site of political contestation and control, regarding everything from birth control to weight to beauty. Feminists have shown that women are subjugated by standards of beauty that demand perpetual youth and unnatural thinness;

they are denied opportunity because of their hormones and bodily functions; and they have long been thought to be driven by their (largely uncontrollable) emotions.

Historically, women and animals have been considered less intelligent than men. Tactics such as objectification and ridicule have been and continue to be used to control and exploit women and animals. Women are called names such as cow, sow, pig, dog, bitch, fox, and hen; by symbolically associating them with animals, they are trivialized. In the fourth century BCE, Aristotle (1977) wrote that because only men possess rationality, it was natural and necessary for men to rule women and animals as well as slaves, all of whom lacked it. Women, slaves, and animals, existed to serve the needs of man.

In the 1970s, anthropologist Sherry Ortner (1974), in attempting to explain why women are universally subordinated to men, created an explanation arguing that women are symbolically associated with nature, and men are associated with culture, and nature occupies a subordinate position to culture. But why would women be seen as closer to nature? Women's bodies, in that they are occupied much of the time with reproductive activities, seem to place them closer to nature, and to animals. Women are, as she put it, the prey of the human species because many of their bodily features and functions do not help with survival but only exist to help the species. In fact, some features may even hinder women, such as large breasts; menstrual discomfort; breast, uterine, and ovarian cancer; and pregnancy—all slow women down. Women are more enslaved by their bodies than men; this leaves men with the freedom to turn to other things, such as creating culture. It is easy to take Ortner's theory and expand it to include animals. In other words, we can argue that one reason why women and animals are subjugated is that they are compared to each other, and women and animals occupy inferior positions vis-à-vis men in society.

Because female humans and female animals are marked by their reproductive abilities, they are also constrained by them—not in a physical sense, but in a cultural sense. Society expects that women will devote a considerable amount of time and energy producing and rearing children, and for that reason, women have faced social and legal barriers throughout history that kept them from achieving many of the accomplishments of men.

Animals, too, are constrained by their bodies, and are often defined primarily by their reproductive abilities. For example, the hormone replacement drug Premarin is made from the urine of pregnant horses. To produce the drug, mares are confined in stalls, repeatedly impregnated, and subjected to invasive procedures throughout their pregnancy (and their lives), only to have

BOX 13.1

ORGANIZATIONAL FOCUS: FEMINISTS FOR ANIMAL RIGHTS

Feminists for Animal Rights was an educational organization dedicated to ending abuse against women, animals, and the earth, and to pointing out the connections between the abuse of women and the abuse of animals. The organization was founded in 1982 by Marti Kheel and Tina Frisco, ecofeminists who recognized that the abuse of women and nature is intimately connected in patriarchal society. From their promotional flyer:

In patriarchal society women and animals are . . .
raped, beaten, hated, enslaved as pets
exploited as wives, sold for money, used
for entertainment, cheap labor, sex experiments . . .

In patriarchal society women and animals are
considered . . .
inferior, "cute," childish, uncontrollable,
emotional, impulsive, instinctive, irrational,
evil, property, objects . . .

In patriarchal society women and animals are
referred to as . . .
chicks, bitches, pussies, foxes, dogs,
cows, beavers, birds, bunnies, kittens,
sows, lambs, shrews, geese, fillies,
bats, crows, heifers, vixens . . .

their colts taken away from them after birth. Ironically, the horses' suffering is used to create a drug that is marketed to menopausal women who are told that their bodies are now diseased. The reality is that not only is menopause *not* a disease but also that drugs such as Premarin that are used to alleviate menopausal symptoms have been linked to cancer, thus causing disease.

Other female animals are likewise exploited for their reproductive abilities. Dairy cows (and goats and sheep) are repeatedly impregnated so that their bodies will produce milk; but instead of that milk being used to feed their own babies, the babies are removed (and male calves are sold to veal production

Figure 13.2. Advertisement for a temporary tattoo encouraging women and girls to stamp themselves "USDA Choice." (Courtesy of Carol Adams.)

facilities) and the mothers are milked to produce dairy products for humans. One of the worst examples of misery is found in the lives of egg-laying hens. They are confined to tiny cages without enough room to spread their wings or lie down, they are kept in the dark at all hours, and their beaks are burnt off to keep them from pecking other birds to death. These animals get no sunlight, no dirt, no grass, and no relaxation, but must lay eggs for human consumption until they die of exhaustion.

Because men have long had more power—political, social, economic, and physical—than women and animals, they have been able to use that power to master both. For instance, in most pastoral societies, only men are able to own animals, and women are the property of fathers and husbands, just like cattle or sheep. The consumption of meat is another good example. Feminist writer Carol Adams (1991) shows that meat eating has long been associated with men and masculinity; we discussed in chapter 7 how meat has been eaten by elites and others with power throughout history, even though women, children, and the poor have eaten what was thought to be second-class food—breads, fruits, and vegetables. Even today, vegetable eating—and especially salad eating—is considered to be weak food, or "rabbit food."

Adams shows how animals and women are objectified, fragmented, and ultimately consumed by men. Animals are treated as objects, killed, dismembered, and consumed as meat, while women are objectified, and, through pornography, dismembered into body parts (breasts, lips, butt, vagina) and then "consumed" through porn or sexual violence. When women say that they "felt like a piece of meat," they are referring to degrading and dehumanizing treatment that is reserved for women and animals, but never men. In addition, we discussed in chapter 12 the links between violence to humans and violence to animals. We saw that many men who control their wives and girlfriends do so not only by using violence and threats of violence against them but also by committing violence against the women's companion animals. Abusing, or threatening to abuse, a woman's companion animal has become yet another way in which some men control and victimize women.

Racism, Slavery, the Holocaust, and Animal Exploitation

Like women, nonwhites have been compared to animals since at least the time of colonialism and certainly since the African slave trade emerged in the seventeenth century. Europeans and Euro-Americans referred to Native Americans and Africans as animals, and many people subscribed to the Christian idea of the great chain of being, which posited that the human "races" were created by God, with whites at the top of the chain. By the nineteenth century, with the rise of evolutionary thought, many people felt that even though all humans may be related to apes, some humans were closer to apes than others; Africans in particular were thought to be the "missing link" between apes and humans. Scientists in the eighteenth and nineteenth centuries came up with all kinds of theories to explain and justify slavery. For instance, scientists and philosophers debated about the origin of the races: One theory, known as **monogenism**, said that all humans derived from the same source but that the nonwhites degenerated over time. Another theory, called **polygenism**, claimed that all humans descended from different sources and indeed made up different species. The latter theory was most popular among slave-keeping countries such as in the Americas. By using animals—especially monkeys and apes—to refer to Africans, and by implying or stating that Africans and African Americans were closer to these animals than whites, whites asserted their human superiority over them.

African Americans were not just thought of as animals; they were treated like animals. Treating a human like an animal is, like calling them an animal, a way to degrade and dehumanize them. African slaves were shackled and muzzled like animals, beaten like animals, branded like animals, and bought and sold like animals; they had their children taken from them like animals and had their humanity and individuality ignored, just as humans do with animals. They were property just as animals were, and could be legally killed by their property owner, just as animals could. Huge numbers of Africans died in transit from Africa to America—a loss of life that was absorbed into the prices of the remaining men and women. Animals that are transported to pet stores in the United States also experience high rates of what the industry calls "shrinkage," and those losses are also figured into the prices of the remaining animals.

Marjorie Spiegel's 1997 book *The Dreaded Comparison* makes clear the similarities in the ways that African slaves were treated, and how animals were then treated, and are still treated today. Further, she shows that whites justified their use and treatment of slaves with many of the same reasons that humans use today to justify their use of animals as food or medical subjects. Blacks were thought to not feel pain or feel love toward their children. They were also thought to be happier under slavery than living on their own. In addition, slavery was an important part of the economy of the American South—how could plantation owners expect to do without it?

We use these same explanations today to justify our use of animals. Animals surely do not feel pain, do not know the meaning of happiness, and do not form bonds with their young, so could not be bothered when their young are taken from them; they are happier and safer living on farms and in factories and zoos than they would be in the wild. And—perhaps most importantly—human economies depend on animal agriculture, and we need animals to experiment on. The difference is that we no longer use African Americans like this, and would be horrified by any society that did. Yet we still maintain these exact practices with regard to animals, and justify them exactly the same way.

In the years after slavery known as the **Jim Crow** period, when American law and social practice continued to prevent African Americans from attaining the same rights as whites, blacks were still thought of as beastly, as unable to control their desires. Whites thought that blacks were more prone to violence and rape (especially rape of white women) than were whites, and lynchings of blacks were common. Blacks were sometimes displayed in zoos and in circus and carnival sideshows alongside wild animals. Unbelievably, even in the twenty-first century, the association of blacks with monkeys and apes still has not disappeared. In the months leading up to the 2008 presidential election in the United States, and in the years following the election of the first African American president, the Internet has been rife with cartoons and images of Barack Obama (or Michelle Obama) represented as a monkey or an ape.

Other groups are animalized too. In the nineteenth century, the Chinese and Japanese were called vermin and were compared to rats by white Americans; during World War II, the Japanese described the Chinese as pigs during their invasion of Manchuria. Even Disney's animated cartoons play a role in this animalization. If we look at Disney cartoons over the years, you can see how animal characters are "raced." Crows, monkeys, and apes are played by African-Americans, Chihuahuas by Latinos, and cats by Asians, and all are stereotyped negatively.

But perhaps most well known is the treatment of the Jews by the Nazis in the 1930s and 1940s. The German word for "race" (*rasse*) is also the word for a purebred animal, demonstrating not only the Nazis' tendency to animalize people but also their concern with maintaining "blood purity" in animals and people. Jews were called vermin, rats, and cockroaches in Nazi speeches and in the German media. They were also known as *untermenschen*, or sub-human. Minister of Propaganda Josef Goebbels said, "It is true that the Jew is a human being, but so is a flea a living being—one that is none too pleasant . . . our duty toward both ourselves and our conscience is to render it harmless. It is the same with the Jews." The Nazi Party manual had the following line, "All good Aryans should squash Jews and members of other 'inferior races' like 'roaches on a dirty wall.'" The Nazi propaganda film *Der Ewige Jude* (1940) included the following line, "Rats . . . have followed men like parasites from the very beginning . . . They are cunning, cowardly and fierce, and usually appear in large packs. In the animal world they represent the element of subterranean destruction . . . not dissimilar to the place that Jews have among men."

But as with slavery, the animalization of Jews went far beyond call-ing. During the **eugenics**, or race purity, movement in the United States and Germany during the early part of the twentieth century, the practice of animal breeding—breeding those with the desirable characteristics and killing and sterilizing the rest—became the inspiration and example for eugenic efforts to upgrade the human population in both countries. (Hitler, for example, although promoting purebred animals and wildlife, mandated that old and sickly animals be killed.) These efforts led to compulsory ster-ilization of the disabled in the United States and compulsory sterilization, euthanasia killings, and, ultimately, genocide in Nazi Germany. During what was called Action T4, from 1939 to 1941, the Germans sterilized some 300,000 Jews, Gypsies, homosexuals, and those with mental and physical disabilities. Starting in 1942, Hitler began implementing the **final solution**, which mandated that all Jews be rounded up and killed—either via the death squads that operated through much of Eastern Europe, or through the concentration camps and death camps.

From 1942 to 1945, European Jews were transported in cattle cars to the camps, some were tattooed with identification numbers as are livestock, and millions were slaughtered en masse, with their humanity and individuality completely extinguished. The camps themselves were modeled on Ameri-can stockyards and slaughterhouses: Nazis borrowed features intended to

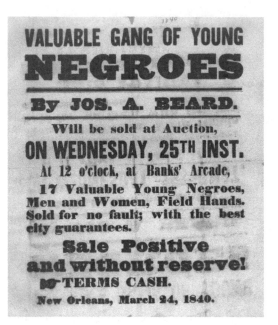

VALUABLE GANG OF YOUNG NEGROES

By JOS. A. BEARD.

Will be sold at Auction,

ON WEDNESDAY, 25TH INST.

At 12 o'clock, at Banks' Arcade,

17 Valuable Young Negroes, Men and Women, Field Hands. Sold for no fault; with the best city guarantees.

Sale Positive and without reserve!

TERMS CASH.

New Orleans, March 24, 1840.

Figure 13.3. Poster advertising a slave auction in 1840 in New Orleans. (Courtesy of Wikimedia Commons.)

make the processing of Jews at the camps as speedy and efficient as possible and to streamline the final part of the operation that took the victims to their deaths. In the gas chambers, Zyklon B, a pesticide normally used on mice, was utilized for the mass killings. Rudolf Hoess, the commandant of Auschwitz, called that camp "the largest human slaughterhouse that history had ever known" (Patterson 2002:122). Jews, Gypsies, and others were also experimented on, in the infamous experiments of Josef Mengele, who called a group of Polish women subjected to grotesque experiments the "rabbit girls."

Many writers have noted the similarity of treatment of animals in slaughterhouses and Jews in the death camps. Jewish author Isaac Bashevis Singer famously wrote, "[F]or the animals, it is an eternal Treblinka" (Singer 1968:26) German philosopher Theodor Adorno once said, "Auschwitz begins whenever someone looks at a slaughterhouse and thinks: they're only animals." And as we have seen, it is no accident that the treatment of animals in agriculture reminds us of the treatment of Jews in the death camps because the latter was modeled on the former.

What Is the Problem with Comparisons?

Because of what appear to be obvious similarities between the treatment of humans by other humans in the context of institutions such as slavery, and the treatment of animals in animal agriculture or biomedicine, animal rights activists and organizations have used these comparisons to draw public attention to the treatment of animals in contemporary society. Yet these comparisons have been, to say the least, controversial. Jewish organizations have been especially horrified at the comparison between animal agriculture and the Holocaust. The Anti-Defamation League (ADL), an organization that fights defamation against Jewish people, for example, has been outspoken in their condemnation of these comparisons and especially the use of Holocaust imagery and language by the People for the Ethical Treatment of Animals (PETA) in their advertisements. For example, Ingrid Newkirk, PETA's president, has said, "six million Jews died in concentration camps, but six billion broiler chickens will die this year in slaughterhouses." PETA has even used photos of concentration camp victims in an advertising campaign called "Holocaust on your Plate" which juxtaposed images of Jews in camps alongside images of factory-farmed animals. The ADL's response is that campaigns such as this trivialize the horrific loss of human life in the Holocaust by comparing humans to chickens or pigs. Even some who are sympathetic with animal rights have condemned these campaigns; Roberta Kalechofsky, president of Jews for Animal Rights, has pointed out that Jews were killed during the Holocaust by people who hated them; humans, even while killing animals, do not hate them. She also writes that, like Jews, animals are trapped by such comparisons "in the symbolism of another group" (Kalechofsky 2003:55). But more importantly, from the perspective of those critical of PETA, is that comparing animal suffering—which most people do not care about—to human suffering reduces the impact of the human suffering. Ironically, the United States Holocaust Memorial Museum's website, in the section entitled "Why teach about the Holocaust?" writes "The Holocaust provides a context for exploring the dangers of remaining silent, apathetic, and indifferent in the face of the oppression of others" (United States Holocaust Memorial Museum). Yet comparing the oppression of people to the oppression of animals is clearly not intended by that statement.

Feminist and animal rights activist Carol Adams takes the opposite perspective, however. She has been highly critical of those who use the language of animal suffering—"slaughter" or "butcher"—to talk about human suffering. When we do so, we erase the dead animals at the root of words such as these; animals become, in Adams's words (1991), the **absent referent**: they are there, but not there. The difference between Adams's take and the ADL's is this: When we use emotionally loaded and historically grounded terms such as Holocaust or slavery to refer to animal suffering, we are aware of what the word means and of the depth of the human suffering behind it. However, if we use butcher or slaughter to refer to human suffering, we do not actually acknowledge the real suffering of the real animals behind them.

Racism and Animal Advocacy

As we have discussed, comparing animals to humans is controversial on a number of levels. It has been common practice in the past to compare minorities and women to animals as a way to dehumanize them. In addition, animal rights supporters have made the comparison between vivisection and the treatment of animals in factory farms, and the treatment of humans in such abhorrent situations as slavery and the Holocaust. In addition, some scholars have pointed out that when animal advocates decry certain activities that have been conducted by non-Americans, or racial minorities in the United States, this too can be controversial, and often invokes claims of racism.

We discussed in chapter 7 the way that laws banning the consumption of dogs in the United States can be seen as racist, and how easy it is for animal advocates to claim that dog eaters are barbaric and cruel, even though those advocates may well condone equally cruel practices that are aimed at cows, pigs, or chickens. Furthermore, by saying that Chinese or Koreans are "dog eaters" or "rat eaters," we dehumanize them, and make them less than us, which is exactly what was done by racists against Jews, Chinese, or African Americans in previous times. This same problem can be seen when Western animal advocates take a stand against practices involving animals that take place outside of the United States. Whether condemning dog eating in Asia, or other Asian practices such as keeping bears in small "crush cages" so that the bile from their liver can be extracted (the juice is then sold as traditional Chinese medicine), it is often easy to slip into racist language such as "barbaric." For instance, French actress and animal rights activist Brigitte Bardot not only has condemned the slaughter of seals in Canada, the hunt of dolphins in Denmark, the killing of rhinos, tigers, and bears in China, and the ritual slaughter of sheep by Muslims, she has also made public statements in which she has complained that France has been "invaded by an overpopulation of foreigners, especially Muslims." Her positions in support of animal rights and against foreigners have intertwined in her public statements and political positions, linking support for animals with antagonism toward foreigners and racism.

As easy as it is to dehumanize groups of people by comparing them to animals, it is just as easy to condemn ethnic groups and nationalities when working to save animals. Calling people who engage in cruel-seeming acts barbarians or savages, especially when ignoring similarly cruel activities in one's own country, not only is hypocritical but also verges on racism and cultural imperialism.

Political scientist Claire Kim (2010) writes about two recent legal and media cases: the controversy surrounding the **live animal markets** in San Francisco's Chinatown in the 1990s, and the Michael Vick dogfighting case of 2007. In both cases, when activists—largely white and middle class—spoke out against what they saw as atrocities to animals (the sale and slaughter of turtles, chickens, frogs, and other animals at Chinatown markets and the dogfighting ring run by professional football player Michael Vick), they were met with charges of racism and cultural imperialism. Even though the situations were on some level very different—the live sale and slaughter of animals in Chinatown were legal but Vick's dogfighting operations were not—the feelings of distrust between the white activists and the Chinese-American and African-American communities were very similar.

In the case of the live animal markets, Chinese merchants argued that their practices are centuries-old, and a deeply embedded part of their culture. Furthermore, who is to say that the methods by which animals are kept and killed in Chinese markets are any more cruel than the methods of U.S. factory farms? The only difference, it would appear, is that in the former case, the killings are done in public; in the latter case, the killings are carried out privately in closely guarded slaughterhouses, and the products of that slaughter are neatly packaged and sold in supermarkets. Kim also notes that it is not thanks to the Chinese that Chinatowns exist. They were created by white Americans as a segregated site where Chinese people were forced to live and work because they were excluded from living and working in white communities. Similarly, when the Chinese were blamed for the spread of avian flu in 2003, or when Mexican slaughterhouse workers were blamed for the swine flu epidemic of 2009, both groups were linked with not just animals but also disease, and they were blamed for conditions of poverty that were not of their choosing.

The Michael Vick saga is, in many ways, more similar to the racially charged debate around the murder trial and acquittal of O. J. Simpson than to the Chinese market controversy. Michael Vick was engaged in an activity—dogfighting—that is illegal and widely condemned in American society. O. J. Simpson was accused of the brutal murder of two people. Both were famous black athletes. But in both cases, members of the African-American community rallied to the defense of these popular celebrities—not because people supported either dogfighting or murder, but because they felt that Vick and Simpson were just scapegoats for a racist justice system that has been historically biased against African Americans. Although whites were quick to claim, in both cases, that race played no role in either the investigations and trials, or in their feelings about the defendants, blacks were just as quick to point out that it is so often wealthy black celebrities who are taken down in this public way, punished for being too rich or uppity, and that these very public trials were really just a modern version of lynchings. In addition, it was difficult to miss the comparisons between Vick and animals, with media commentators and the public alike suggesting that

Vick was the "real animal" for making other animals fight to the death. Here again we draw, without perhaps even recognizing it, on the old comparisons between African Americans and beasts. And finally, many Vick supporters asked why so many whites were outraged that Vick killed dogs, yet have no outrage for the many African Americans who are killed every year, sometimes by the police.

Unfortunately, for (white and American) animal advocates, charges of racism mean that there is often no safe way to protest animal practices conducted by racial minorities or outside of the United States. And even more troubling, these controversies bring up one final point: The way that people accuse animal activists of caring for animals more than people. African Americans in particular can take the position that for years many white Americans (and American society in general) cared more about dogs and treated them better than they did African Americans. This is not an unreasonable position to take.

It is all too easy for people opposed to animal rights to trivialize animal activism by bringing up some human travesty—hunger, poverty, child abuse, police brutality, or homelessness. If someone is working on behalf of animals, rather than on behalf of the homeless or children, then that must mean that they care less about people than about animals. Unfortunately, this strategy serves to trivialize animal suffering by constantly comparing it to human suffering. But many animal advocates would argue that we should pay attention to wrongs against humans *and* wrongs against animals. One does not have to elevate one type of wrong above the other. Instead, as we have discussed in this chapter, these two types of wrong are often linked, and that, for those of us who care at all about righting society's wrongs, we can in fact tackle both.

Capitalism and the Expansion of Oppression

We have argued that human oppression and animal oppression emerged with the domestication of plants and animals, and the rise of state-level civilizations. Many scholars go further to suggest that human and animal oppression have escalated with the rise of **capitalism** as an economic system.

Capitalism is an economic system based on private ownership of property, competition in the production and distribution of goods and services, and the maximization of profits for those who own the **means of production** (the factories, land, machines, and tools used to make products). In a capitalist system, profit is produced by limiting the cost of the **factors of**

production—labor, resources, or technology. In other words, by underpaying workers and controlling other production costs, profit is increased for the property owners. In Karl Marx's analysis, although the capitalists own the means of production, the workers actually produce the products. But the workers do not benefit from their production or distribution. Profit is thus, in this interpretation, produced through the exploitation of the worker.

In capitalism, animals too are exploited. (Because they are considered resources, not labor, they are deemed a factor of production. Limiting the costs associated with animals in an industry such as animal agriculture means profits can be increased.) As we discussed in chapter 7, that means that animals are now tightly crowded into factory farms or confined animal feeding operations, with feeding and watering largely automated. They are not provided the opportunity to get fresh air, sunlight, or exercise, and are genetically modified so that their bodies produce the maximum amount of growth in the shortest possible time, and they can be slaughtered as soon as possible. Natural behaviors such as grooming, playing, socializing, and even laying down are prohibited because they do not promote profits. Capitalism, then, exploits some people (workers) and all resources (animals) so that some other people (owners) can benefit.

Why do we continue to support a system that is so exploitative? Those who are at the top of the system—the business owners—do so for obvious reasons: because they benefit from it. The rest of us hope to benefit from it someday, and also do not tend to see capitalism as exploitative. But on top of that, all humans—rich or poor—benefit from the bodies of animals. Even here, though, some people benefit from the exploitation of animals much more than others. Those who profit the most from animal agriculture, the company owners in the biomedical or the pet-breeding and -selling industries, will work extra hard to ensure that their industries remain laxly regulated and that the public remains uninformed about their practices, so that their profits will not be reduced.

And those humans at the bottom of the system must do the brutal work to ensure that the system continues; they work in the slaughterhouses, on the factory farms, and in the packinghouses, where they too are treated "like animals." Work here is dangerous and employees are found in communities made up of the poor and of minorities. Meat packing plants even advertise in Mexico to attract the poorest and most vulnerable populations to work in their factories.

Sociologist David Nibert's work (2002) has involved showing that all oppressed groups are victims of the same overwhelming forces of capitalist society. Profit prevails over everything else, including fair worker treatment and humane treatment of animals. Suffering in this analysis is not accidental; it is the logical outcome of a system that demands profit above all else. And this extends from the economic system into the political system. As we saw in chapter 7, wealthy businesses and industries not only lobby the government to ensure that laws are favorable to them, but also get themselves appointed to the very same federal agencies established to regulate their industries. Thus we see close links between the agricultural industry and the USDA, which results in the USDA's intensive promotion of meat consumption to

BOX 13.2

THE BP OIL SPILL

In April 2010, an oil rig exploded in the Gulf of Mexico, resulting in the deaths of eleven workers; it was the greatest environmental catastrophe in U.S. history. Animals that live in the area and that have been impacted by the spill include sea turtles, whales, porpoises, dolphins, tuna, sharks, pelicans, oysters, and crabs and a variety of migratory shorebirds and songbirds. British Petroleum, the company that operated the oil rig, found that "mistakes were made" that led to this deadly disaster, and that these mistakes should not reflect negatively on the company.

Since the spill, however, journalists have uncovered the ways that British Petroleum *routinely* engages in practices that put humans, animals, and the environment at risk. Although oil rig explosions may be accidental, in the sense that no one intends for these tragedies to occur, these occurrences are not aberrations; they are standard and expected occurrences for industries that put profits above worker and animal safety. In May 2010, documents surfaced that showed that British Petroleum used a cost-benefit analysis in 2002 in order to help them decide what kind of housing to build for its workers at a Texas refinery—inexpensive trailer homes that would have no chance of surviving a refinery blast, or concrete and steel housing that would cost ten times as much, but could withstand such a blast. The document, which used the "three little pigs" fairy tale (and was even illustrated with drawings of three pigs) as an analogy, recommended the cheaper housing. Another BP document put a $10 million value on the workers' lives (based on estimated costs that would be incurred in possible lawsuits). Even with that figure, the cost of cheaper housing combined with the potential lawsuits were still recommended over the more expensive housing. Three years later, the refinery caught fire and fifteen workers (most of whom were in the trailers) were killed; 170 others were injured.

Figure 13.4. Mercy for Animals protesters march at the Marriage Equality March in New York City, September 2010. (Photograph courtesy of Mercy for Animals.)

American consumers. The result? More animals die—billions per year—and, ironically, more people suffer from hunger, as grains and water are diverted into meat production.

Today, sociology textbooks and courses are beginning to include sections on the relationships among class, race, and the environment. Social scientists are becoming increasingly aware that communities in which the poor and minorities live are targeted by industry for toxic industries and for waste disposal. Known as **environmental racism**, this includes the longstanding practice of burying radioactive waste on Native American lands, drilling and mining for minerals and fossil fuels in poor communities of color, locating nuclear and other power plants on Native American lands, and exporting pesticides and other chemicals that are banned in the United States to third world countries. It also includes locating slaughterhouses and pig factories in poor communities, where the neighbors do not have the political clout to effectively complain about the reduced air and water quality. But most social scientists have still not recognized the connection between these practices and those that exploit animals, even though the results of animal exploitation harm more than just animals. As we have seen, people and the environment suffer as well, and those harms increase the further one moves down the socioeconomic ladder.

Suggested Additional Readings

Adams, Carol and Josephine Donovan, eds. 1994. *Animals and Women: Feminist Theoretical Explorations*. Durham, NC: Duke University Press.

Baker, Lee. 1998. *From Savage to Negro: Anthropology and the Construction of Race, 1896–1954*. Berkeley: University of California Press.

Birke, Lynda. 1994. *Feminism, Animals and Science: The Naming of the Shrew*. Buckingham: Open University Press.

Davis, Karen. 2005. *The Holocaust and the Henmaid's Tale: A Case for Comparing Atrocities*. New York: Lantern.

Donovan, Josephine and Carol J. Adams, eds. 2007. *The Feminist Care Tradition in Animal Ethics: A Reader*. New York: Columbia University Press.

Gaard, Greta, ed. 1993. *Ecofeminism: Women, Animals, Nature*. Philadelphia: Temple University Press.

Haraway, Donna. 1989. *Primate Visions: Gender, Race, and Nature in the World of Modern Science*. New York: Routledge.

Kalechofsky, Roberta. 2003. *Animal Suffering and the Holocaust: The Problem with Comparisons*. Marblehead, MA: Micah Publications.

Nibert, David. 2002. *Animal Rights; Human Rights: Entanglements of Oppression and Domination*. Lanham, MD: Rowman & Littlefield.

Patterson, Charles. 2002. *Eternal Treblinka: Our Treatment of Animals and the Holocaust*. New York: Lantern Books.

Spiegel, Marjorie. 1988/1997. *The Dreaded Comparison: Human and Animal Slavery*. New York: Mirror Books/I.D.E.A.

Connecting the Dots: Legitimating Oppressions
DAVID NIBERT
Wittenberg University

Like most people, I was raised to believe that eating other animals and using them in various other ways was normal and natural. I never gave it much thought; no one suggested I should. When I went to college I began to learn about racism, sexism, classism, and other forms of injustice—and how they were deeply grounded in the working of the capitalist system. Increasingly, I became active in community politics, marched against racism and sexism, and worked as a tenant organizer.

Then, on a visit to Madison, Wisconsin, in 1983 I happened upon a large demonstration for animal rights. I sat down on a hillside with thousands of other people and listened to the speakers. For the first time in my life I heard the argument that human exploitation of other animals—in all forms—is wrong. Speakers talked about the sentience of other animals and how they were subjected to horrific treatment and unimaginable suffering by the hundreds of millions.

As I traveled home from Madison, I reflected on my lack of awareness of the treatment and experiences of other animals. Eventually, I began to realize that the same processes underlying human oppression also were responsible for the oppression of other animals. That is, the oppression largely was motivated by economic gain, protected by the power of the state, and thoroughly rationalized and legitimized to secure public acceptance.

Soon I began to participate in demonstrations against hunting, factory farms, and various other forms of oppression of other animals, and I marched for animal rights. However, many of my friends in the movement were not receptive to my critique of capitalism. Most were only concerned with stopping the mistreatment of animals, a problem that they believed stemmed from prejudice and a lack of empathy.

As I moved from the life of a community activist to that of a sociology professor, I began to teach about the oppression of animals and to speak on the subject at different university events. Some members of the faculty offered condescending congratulations on my heightened level of "sensitivity" for animals, but admonished me to turn my attention to the "more important"—and, in their minds, apparently, unrelated—issues affecting so many fellow humans. As I pondered this anthropocentric critique, I became aware that the varieties of oppression of humans and other animals were not only similar but also entangled and could only be considered, and addressed, together. My developing awareness of this intertwined oppression was grounded in no small part in the work being done by feminist scholars who were highlighting the connections between sexism and speciesism.

Such a position was difficult to advocate in the sociological community, as existing paradigms were not conducive to comparing human oppression to that of other animals. I sought to introduce such a framework in my book Animal Rights/Human Rights: Entanglements of Oppression and Liberation *(2002). In that work, I argued that the old, problematic sociological term,* minority group, *was a euphemism that largely masked the reality of oppression; it also entirely excluded other animals from the analysis of the uses of power by one group against another. I proposed a more appropriate term,* oppressed group, *defined in a way that includes other animals.*

Next I suggested that the word speciesism, frequently used by animal advocates to mean injustice caused by prejudice and flawed moral reasoning, should be defined in a manner similar to racism, sexism, classism, and related words. All of these are ideologies, or systems of belief, that were created to legitimate oppressive arrangements and practices. This insight permitted a theoretical comparison that showed the ways in which speciesism as an ideology—such as racism, classism, sexism, homophobia, ageism, and other supports for oppression—was motivated by the workings of capitalism and had deep roots in the existing, powerful institutional arrangements. Therefore, the abolition of all these types of oppression requires a challenge not just to prejudice but also to the economic, political, and social institutions that perpetuate the injustices.

Using this theoretical framework, which highlighted the economic, state power, and ideological dimensions of animal exploitation and related injustices, I reviewed the history of entangled oppression; from the inception of women's subordination that accompanied the advent of men's stalking and killing of free-living animals, to the displacement of indigenous humans for the expansion of oppressive ranching operations, the history of entangled oppression of humans and other animals is long and tragic.

My current project builds on the ideas developed in Animal Rights/Human Rights, *particularly the concepts of the entangled oppression of human and other animals and the economic basis for oppression—particularly with the emergence of capitalism. The specific focus of this new work is the intertwined oppression that resulted from the process referred to as animal "domestication," in which cows, pigs, goats, camels, horses, and other animals were captured and exploited on a large scale, ostensibly to serve human interests. Although much of the history of "domestication" is portrayed in mainstream scholarship and media as a benign partnership between humans and other animals, yielding benefits for both, in fact the social gains for humans from the exploitation of animals overwhelmingly have been surpassed by the large-scale violence that ensued.*

For example, beginning roughly 8,000 years ago throughout Eurasia, where animal exploitation was at the time conducted on the largest scale, powerful societies of nomadic pastoralists emerged. Led by warrior elites, such as Genghis Khan and Attila the Hun, they ravaged and terrorized much of Eurasia for thousands of years. Such enormous levels of human violence, which became indelibly stamped in the development of human societies, were made possible only through the exploitation of and violence against other animals that were used as instruments of warfare, of laborers, and for rations. Such wide-scale violence did not occur in areas without populations of large, social mammals that could be exploited in such ways. Oppressive uses of animals also enabled Roman imperialism and built other aggressive civilizations, which in turn developed significant stratification, human enslavement, and deprivation for many, to benefit a relative few. Later, animal oppression was deeply entangled with the cruelty, tyranny, and monumental loss of life that accompanied Western colonialism, imperialist expansion, and the development of capitalism for the past several centuries.

After considering the appalling consequences of such entangled oppression on all continents over the centuries, I am now turning my attention to the growing forms of entangled oppression that characterize the twenty-first century. Today, the expropriation of arable land and scarce fresh-water resources, especially in the Third World, is necessary to raise the billions of animals that spend their last dreadful hours in the slaughterhouse, in the drive of agribusiness and fast-food corporations to expand the global consumption of products derived from animals. The fates of humans displaced from the land to make room for pasture and feed crop production, who experience water shortages, hunger, and other deprivations, are deeply entangled with the growing numbers of sentient animals now treated merely as "biomachines." The massive violence of the past, which was furthered by the promise of wealth that accrued to those with control over large numbers of animals, continues in the twenty-first century—and, indeed, is increasing through the work of a powerful and profitable animal industrial complex. We can expect to see only more conflict and large-scale violence as capitalism's imperative for expansion and growing profit levels is creating global warming, environmental devastation, and growing scarcities of water, arable land, and oil. All the while, deadly diseases linked to the consumption of products derived from animals, such as cardiovascular disease and many forms of cancer, are growing rapidly around the world.

It is my hope that my work emphasizing the entangled oppression of humans and other animals can be used to counter anthropocentric priorities and to help illuminate the commonalities among oppressed groups—and the fundamental role of capitalism in the various forms of oppression. Ultimately, it should illustrate that abolishing the exploitation of animals is imperative if we are to eliminate the oppression of humans as well, to develop a more just economic system and a more peaceful future for all.

IMAGINING ANIMALS

Animals as Symbols

14

Animals in Human Thought

In the beginning before there were people, before there were animals, a lone woman lived in a cave. She lived on the roots and berries of the plants. One night a magical dog crept into her cave and stretched out on her bed beside her. As the night grew long the dog began to change. His body became smooth and almost hairless. His limbs grew long and straight. His features changed into those of a handsome warrior. Nine months later the woman birthed a child. He was the first Chippewa male and through him came the Chippewa peoples.
—CHIPPEWA CREATION MYTH

UP UNTIL NOW, THIS textbook has covered the role that real animals play in human lives, and the variety of relationships that have been formed between humans and other animals. But one of the most important ways that animals play a role in human cultures is through their representations. Animals have been portrayed in the art, literature, folklore, religion, and language of human cultures for millennia. As such, they are important **symbols** that humans use to make sense of our world and ourselves. Biologist Edward Wilson wrote that animals are "agents of nature translated into the symbols of culture" (1984:97). But how animals are represented by humans is not just a metaphorical question. The ways in which we paint, worship, and tell stories about animals also shape how we treat them in turn. In addition, for many people, the real relationships that humans once had with animals have been largely supplanted by symbolic representations, with important implications for people and animals. Finally, as we shall see in this chapter, real animals and metaphorical, artistic, mythic, and virtual animals overlap in many ways.

The Use of Animals in Human Language

Language is more than a system of communication. Language reflects, and also shapes, how we see the world. On one level, language is just a system of categories. We create words to categorize the world around us—words such as furniture refer to a class of objects such as tables and chairs that are used in a particular way. Without linguistic categories such as these, it would be difficult to teach our children about the world that we live in.

Words about animals shape our understanding of animals. We have already discussed how terms such as "pet" and "livestock" reflect a particular understanding of animals, and then shape our treatment of them. In our culture, once an animal has been classified as a pet it would be difficult to turn that animal into meat. Similarly, we do not think twice about consuming animals that have already been classified as livestock.

Language then reflects cultural values and practices, and shapes those very same values and practices. This view is found in an anthropological theory known as the **Sapir-Whorf hypothesis**, which states that language helps to define the world view of a people, and thus the culture, of its speakers. It does so by providing labels for certain kinds of phenomena that different languages define according to different criteria. In other words, the grammatical categories of different languages may lead their speakers to think about things in particular ways.

So if language shapes how we see the world and teaches us cultural values and norms, what does it say about how we see the world when we use words such as dog, cow, or fox to describe people, or when we call someone bitchy, catty, or chicken?

There are a number of related issues at play here. First is the way in which humans so easily draw on animals to make sense of human realities. This is the case when we call people by animal terms such as bitch or sow, and also when we use phrases and idioms such as "flog a dead horse" or "skin a dead cat." Why are we so quick to draw on animals in this way?

Second, we need to look at the implications of using animals this way. What are the implications of calling people by animal names, especially when those names are used as pejoratives? **Animal pejoratives** reinforce attitudes toward marginalized humans by comparing them with another marginalized group: animals. Negative animal imagery is most often used to disparage women and minorities. Words such as bitch, cow, and sow almost exclusively refer to women and girls only and imply a number of negative traits such as stupidity, passivity, moodiness, and obesity. Calling a woman a dog or a pig not only implies that that woman is ugly or fat, it also says something about all women as well: that they are to be judged primarily on the basis of their physical appearance, and that if they fail in achieving the cultural standard of beauty, then they deserve to be called an animal. In other words, the huge number of expressions linking negative traits in women with animals demonstrates the disdain not just for the animal but for women as well.

Third, there are implications for animals as well. How does using animal idioms and animal pejoratives impact the real-world treatment of animals? Language is never neutral—it shapes behavior. According to Joan Dunayer (1995), animal pejoratives denigrate certain categories of people—women and minorities, primarily—but they denigrate animals as well. Anthropologist Edmund Leach (1964) once noted that by creating pejoratives from the names of animals, humans establish distance between themselves and the animals they regularly abuse. The fact that dog, cow, pig, sow, bitch, and chick are all thought of as negative terms means that those animals are thought of negatively. A woman who is called a cow is thought to be fat and dull; likewise, the animal, the cow, is thought of as dull. A dull animal is an animal that, ultimately, does not deserve to live. In her analysis of the history and mythology of the turkey, Karen Davis (2001) makes the same claim: When we are determined to do violence to an animal, we must first turn the victim into a despicable "thing" that deserves such treatment.

In addition, idioms such as "skin a dead cat" contribute to a permissive social attitude toward the abuse of animals. Language influences the ways in which animals are socially constructed and therefore treated in human society. Negative animal idioms normalize or trivialize violence toward animals. When sayings such as "flog a dead horse" are used and become a normal part of our vocabulary, we can no longer "see" the implications of human violence against animals. These expressions mask the real violence within them and demonstrate human power over animals. Further, what does it mean when we use phrases such as "slaughter" or "butcher" or "hunt" to discuss the killing of humans? It makes killing humans vivid and awful, yet we do not consider that when we use those words to discuss animals; they are normal, and happen every day.

In addition, we have discussed in this text the ways that the language used in animal industries serves to depersonalize and de-animalize animals. Terms such as "breeding stock," "meat," and "research tool" serve as absent referents, hiding the animal underneath the term. We have also talked about scientific writing, and how the use of third person passive voice ("the animals were euthanized") makes the animals into objects and takes away all human culpability for their deaths. We see this same phenomenon at play when celebrities are forced to apologize for crimes that they have committed, and yet do not want to actually accept responsibility for perpetrating those crimes. In 2009, R&B singer Chris Brown was charged with domestic violence against his-girlfriend, at the time singer Rhianna. In his public apology, Brown said: "I regret the incident," as if he himself did not cause "the incident."

Another example of the use of language to reinforce attitudes surrounding the exploitation of animals is the use of different terms for describing the killing of humans and nonhumans. We routinely use words such as murder or execution when talking about a person being killed, but we never use those words when referring to animals; instead, we use slaughter or euthanize, both of which have very specific meanings, and neither of which include the sense of culpability (or illegality) that murder conveys. On the other hand, we often use words intended for animals, such as slaughter, to refer to humans. Using slaughter in this way implies a particularly brutal murder, yet when we use it to refer to animals, most of us never think about the brutality of the killings behind it.

Most negative animal idioms about companion animals are about cats. This reflects a long history of cats being mistreated. Cats really were at one time stuffed in bags and drowned, skinned, and swung around. Even though we no longer skin cats, cats and other pets are still routinely abused, neglected, and abandoned. But when we use an expression such as "swing a dead cat," it seems funny and unreal. The rabbit is another animal that is routinely trivialized, yet subject to horrific abuse. Terms such as "dumb bunny," "bunny slope," "snow bunnies," and "ski bunnies," all of which are aimed at women, paint women and rabbits as being dumb and childish. Most terms associated with rabbits are derogatory, including "harebrained" (meaning frivolous or stupid), "rabbity" (meaning small, cowardly or rabbit-like), and "rabbit shouldered" (meaning slumped in the shoulders).

The treatment of animals is shaped by another linguistic practice as well. It is linguistic convention in popular media to refer to animals as "it" unless the writer or speaker knows that the animal in question is male or female. Even then, animals may still be referred to as "it." I once wrote an article about rabbits for a well-known rabbit publication, and was shocked when the magazine came out and all of my "he or she" and "his or her" were transformed into "it" and "its." I complained loudly to the editor of the magazine, but she told me that it was standard practice in the publication. I was shocked that in a magazine about animals, ostensibly for animal lovers, that this could be the case. Likewise, addressing an animal directly in print is almost never done. In the introduction to Dave Aftandilian's *What Are the Animals to Us?* (2007),

ANIMAL IDIOMS

All of these popular idioms have violence toward animals at their roots:

Bleeding like a stuck pig
Cook someone's goose
Dead as a dodo
Got bigger fish to fry
Killing two birds with one stone
Like a chicken with its head cut off
Like a lamb to the slaughter
More than one way to skin a cat
Not enough room to swing a cat
Shooting fish in a barrel
So hungry I could eat a horse
Take the bull by the horns
There's no sense beating a dead horse
To let the cat out of the bag
To shoot the bull

the book was devoted to animals: "It is you that we have written this book about and for" (p. xvii). Even though I myself have dedicated a couple of my books, at least in part, to the animals in my life, seeing this was still somewhat shocking, as it is so rarely the case that animals are directly addressed as subjects.

Animals as Symbols

Throughout history, and all around the world, various cultures have used animals as symbols. Symbols are things to which people give meaning, things that stand for, or represent, something else. The meaning of symbols depends on the cultural context in which they appear, and shape how we think about things. The cross, the American flag, and the swastika are all symbols that are imbued with different meanings in different cultural contexts. There is no inherent meaning in the symbols themselves. Many symbols are also "polyvocalic"; that is, they can signify multiple, abstract concepts all at once, not all of which can be easily articulated in words.

Animals are used to symbolize a whole host of characteristics that we see in ourselves, or want to project onto others, but that may be dangerous or foreign to us. Thus animals can be lustful, deceitful, murderous, or promiscuous. They can also symbolize more positive qualities such as love, altruism, and sacrifice. In the Middle Ages, for example, books called **bestiaries** were extremely popular. They were filled with images of animals, a bit about their natural history, and what the animals symbolized, and were accompanied by a moral lesson.

Animals are *like us*, but also *unlike us*. Because of this ambiguity, they are a perfect vehicle for expressing information about ourselves, to ourselves. As we saw with our discussion of language, we bestialize people (who we call bitches, cows, or foxes, or, in the case of whole groups of people, beasts or vermin) and humanize animals (that we anthropomorphize). And although we can use animals to highlight a person's good qualities (brave like a lion), we more commonly use animals negatively (cunning like a fox), especially to denigrate racial minorities.

Zodiacs, too, in cultures around the world, feature animals. The term zodiac, in fact, means "circle of little animals" in Greek. Zodiacs use animals to predict the future and to tell stories about the present and the past. In Western astrology, for example, an individual's personality can be predicted from the relationship at the time of the person's birth between the planets and the constellations, most of which are named after animals. So a person who was born when the sun was in the constellation Taurus will have the characteristics associated with the bull: steady, stubborn, and stable. Similar to the Western zodiac, the Chinese zodiac consists of twelve signs, all of which are represented by animals. But these animals do not represent the constellations; rather they represent the year in which a person was born. Thus a person who was born in the year of the rabbit is thought to share qualities with the rabbit; they are articulate, talented, and ambitious.

Why are animals so commonly used as symbols? Anthropologist Claude Lévi-Strauss, in his work on totemism (1963), said that animals are chosen as totems not because they are good to eat, but because they are "good to think (with)." In other words, animals have great metaphorical potential, and, for Lévi-Strauss, they are especially useful for representing social classifications such as clans and other aspects of kinship systems. Ancient and traditional cultures saw animals as closely related to humans, so it made sense that they would be incorporated into human symbolic orders, and used to represent human behaviors, desires, and dreams.

But why do we choose *some* animals to symbolize specific things? On the one hand, symbols are arbitrary—there is nothing in the stars and stripes of the American flag that indicates freedom or democracy. But because of the flag's history, it has been imbued with those symbolic associations so that today, they seem natural. In the case of animals, there is generally some kernel of truth at the root of the animal's symbolic association—some aspect of its biology or behavior that make people interpret it in a particular way.

Cats, for instance, can be symbolic of bad luck, evil, or witchcraft in Western cultures. Why the cat was chosen, and not, for example, the dog, tells us something about cats (they famously see in the dark, are solitary, and have a blood-curdling cry when fighting) and about the people in the cultures that use cats in this way. Often what we see in an individual animal is obvious: Ants often symbolize teamwork and discipline, the deer grace and sensitivity, the snake—because of its ability to shed its skin—regeneration, and the rabbit quickness. Some animals, such as bulls, goats, and apes, are especially thought to represent sexuality because of their behavior or the size of their genitals; rabbits and hares represent female sexuality (because of their well-known fertility).

In many Native American cultures, bears are thought to most closely represent man because they walk on two feet and are said to have uncontrolled emotions, as do humans. Bears are also used in Native American cultures to symbolize introspection because they are solitary and hibernate alone in the winter. Other animals important to Native peoples gained symbolic importance because of their economic importance. For instance, the bison was used by Plains Indians for food, shelter, clothing, trade items, and other objects, and thus represented life to them.

Sociologist Colin Jerolmack, in a discussion of the metaphorical qualities of the pigeon (2007), notes that pigeons tend to mate for life. This makes pigeons an excellent symbol of monogamy and enduring love; thus they are a popular animal used in fertility magic. Pigeons or doves (which are really the same bird) once represented Aphrodite, the Greek goddess of love, and Kamadeva, the Hindu goddess of love. Pigeons and doves have been used to deliver messages for thousands of years. Because of this use, the biblical story of the flood told in the Book of Genesis has a dove delivering the message to Noah (in the form of an olive branch) that the flood is over and that dry land has appeared. Since then, the dove has represented deliverance and God's forgiveness to Christians. Because of their gentle natures and white plumage, they also are used to symbolize peace. Ironically, Jerolmack points out that today, doves have inherited the positive symbolic associations mentioned here, although pigeons now represent filth because of their tendency to live in cities and eat human leftovers.

Another bird, the sparrow, has been subject to a great deal of symbolic associations. The English sparrow was imported to the East Coast of the United States in the mid- to late nineteenth century in order to control a species of worm that had begun infesting American trees. The sparrows were so successful that they began to out-compete some local birds, and were perceived as a menace by some scientists, although they were defended by bird lovers. A number of jurisdictions passed laws encouraging the killing of the sparrows, and even establishing bounty programs, which encouraged children to kill them in great numbers. Sociologists Gary Alan Fine and Lazaros Christoforides (1991) claim in their article on what were called the "great sparrow wars" that one reason for the issue becoming so important was that it occurred during a time when Americans were growing increasingly concerned about high rates of immigrants arriving in the United States from southern and eastern Europe, as well as from Asia. These new immigrants threatened native-born Americans and the anti-immigrant rhetoric of

Figure 14.1. Black cats like Sniffles are often considered to be bad luck in the United States. (Photograph courtesy of Anita Carswell.)

the time was used to fuel antisparrow sentiment, with the sparrows acting as stand-ins for foreign and unruly outlaws.

Like pigeons and doves, rabbits (and hares) are used to represent fertility and sexuality, but certainly not monogamy. Rabbits are linked to sexuality—and particularly female sexuality and fertility—in the rituals, myths, and symbols of ancient Greece and Rome, and were related as well to Diana, goddess of childbirth. Rabbits were often used in love spells, as aphrodisiacs, and to aid in fertility. In Eastern cultures as well as in Mesoamerica, ancient Europe, the Arab cultures, and Africa, the rabbit is also linked to the moon, which itself is linked to fertility and childbearing, with lunar goddesses commonly represented with or dressed as a rabbit or hare. Finally, thanks to the rabbit's fertility, they are also used in myths and rituals to signify rebirth, explaining the rabbit's role in the symbols of Easter.

Animals in Artwork

Animals' symbolic attributes also make them fine subjects for artist representation. They have been used by artists as a decorative motif, to represent the real place of animals in society, and also symbolically—to symbolize the same qualities that they represent elsewhere in the culture. Images

Figure 14.2. Sculpture of a monkey from the Chinese Zodiac on display in Moganshan National Park in China. (Photograph courtesy of Jakub Hałun, Wikimedia Commons.)

involving animals date back to the earliest forms of artwork from the Paleolithic era. The European cave paintings from 30,000 years ago featured animals more often than they did humans, and included large animals that played a major role in human lives—as predators and prey. Later, with the rise of agriculture in the Neolithic, domesticated animals began to displace wild animals as subjects in artwork. Cattle, for instance, were important animals in terms of labor, food, and religious ritual, and were found on jewelry, on musical instruments, on pottery, and on the walls of homes, tombs, and temples in the great classical civilizations.

In medieval art, animals were represented in different ways. They commonly appeared on jewelry, paintings, and utilitarian objects and were used to illustrate Bibles and other religious texts. When lambs were used, they generally represented purity and Jesus' sacrifice for humanity; doves were used to show love, and griffins were depicted as guardians for the dead. Medieval English coins were often decorated with animals whose symbolic meaning would have been clear to those using the coins. English artists depicted animals—such as lions or peacocks—that were unknown in England except through artwork and fable, but common animals were included as well, such as chickens or doves. Hens were used to represent the Mother Church; the peacock represented immortality and the Resurrection of Christ. Lions, which were often depicted fantastically because the artists had never before seen one, symbolized Jesus, and also the watchfulness of God. Snakes in the Christian tradition represented evil, but they were also seen as having protective capabilities, which was probably their function on medieval coins.

With the rise of the Renaissance in Europe, animals were represented more naturalistically and less symbolically than in the Middle Ages. Artists such as Leonardo da Vinci and Albrecht Durer drew very realistic portraits of animals, and other artists showed dead animals as food, surrounded by fruits, vegetables, wine, and bread. As in earlier times, hunting remained a prominent theme of Renaissance paintings.

The Japanese have used animals in their artwork for centuries. In ancient Japanese art, noblemen and warriors enjoyed images of hawks and eagles, which represented power and strength. Japanese carp, or koi, that live a great many years, symbolized persistence and longevity; peacocks, because of their beautiful, extravagant tails, were used to represent pride and service to the gods. Cranes have long represented good luck and are used in artwork and functional objects such as screens to divide rooms in the house. Today, a gift of a thousand paper origami cranes is given to someone as a gesture of good luck or healing in the case of illness. Other animals often represented in Japanese art, which borrowed heavily from Chinese art thanks to the strong influence of China until the ninth century, are the twelve animals of the Chinese zodiac, all of which share their symbolic qualities with people born in the year of that animal. For example, the horse is thought to be intelligent, independent, and free-spirited, and those qualities are shared by those who are born in the year of the horse.

One of the most popular animals in Japanese art is the monkey. Monkeys are sometimes depicted realistically, but are often shown in human clothing and engaged in human activities. Monkeys are particularly useful to represent human characteristics because unlike horses or cranes or carp, they really do look like, and often behave like, humans. For that reason, Japanese macaques are a frequent subject in Japanese art and myth. They are thought to be messengers of the Shinto gods and are thus revered. Foxes, too, were associated with the gods. In particular, the fox was considered to be the messenger of the goddess of the harvest, Inari, because foxes killed mice that destroyed the rice crops. Later, foxes took on a darker symbolism, and became cunning tricksters that could bewitch humans, and even shape-shift into human form.

For years, African traditional art has also depicted animals, in paintings, sculpture, masks, jewelry, and even rock art. As in other cultures, in African art, animals are chosen for their symbolic qualities as well as their beauty or distinctiveness in appearance, sound, or behavior. For instance, the sankofa bird is known for its odd habit of looking backward; for the Fante people, the bird is used in art to symbolize wisdom and learning from the past. The spider is a popular African trickster animal, and symbolizes cleverness in Akan art. Animals such as tortoises and snails are important food sources to many African tribes, and because they are gathered by hand, rather than shot with a weapon, they represent peace when depicted in artwork. As in other cultures, the snake, because of its ability to shed its skin, is used as a symbol of transformation. In African art, the pattern of a snakeskin can be used to decorate an item, with the same symbolic connotation as when the snake itself is used.

In contemporary art, animals, if they appear, are used in very different ways than in the past. In particular, it is rare to see animals depicted sentimentally or as symbols anymore. British artist Damien Hirst, in the 1990s, created pieces that do not just represent animals, but that are composed of the parts of dead animals. His first major "animal installation" was called *A Thousand Years* (1990) and consisted of a cow's head rotting inside of a glass case, complete with flies and maggots. Another piece, entitled *Mother and Child Divided* (1993), consisted of a bisected and pickled cow and her calf. His work has been controversial among gallery owners (who have been concerned about patrons vomiting when seeing his work), some art critics who considered his work overrated, and animal rights activists, who deplored the use of dead animals. "Meat art" is also a popular art form in some circles today. In 2010, singer Lady Gaga famously wore a dress made of meat for the MTV Video Music Awards Awards, saying that she wore it to show that she is not a "piece of meat" and that if people do not stand up for their rights, they will soon have as much rights as the "meat on our bones." Meat as metaphor is also the foundation of artists who make pieces out of meat, or artists such as Nicolas Lampert, whose art represents meat (but does not use it).

Other artists have used and sometimes killed animals in order to provoke reactions from the public. In 2003, Chilean artist Marco Evarisitti created a piece, for example, that involved live goldfish swimming in blenders full of water. Patrons were given the opportunity to press the buttons on any of the blenders, killing the goldfish within, which the artist said was an invitation for the public to "do battle with their conscience." More controversial was Swedish photographer Nathalia Edenmont, whose work involved actually killing rabbits, mice, chickens, and cats, and then photographing their chopped up bodies alongside flowers, fruit, and other objects. Like Evarisitti, Edenmont has said that her work was intended to challenge the public, and claimed that those who opposed it were hypocrites for not opposing the killing of animals for makeup or for food.

Live animals feature in the work of a number of other artists as well. In one of the earliest pieces involving live animals, Greek artist Jannis Kounellis's 1969 piece, *Untitled (12 Horses)*, was an installation in a Roman art gallery of twelve live horses. In 1974, performance artist Joseph Beuys created a piece called *I Like America and America Likes Me* in which he lived in a room with a live coyote for three days; Beuys said that the piece demonstrated

Euro-America's relationship with coyotes and with the indigenous peoples of America. In 2007, Costa Rican artist Guillermo Vargas displayed a starving dog, tied to a gallery wall, in a piece he called *Exposición N° 1*. On the wall behind the dog was written *eres lo que lees* (you are what you read) in dog food. It was alleged (but never proven) that the dog starved to death in the course of the gallery show. Vargas later said that the show was intended to uncover the public's hypocrisy, because dogs starve all the time on the street, and no one takes an interest. American artist Eduardo Kac commissioned the National Institute of Agronomic Research (INRA) in Paris to create for him a glowing rabbit, made by splicing a rabbit's DNA with that of a jellyfish, as a conceptual art piece in 2000. Kac claims he and the INRA had an agreement that the rabbit, Alba, would go home and live with Kac's family in Chicago, but the INRA claims that no such agreement ever existed. Even though Kac never brought the rabbit home, the image of the green glowing bunny that he created became one of Kac's most popular works. Another artist who famously uses live animals in his work is Mark Dion, whose installations often include live birds in a museum context, and whose intent is to sensitize the public about birds.

Other artists today use their work to directly challenge cultural views about animals, and take a strong position in favor of animal rights. For example, Tasmanian artist Yvette Watt creates paintings of animals as well as animal-human hybrids as a commentary on the changing relationships between human and (other) animal and on the objectification of animals in society. She also depicts farm animals in unusual situations in order to force the viewer to think about our close (we eat them) and yet distant (they are kept far away from us) relationship to them. Her work also reflects an interest in the relationship between how nonhuman animals are depicted and what this might have to say about how these animals are thought about and treated. Photographer Jo-Ann McArthur, in a project titled "We Animals," photographs animals in human environments, in order to demonstrate that "humans are as much animal as the sentient beings we use for food, clothing, research, experimentation, work, entertainment, slavery, and companionship." Her aim is to challenge the human-animal border that allows for so much exploitation of nonhuman animals. Britta Jaschinski is another photographer whose work focuses on animals and animal suffering. Her project "Zoo" depicted animals in captivity, and captured their boredom, frustration, and unhappiness. Sue Coe is an American artist whose etchings, paintings, and lithographs document graphic animal (and human) suffering in places such as factory farms and meat-packing plants. And British artists Olly and Suzy travel around the world, painting and drawing endangered

Figure 14.3. Miriam from *Offering #2* by Yvette Watt. The image of Miriam was painted in the artist's own blood. (Courtesy of Yvette Watt.)

animals in the wild, getting as close as they possibly can to the animals themselves. They often let the animals interact with their work, as they did in 1997 when they painted great white sharks underwater and then photographed one shark taking a bite out of their painting. They then use their work to educate the public about the tenuous status of the animals that they paint.

New Zealand artist Angela Singer uses old taxidermied animals to show how animals are abused, not just by hunting and taxidermy, but by society as a whole. She noted in an interview that she was always shocked at how dead animal heads could decorate the walls of homes and businesses, and their presence typically elicits no commentary at all. Her work confronts and makes public the violence that was done to these animals, and asks the viewer to question the human superiority over animals. In addition, she intends to honor those animals whose lives have been taken from them for a trophy. Art critic John Berger notes that animals are already becoming lost to the contemporary world, and that much of today's art involving animals can be seen as a sort of memorial to their loss. Certainly the work of artists such as Olly and Suzy, who focus primarily on endangered animals, or Britta Jaschinski, whose latest work featured disappearing wildlife, is not just a memorial to this loss but also a cry to stop it.

Artist Yvette Watt writes of the way that other artists use animals as symbols or metaphors for the human condition:

> I propose that this results in the animals becoming marginalized, allowing the artists to avoid addressing the broader ethical issues surrounding the ways humans interact with animals. This avoidance of the politics of animal representation in the visual arts is at odds with a rethinking of animals and human-animal relationships in other disciplines where there is an increasing emphasis on the importance of foregrounding the ethical and political issues surrounding human-animal relationships.
> (2010:111)

Mirrors for Human Identities

What animals ultimately do, whether used in art, language, or, literature (which we will discuss in chapter 16), is allow humans to express ideas about human identity. What does our depiction of animals, as well as our use of animals, tell us about ourselves?

Anthropologists have long analyzed cultural practices involving animals as being, at least in part, a commentary about human culture, values, and practices. For instance, anthropologist Garry Marvin (1994) argues that bull-fighting is symbolic of the opposition between nature and civilization, and force and intelligence, with the bullfighter (who represents culture) conquering the bull (signifying untamed nature) in the ring (a cultural space) by the use of his intelligence over the bull's raw force. Anthropologist Clifford Geertz, in his classic explanation of the Balinese cockfight (1994), says that the cockfight is a "metasocial commentary" about the human relationships in Balinese society. Geertz suggests that Balinese cockfighters see their birds as the symbolic representation of the men such that it is really the men fighting for their own honor in the ring. For Geertz, the cockfight is a story the Balinese "tell themselves about themselves." And veterinarian and rodeo scholar Elizabeth Lawrence called rodeos "ritual events," which serve to "express, reaffirm, and perpetuate certain values and attitudes characteristic of the cattle herders' way of life" (1994:211). Emiko Ohnuki-Tierney's 1987 discussion of the use of monkey as metaphor in Japanese society addresses the monkey's closeness to humans, and the various implications of that closeness, including its use as a deity, or mediator between the gods and humans, and also its use as a scapegoat, mocked for its attempts at being human. Ultimately, like the other works mentioned here, the monkey serves to help the Japanese understand themselves and Japanese culture. As Japanese culture changes in the modern era, so do interpretations of the monkey.

Suggested Additional Readings

Baker, Steve. 1993. *Picturing the Beast: Animals, Identity and Representation*. Manchester, UK: Manchester University Press.

Jaschinski, Britta. 1996. *Zoo*. London: Phaidon.

Jaschinski, Britta. 2003. *Wild Things*. London: Phaidon.

Malamud, Randy. 2003. *Poetic Animals and Animal Souls*. New York: Palgrave MacMillan.

Noelker, Frank. 2004. *Captive Beauty*. Urbana: University of Illinois Press.

Palmeri, Frank. 2006. *Humans and Other Animals in Eighteenth-Century British Culture: Representation, Hybridity, Ethics*. Burlington, VT: Ashgate.

Pollock, Mary Sanders and Catherine Rainwater, eds. 2005. *Figuring Animals: Essays on Animal Images in Art, Literature, Philosophy, and Popular Culture*. New York: Palgrave MacMillan.

Shepard, Paul. 1998. *Thinking Animals: Animals and the Development of Human Intelligence*. Athens: University of Georgia Press.

Wolfe, Cary. 2003. *Animal Rites: American Culture, the Discourse of Species, and Posthumanist Theory*. Chicago: University of Chicago Press.

Wolfe, Cary, ed. 2003. *Zoontologies: The Question of the Animal*. Minneapolis: University of Minnesota Press.

Animals and the Creative Arts

CAROL GIGLIOTTI
Emily Carr University

My first memory of being aware that I was an animal came at the age of three and a half. I was in a Saturday arts class in a small public elementary school near my home. That I was there at all was unusual, since I was inevitably bound for the Catholic grade school across the valley. I think of this particular experience fondly since it seems to be my first "aha!" moment. The pertinent activity was led by a young drama student who placed all her little charges in a circle and encouraged them to act like a particular animal. I "became" an elephant. I still remember how physically acting like an elephant, putting my arm in front of my head and swinging it back and forth, feeling my now, suddenly, large ears flapping on either side of my head and bringing each of my now enormous feet down slowly as I ponderously made my gigantic way around the circle, allowed me to be an elephant. In hindsight, that day I learned two things that have had an enormous impact on my life: I can feel something of what others feel through immersing myself in their way of moving through the world and there is no species barrier when it comes to empathy.

Though I was an actress during my high school and college years, I realized early on that acting left me little chance to communicate what I was most interested in and passionate about, animals and the natural world. Though I had always drawn and painted, it was not until I became a printmaker that I saw a way to communicate that interest and passion though a medium that had always been associated with social and political change. During the 1970s and 1980s and early 1990s, my goal as a practicing artist was to bring ideas of animal rights and animal liberation to my audience through first printmaking, and then, large mixed-media narratives that took on the topic of factory farming and animal experimentation. I initially became involved with digital interactive technologies and animation for the purpose of communicating these ideas through animation. While I was learning high-end animation at the Advanced Computing Center for the Arts and Design at The Ohio State University, I was also writing my dissertation on the Ethical Issues in Interactive Technological Design, *specifically looking at virtual environments. Through this research I began to understand how ubiquitous interactive media would become, and how important it was to question what motives and consequences the widespread use of it would bring.*

I turned all my efforts to writing about the ethics of these emerging technologies and became immersed in the new media arts world. My publishing, speaking, and involvement in this area focused on the necessity of considering the natural world and animals, as well as social justice issues, in the development of these technologies by artists, designers, computer scientists, and engineers, a number of whom I found to be lacking in knowledge or interest about these concerns. My ethical perspective was informed not only by my long involvement in the philosophies and practices of animal rights but also the emerging science of cognitive ethology (the study of animals in their natural surroundings).

The introduction of the genre of "bioart" by artists such as Eduardo Kac, SymbioticA, and Joe Davis in the early years of this decade was a revealing example of how thinking in technological research and development in the sciences and the arts appeared to be far from any consideration of nonhuman animals as having value in and of themselves. The goals of creativity and thinking about these technologies continued to be based on anthropocentric goals, that is, goals based on only human needs and desires.

> While researchers in comparative ethology are contributing to comprehension of the cognitive and emotional lives of other beings, much of the work in biotechnologies is reinforcing an understanding of animals as suited to act as a material language, a symbolic technique, without concern for their intrinsic value as beings with whom we share this planet. Animals have been conscripted into these technologies to further an agenda of controlling the creation of all life through the manipulation of various manifestations of code. *In today's biotechnologies, animals have become code.*
> (GIGLIOTTI 2009:XI–XXVIII)

In other words, animals, already redefined as such by centuries of use in human food and labor, are now approached by the life sciences and medical practices as data warehouses of information. As information, animals are now able to be reconfigured, recoded, and most importantly redesigned for commercial enterprises: food, health, military, and even "eco-friendly" or "sustainable" undertakings.

This research, combined with research on species extinction, global warming, and related environmental degradation, has led me to consider how our technological world rests solely on the existence of a free, self-willed, self-replicating wild agency. The ongoing suppression and destruction of this creative wild agency, often through the medium of technology, have corrosive effects on sustained human creativity, as well as the ongoing survival of nonhumans and humans alike. I have begun to investigate and attempt to articulate the deep connections between the destruction of wildness and the dilution of human creativity. Within this new body of research, I hope to advance arguments for shifting human creative thought and practice toward a central ethic of creativity in which wild agency is understood as indispensible.

My pathway through the arts and technologies has been guided by my passionate interest in, and commitment to, shifting our attitudes and relationships toward animals. In looking back, I have to thank that young drama teacher for giving me the gift, through a creative process, of imagining what it might be like to be *an elephant.*

15

Animals in Religion and Folklore

When Buddha lived on earth as a hermit, he became lost and came across a rabbit. The rabbit asked if he could help the hermit to find his way out of the forest, but the hermit replied that he was poor and hungry and could not repay the rabbit for his kindness. The rabbit instead told the hermit that if he was hungry, he should light a fire, roast, and eat the rabbit. The hermit then lit a fire and the rabbit immediately sprang into the fire and cooked his own flesh. The hermit then manifested himself as the Buddha and pulled the rabbit out of the flames, and, to thank him for his sacrifice, sent him to live in the moon palace, where he became the Jade Rabbit. And that is why the rabbit lives on the moon, where he continues to be seen to this day.
—CHINESE TALE

Animals in Religious Thought

Animals play an important role in many of the world's religions. As symbols, animals help us to understand important religious concepts such as purity, sacrifice, morality, and creation. As such, they play important roles in the myths of cultures around the world. Some religions hold certain animals to be sacred, and some religions worship certain animals, or taboo certain animals for religious purposes. Other cultures hold that some animals are the mythic founders of a group or clan, and have set those animals aside as totems. Finally, religious rituals often make use of animals directly, often as a sacrifice to the gods or ancestors.

Religions, through their myths and practices, encode a worldview that is specific to that religion. In this sense, a worldview is an orientation to the world that includes assumptions about the natural world and our place in it, our relationship to the sacred realm, and notions about the construction of reality. In a worldview is often, either explicitly or implicitly, a set of assumptions about humanity's relationship with nature and with other animals.

For instance, the **monotheistic** religions, Judaism, Christianity, and Islam, often called the "religions of the book," because they all share the Old Testament as a source of God's wisdom, have some similarities in their views toward animals, as well as some differences. In the Old Testament, God created man in his own image, and the animals and plants of the world, over which man was given dominion (Genesis 1:26). Even though this can be interpreted to mean that humans were given rights of dominance over animals, some interpret this to mean stewardship, in which humans have fundamental obligations toward animals. In any case, it is clear that in the Old Testament, animals and humans are separate creations, with the latter being given authority over the former. There are other passages in the Old Testament that give further information about how it is that humans should treat animals. For example, Proverbs 12:10 says, "The righteous person regards the life of his beast," and there are other passages that caution readers to take care not to cause unnecessary pain to animals (Deuteronomy 12:4), to feed them before oneself (Deuteronomy 11:15), and to let them rest on the Sabbath (Deuteronomy 5:14). Yet animals are still placed on earth for man's use, whether for food, clothing, or labor. In fact, most of the provisions in Deuteronomy and Leviticus that regard the proper treatment of animals only deal with domesticated animals and their proper treatment, as well as proper methods for animal sacrifice.

Christianity, because it shares the Old Testament with Judaism, shares the idea of humans having been created separately from animals, with the power and responsibility of dominion over them. In the New Testament, the book of the Christians, there is very little written about animals. Most of what we know about Christian views toward animals comes later, starting in the fourth century with the writings of St. Augustine (Augustinius 2007), who wrote that only man has a rational soul created by the breath of God, as opposed to animals, which only have sensation and cannot attain eternal life. It is Augustine's view (which itself borrows on Greek thought about the inferiority of animals), later expanded by the thirteenth-century theologian Thomas Aquinas (Aquinas 2007) who held that man had no moral obligations toward animals, that sets out what would become the dominant Christian view toward animals. In this view, because only humans were created in God's image, only humans have an immortal soul, and because God became a human being (in the body of Jesus), animals and humans belong on different **ontological** levels—in other words, they are radically different beings that do not deserve the same consideration as humans. For example, the Papal Encyclical "Evangelium Vitae," although opposing needless suffering for animals, does so because it is contrary to human dignity, not because it is harmful toward animals. It also says that humans should not spend money on animals that should instead go to relieving human suffering. And the 1994 Catechism of the Catholic Church tells us that animals are "destined for the common good of past, present, and future humanity" (1994: par. 2,415, p. 516).

On the other hand, many Christians take a different view, arguing that Jesus preached compassion and mercy. These Christians recognize that animals do have intrinsic (rather than utilitarian) value and have a place in God's kingdom. For them, Christian values demand careful stewardship over animals that are God's creatures, rather than dominion over them. This is the view encapsulated by St. Francis of Assisi who was said to preach to animals and humans alike. Some even think that Jesus was a vegetarian, basing their views not only on what he preached but also on the Gnostic Gospels, an alternative set of writings about the life of Jesus that were not included in the Canonical Gospels of the New Testament. Others see the fact that the Garden of Eden was vegetarian (Genesis 1:29–30 tells us that humans were told to eat of the seeds and fruits of earth in Eden; they were given permission to eat animals only after the fall of Adam and Eve), and that the Kingdom of God may be vegetarian as well, as indicative of a much more peaceful and animal-friendly Christianity.

In Islam, the third religion of the book, all creatures were made by Allah, who loves all creatures. The prophet Muhammad, too, was said to love animals, and he features in a number of stories which relate his caring. The Quran explicitly states, however, that animals can be used for human benefit. "It is God who provided for you all manner of livestock, that you may ride on some of them and from some you may derive your food. And other uses in them for you to satisfy your heart's desires. It is on them, as on ships, that you make your journeys" (Quran 40:79–80). In the previous traditions, however, even though animals should be treated with kindness and compassion, they exist for the benefit of human beings. So although a number of animal practices are condemned in Islam, such as sport hunting and animal fighting, Muslims certainly do use animals for food, labor, clothing, and other purposes. Animal sacrifice, which predates Islam, also remains an important part of the religion. In particular, once a year, at the Feast of Sacrifice, which commemorates Abraham's willingness to sacrifice his son Isaac, animals are sacrificed by the faithful. Animals may also be sacrificed seven days after the birth of a child, to fulfill vows, and to atone for sins committed during the pilgrimage to Mecca.

Buddhism and Hinduism, the primary religions of India, differ substantially from Judaism, Christianity, and Islam in their worldview regarding animals. Hinduism is a **polytheistic** religion with a great many traditions within it, and thus has no single view regarding animals. However, many Hindu gods take on animal forms, such as Ganesh, who has the head of an elephant, or Hanuman, who takes the shape of a monkey, and cows are sacred to Hindus. The god Vishnu was incarnated as a fish, a tortoise, a boar, and a half-lion/half-man; he is worshipped today in all of those forms. Because of these beliefs, Hindus do not eat cows, and villagers will not harm monkeys; both animals are protected by Indian law. Other Hindus are vegetarian because of the concept of *ahimsa*, which means avoidance of harm. And finally, Hindus believe in **reincarnation**, which means that one is reborn into another life after death. Because animals and humans both have souls, and together exist in the cycle of rebirth, humans can be reborn as animals, and vice versa. But even though humans and animals exist in the same continuum of life, humans are considered to be the apex of what life should be; humans are thus superior to animals, and accruing negative **karma** means that one will be reborn as an animal, which is considered to be an unhappy fate. Although many Hindu texts and beliefs promote the humane treatment of animals, and the coexistence of humans and animals, many Hindu practices are, nevertheless, harmful toward animals.

Buddhists also share with Hindus the desire to do no harm, and the concept of reincarnation (or **samsara**, through which animal and human lives are recycled) and karma; cruelty of all kinds should be avoided for Buddhists. Like Hindus, many Buddhists are vegetarians. Finally, animals and humans both have "Buddha nature," which is the potential for reaching enlightenment. However, in Buddhism, as in Hinduism, being reborn as an animal is seen as negative because animals are not seen as having the ability to improve themselves, making it harder to work off bad karma and ultimately reach enlightenment. Clearly, Hinduism and Buddhism see animals in a complex way.

Finally, although there are hundreds of Native American traditions, we can generalize from them to gain a glimpse of how animals are seen within many Native American religions. Unlike the book of Genesis in the Old Testament, which teaches that God created animals and plants, and separately created humanity, in many Native traditions, humans, animals, and plants are created together; they share a spiritual kinship. Humans are not separate from the natural world, but are part of it. In fact, many Native **creation myths** have animals as creator gods, giving birth to humans and animals. Although no Native American religion prohibits the killing of animals, they

Figure 15.1. Recognizable by its elephant head, this statue depicts the Hindu god Ganesh. (Photograph courtesy of Sujit Kumar, Wikimedia Commons.)

typically do require that animals only be killed when necessary, and that it be done with respect and gratitude toward the animal.

Of all the world's religions, only one mandates total nonviolence toward animal life. Jainism is a religion of India that dates back perhaps 5,000 years. Like Buddhists, Jains believe in reincarnation and the cycle of life, death, and rebirth. But for Jains, the only way to escape the cycle is to adhere closely to the principle of nonviolence. This means, in practice, not only refraining from eating animals but also from accidentally harming even the smallest of insects, which leads some practitioners to wear masks so that they do not accidentally inhale tiny bugs. Different religions clearly shape different cultures' views of, and treatment toward, animals.

Animal Tales

Animals play a major role in the folklore of people around the world. Legends, folk tales, fables, and proverbs are filled with animals that speak, animal-human hybrids, and magical animals. Many contemporary folk tales contain fragments of ancient myths, often from religions that disappeared long ago. Others are educational, in that they explain why certain things are they way they are. Still others, such as the fable, impart a moral lesson, and all serve to entertain as well.

Folk tales that use animals as the main actors are called **animal tales** by folklorists. Such stories exist in almost every culture in the world. One common type of animal tale is the "just so" or explanatory story. Native American tales include a number of **just so stories** that explain how a particular animal got to be the way he is. The Sioux legend "How the Rabbit Lost His Tail," for instance, tells why rabbits have such stubby tails. According to this story, the formerly long-tailed rabbit, arrogantly overestimating his own strength, interlocked his tail with the tails of three wolves that pulled so hard that they broke the rabbit's tail in half. Other animal tales explain features of the natural world, such as the Creek tale "How Rabbit Brought Fire to the People." According to this story, in the beginning, the Weasels were the only creatures that had fire, and the people asked the animals if they could get some. Rabbit was the only animal brave enough to steal it, using his cleverness and speed, thus bringing fire to all the people. Tales such as this one also illustrate the interconnectedness often found in Native American worldviews: weasels, rabbits, and humans all live and communicate together. Other just so stories are in fact creation myths, explaining how humans and animals came to be in this world; often, an animal serves as the creator god.

The animal tales that we know best in the West include literary fables, medieval animal tales, and oral legends, all of which have become intertwined throughout history. Even though most of us have become familiar with such tales through the collections of Aesop or the Brothers Grimm, a large portion of European folk tales are not European in origin at all. Instead, they probably descended from the twelfth-century Hindu collection of stories known as *Panchatantra* and the *Jataka*, a collection of Buddhist tales from the fifth century BCE. The original version of "The Tarbaby and the Rabbit" for instance, comes from the *Jataka*, and has been found in 250 versions around the world. The best-known version is part of the African-American Brer Rabbit series popularized by Joel Chandler Harris.

Animals that play a leading role in animal tales are usually cast because they have a certain set of characteristics that other animals do not possess. For example, in the European tradition, foxes are trickster; rabbits are foolish, cowardly, or arrogant; and lions are brave. In the African and African-American traditions, however, hares are tricksters and spiders are clever, whereas for Native Americans, coyotes are tricksters and badgers are courageous. These tales rely on anthropomorphism—the attribution of human traits, such as speaking, to nonhuman animals—in order to tell the listener something about humans, or impart a moral lesson. Some animals are gendered female, for example, such as cats and rabbits, but others are gendered male, such as horses and dogs. Of course, what appear to be real behaviors of the animals are also used in their characterization; lions really do seem to be brave, dogs really do behave as if they are loyal, and foxes really do appear to be sly.

Animals also serve as "types" in folklore. The trickster is a common folklore motif; he uses his wits to trick stronger animals into doing what he wants, or to escape trouble. African trickster animals include the spider, rabbit, and tortoise, although the European trickster is generally the fox, and the coyote is the most common trickster in Native American tales. Animal brides or bridegrooms that marry humans are another common motif, as are humans who have been bewitched and are now in animal form. Animal helpers are another popular motif; in tales involving animal helpers, often the human who receives the help is ungrateful, and pays a price at the end of the tale for his lack of graciousness.

Folktales generally exist to impart moral or educational information to listeners, and often use animals as stand-ins for humans, and thus should not be seen as a literal representation of animals. However, they can be read as a glimpse into the worldview of the people who created them. Animals can be helpful and harmful to humans—often both qualities appear in the same

Figure 15.2. Brer Rabbit swinging with Mr. Fox. Drawing by Palmer Fox for *Harper's Young People*, June 8, 1880. (Courtesy of Project Gutenberg.)

tale, indicating the ambiguous relationship between humans and animals. Because animals and humans cooperate, trick each other, fight with each other, talk to each other, and even marry each other in such tales, we can say that in the traditional and ancient cultures in which the tales were created, the human-animal boundary which emerged in the West had not yet formed. Animals could serve as helpers to people, guardians, inspirations, or sources of wisdom. Animals and humans still shared the same world, and animals played a pivotal role in that world.

Animal-Human Transformations

One of the most common themes in animal tales and myths is the human-animal hybrid. Humans have been transforming into animals, and vice versa, for thousands of years, at least within the stories and songs of cultures around the world. A number of religious traditions have animal deities, or deities which are part human, part animal. For instance, the Egyptian god Anubis has a dog head, Sebek has an alligator head, and Sekhmet has a lioness head; the Hindu god Ganesh has an elephant head.

The *tanuki* is a Japanese animal that resembles a racoon or a badger, but is a member of the dog family. The tanuki is often featured in Japanese folktales as a shape-shifter—a creature that can turn into humans (or even inanimate objects) to get what he wants. According to the tales, tanuki often transform into Buddhist monks in order to beg for food. Another Japanese creature is known as the *kitsune*, which is a fox that can transform into people.

Werewolves are an example of the transformation from human to animal and back again. Werewolf stories were common in Slavic and Chinese cultures, and often involve a curse that causes the person to transform into a beast. Shamanism, on the other hand, is a voluntary, ritual transformation of a person into an animal, usually in order to heal someone (alternatively, it involves the use of animal spirits to help the shaman). Some shamans of the Amazon region in South America, for example, take hallucinogenic drugs in order to transform themselves into jaguars.

Throughout the world, witches are seen as having the ability to transform into animals. The first recording of such a transformation was in 1211 when Gervase of Tilbury claimed that he personally witnessed witches in the guise of cats. A number of Scandinavian tales focus on the milk hare, a spirit that witches use to steal their neighbors' milk from their cows. Milk hares were known throughout Sweden, Norway, and Iceland, and the tradition dates back to at least the fifteenth century. In the Irish tale "The Old Hare," for example, an old woman rumored to be a witch lives in a cabin by herself. Nearly every week some of her neighbors' milk is gone. When a suspicious neighbor sees a hare come out of the old woman's cabin, he shoots the hare, hitting it in the shoulder. Upon entering the woman's cabin the next morning, the neighbor finds the old woman, her bloody shoulder wrapped in cloth.

Many medieval witch tales told of witches who did not transform into animals, but instead had animal familiars that did their bidding for them. According to Boria Sax (2009), the first person charged with having such a familiar was an Irish woman named Alice Kyteler, who was alleged in 1324 to have a spirit that would appear as either a dog or a cat and do her bidding. The two issues may be related—witches who transform into animals and witches who keep animals as familiars. Shamanic traditions in many cultures feature the belief that shamans utilize animal spirits that help them to gain access to separate worlds; to access those spirits the shaman often dresses up as the animal. These shamanic practices may be at the root of medieval witch beliefs. Another related concept is the classical practice of associating animals with Greek and Roman gods, which may have morphed into the animal mascots of Catholic saints. Finally, witches' familiars may also derive from beliefs in many cultures about pre-Christian mischievous spirits such as fairies or elves that may have been transformed into demons and demonic animal spirits with the rise of Christianity in Europe.

Figure 15.3. Inspired by the myths of human-animal gods, this sculpture of a rabbit-human hybrid is from Jonnie Russell's mixed species series. (Photograph courtesy of Jonnie Russell.)

A related folkloric tradition is the marriage between humans and animals. For instance, in the Korean folktale "Silkworm," a princess whose kingdom is at war promises her father that she will marry the first man who kills the enemy's commander. When it appears that a white stallion has killed the commander, thereby winning the war, the princess declares that she will marry the horse. The king, on the other hand, is horrified at the prospect, and orders that the horse be killed and skinned. The princess, however, is filled with grief, and wraps herself in the dead horse's hide, whereby she transforms herself into a silkworm. The king takes the worm and passes her eggs out to all the people, who benefit from the beautiful silk thread that they produce. A number of Native American legends discuss the marriage of girls and dogs, resulting in offspring that are sometimes hybrids, sometimes human, and in one Cheyenne tale, "The Girl Who Married the Dog," the puppies become the stars of the constellation Pleides.

Often, the animal that the human marries is really a human who has been cursed, and is forced to take on the guise of the animal. (Such curses are found in the myths and folktales of peoples around the world. For instance, in Greek mythology, Arachne was a weaver who boasted that her skills were greater than those of the goddess Minerva; upon hearing this boast, Minerva transformed Arachne into a spider.) Sometimes upon marriage, or upon a kiss, a cursed animal can transform back into the human as in the Grimm's fairy tale "The Frog Prince." (The frog is a common animal in transformation stories, probably because it transforms during its own lifetime from tadpole into frog.) In Scottish lore, selkies are seals which can, upon removal of their skin, take on human form and marry humans. The human generally is unaware of the real identity of their spouse, until they find the seal skin. Discarding the skin means the selkie will never be able to return to the sea. In China and Japan, foxes commonly transform into humans who then seduce men into marriage. These fox-maidens, however, can be detected by other animals, and when they are caught, must transform back into foxes and leave their human husbands. A similar idea is behind the Japanese tale, "The Crane Wife." This tale tells of the marriage between a poor fisherman and his wife shortly after the man cares for and eventually releases an injured crane. The wife is able to magically weave sails for the husband to sell, saving them from poverty, but warns her husband to never look behind the curtain while she is weaving. One day he looks behind the curtain and startles a crane that flies away, leaving the fisherman all alone.

Religious Symbolism

Religious myths rely heavily on symbols to convey meaning and often animals are used to convey important religious concepts such as purity or sacrifice. For example, Christianity uses a number of animal symbols such as the fish, which represents Jesus, the apostles, and Christianity; the peacock, which represents the resurrection of Christ; the dove which represents the Holy Spirit, purity, and peace; the eagle which signifies God's omniscience; and the lamb which represents Christ, purity, virginity, the twelve apostles, and resurrection. In fact, the first crosses of the early Catholic Church had lambs on them, rather than Jesus. Many animals in Christianity have negative associations as well. For example, ravens represent ignorance and false thoughts, and were also once thought to carry off the souls of the damned to hell. In Genesis 8:7, Noah originally sent a white raven to see whether or not the flood waters had subsided, but the raven never returned (and, ultimately, a dove did bring back the proper message). In punishment, God turned the raven black and condemned it to eat carrion.

Because of their historical closeness with humans, dogs tend to be represented positively in religions around the world. For example, in Native American traditions, dogs symbolize friendship and loyalty. But dogs also are often seen as guardians of the dead. In Greek and Roman mythology, Cerberus, the three-headed dog, guards the gates of the underworld. In the ancient Zoroastrian religion of Persia, dogs were intermediaries between the profane and the sacred realms, and were especially important in rituals surrounding death. If a person died, for example, dogs were brought in to witness the body, and they played a role in funerals as well. After death, the soul is accompanied to the afterlife by two dogs. Many of the cultures of ancient Mesoamerica such as the Olmecs, Toltecs, Mayans, and Aztecs, saw the hairless dog as the guardian of the dead. Some were kept as sacrificial food animals and others were used to protect homes from evil spirits or to heal the sick. In addition, the dog was thought to accompany the souls of the dead to the underworld; for this reason, mummified dogs were buried in the tombs of the deceased throughout Mesoamerica. Even today, small dogs are thought to ward off ailments in parts of Central America. Also, because dogs are scavengers and often eat the bodies of the dead, they are seen as filthy in some traditions, such as in Islam.

Cats were famously worshipped in ancient Egypt, as they represented the goddess Bastet, but under Christianity, the reputation of cats suffered. Cats were denounced because of their high status in pre-Christian cultures, but also because of their nocturnal abilities which associated them with evil. Cats could also bring good luck, although they often had to die to do so; in medieval Europe, cats were buried under the fields to ensure a good harvest, and were buried in walls to protect houses from evil spirits. Cats, like many other animals, were thought to be witches' familiars and were often executed alongside of accused witches.

Goats are another animal with negative symbolic connotations. In Christianity, goats represent the damned and sheep represent the saved. Goats are typically gendered male, and often are said to represent uncontrolled male sexuality. Medieval European depictions of the goat represented it as the devil, and goats were thought to be witches' familiars, and to have sex with witches. The Greek half-man/half-goat god Pan was probably an early inspiration for the later Christian image of the devil as a half-man/half-goat. In ancient Jewish tradition, the annual Day of Atonement involved two young goats that were made to atone for the sins of Israel; one was sacrificed, and the other was banished into the wilderness (or pushed off a cliff), acting as a

---

BOX 15.1

ANIMAL MUMMIES

In ancient Egypt, a variety of animals were mummified. Many families mummified their pet cats and buried them in the family vault, in the hope that they would be reunited together in the afterlife. Cats and other sacred animals were also mummified because they were thought to be incarnations of gods, so they were buried in an appropriately respectful fashion. Some animals were sacrificed to other gods, and mummified and buried, and still other animals were mummified and buried as food to be eaten by humans in the afterlife.

"scapegoat" for the community. Like the lamb, the goat was one of the most commonly sacrificed animals in ancient Israel, and was a popular animal to be given as **bride-price**.

Animal Cults

Outside of the major world religions, animals played a major role in a number of religious traditions around the world, sometimes as deities, sometimes as totems, sometimes as the abodes of the souls of the dead, and sometimes as sacrificial offerings. Animal cults refer to religions in which deities are represented by animals. They are found in pastoral societies in which certain species of animals are highly revered, but not worshipped, in hunting and gathering societies in which animals are honored before they are killed, in tribal societies in which clans trace their ancestry back to an animal totem, in shamanistic cults where animals assist shamans or where shamans transform into animals, and in agricultural societies which may either have animal deities, or deities which are sometimes represented as an animal.

One of the most sacred of all animals in agricultural societies is the cow, probably because of its economic importance to those cultures. Besides Hinduism, which regards the cow as sacred and prohibits its slaughter and consumption, there were once cults which revered cattle in ancient Egypt, Greece, and Rome and in Zoroastrianism. For example, Apis was an Egyptian bull god who was carefully chosen and revered when alive; after death, he was buried, and a new bull found to replace him. The ancient Hebrews worshipped the golden calf, which they probably borrowed from the Egyptian god Apis, until it was prohibited by Moses. Cattle are also highly revered in pastoral societies which herd cattle.

Wild animals such as bears or wolves tend to be worshipped in hunter-gatherer cultures. For example, the Ainu, the indigenous people of Japan, believe that a spirit called *kamuy* runs through all living beings, but that bears have more than other creatures, and that when the gods appear to humankind, they do so in the guise of bears. Generally, those hunting cultures that depend on one animal above all else tend to ritually honor that animal, and may hold festivals in their honor, as do Alaskan tribes for the whale. Fishing cultures, too, may have gods that reside in fish, as with the Japanese deity Ebisu-gami, that took on the guise of either a fisherman or a shark, and was revered among fishing peoples.

Cultures that believe that humans can be reincarnated into specific animals, or that specific animals hold the souls of the dead, will likewise treat

BOX 15.2

THE SACRED COW

The cow is sacred to Hindus throughout southern Asia. As such, they may not be slaughtered or consumed by Hindus (although Muslims in India and Bangladesh do eat them), and it is illegal to kill healthy cows in most Indian states. Cows roam free in India, and some temples and shelters care for homeless cows. Although they are not worshipped as deities, there are specific temples and shrines where Hindus can honor them, and they are the object of celebration during certain festivals, such as Bail Pola in August or Gopashtami, usually held in November, during which cows are bathed, decorated, walked through town, and fed sweet dishes that were made especially for them. In addition, Hindu rituals that require sacrificial offerings demand that ghee, or clarified butter made from cow milk, be used.

Most scholars explain the Hindu veneration for cattle in economic terms. Cows are India's most valuable animal because of the importance of dairy in the diet of Indians, the importance of cow dung as fuel and fertilizer, and the role that cattle have historically played in agriculture as draft animals. Hindu religious texts dating back thousands of years include references to the importance of cattle, including the Rig Veda and the Puranas, and by Gandhi's time, cow veneration was a central aspect of Hinduism. In fact, the Indian Constitution mandates that cows be protected in the country. Even with that constitutional protection, however, cows are illegally marched or shipped across state borders for slaughter in those states that allow cow slaughter (Kerala, West Bengal, and the far northeast of the country allow it), and even in states where it is banned, there is a thriving underground industry that slaughters cows for local consumption and for the export of leather.

that animal with respect. In Thailand, for example, Buddhists believe that the souls of the ancestors are located in the bodies of white elephants; consequently, they cannot be killed and are celebrated. The same holds true for parts of Africa, where snakes are thought to be incarnated ancestors. Similarly, cultures that believe that an animal guards the souls of the dead, or accompanies them to the afterlife, will revere that specific animal. Sometimes that means refraining from killing them, as with Nepalese Hindus, who worship dogs at an annual festival called Kukur Puja; sometimes it means sacrificing them, as with the hairless dog of pre-Columbian Mesoamerica.

Figure 15.4. Sacred to Hindus, a cow walks along the streets of Delhi in India. (Photograph courtesy of John Hill, Wikimedia Commons.)

Totemic cults are among the most common forms of animal cult, and are found in tribal cultures around the world, although they are most well-known in Native American cultures. The Tlingit, for example, see the raven as the creator god of the people, and as the literal ancestor to one clan. In totemic societies, the totem animal is only sacred to a specific clan; to all others in the society, the animal can be killed or eaten. The clan that holds the totem, however, will generally treat that animal with reverence, although they may sacrifice it on very special occasions.

In shamanistic cults, found in foraging and pastoral societies, shamans are religious intermediaries who generally act as healers through accessing the spiritual realm. They communicate with the gods or ancestors via trance and spirit possession, and often use animal assistants to gain access to that realm. Through trance, the shaman, who often wears the feathers or fur of a particular animal, can be temporarily transformed into the animal spirit, which helps him to reach his goal. Mesoamerican shamans, for example, used jaguar spirits as companions to protect them from evil spirits while they journeyed to the spirit realm. Another way in which animals play a role in shamanism has to do with hunting rituals. In these rituals, the shaman wears an animal mask in order to draw on the spirit of the animal, and either appease them or take their power, generally to bring success in an upcoming hunt. A different type of ritual associated with hunting concerns that of dangerous animals such as leopards that are sometimes thought to take vengeance on those who hunted them. In this case, there may be a prohibition on hunting that particular animal.

Sacrificial Lambs

To some readers, it may seem odd that religious traditions that revere or even worship animals may also kill those very same animals. Animal sacrifice has been a part of religious traditions for thousands of years, and is generally used as an offering to the gods or ancestors in exchange for favorable treatment. Anthropologist Marcel Mauss (1925) also claimed that because the animal, at the moment of its death, straddled the human and sacred realms, the participants in the ritual too would be able to access the sacred realm, through the aid of the animal. Sacrifice is practiced in agricultural societies, and domesticated animals are always used. Commonly sacrificed animals include chickens, lambs and sheep, goats, and cattle. Generally, when an animal is sacrificed, its throat is slit and the blood is allowed to run out onto the ground or altar; the blood itself is a necessary part of the offering. The animal is then usually eaten by ritual participants.

In ancient Greece, the animal to be sacrificed had to appear to assent to its killing; sprinkling water on the animal's head prior to sacrifice was done to make the animal "nod" in assent. After the animal was bled out, the priests would remove the entrails to inspect them to see whether the gods accepted the sacrifice. The meat would then be cooked and eaten by the participants, with the long bones reserved for the gods.

In ancient Israel, prior to the destruction of the Second Temple by the Romans in 70 CE, sacrifice was an important part of Jewish ritual. For the Jews, sacrifice was part of the covenant between the Jewish people and God, made when Abram sacrificed a cow, a goat, a ram, a turtle dove, and a pigeon, following God's command. God next requires that he sacrifice his own flesh, circumcising himself and his descendents. Long after this time, the descendents of Abraham (whose name changed from Abram to Abraham after his circumcision) would use sacrifice in order to cement this relationship with God. For example, Noah, after the end of the great flood, sacrificed one of every "clean" animal and bird and offered them as a burnt offering to God in thanks. Genesis 8:20–21 says, "And when the Lord smelt the pleasing odor, the Lord said in his heart, 'I will never again curse the ground because of humankind, for the inclination of the human heart is evil from youth; nor will I ever again destroy every living creature as I have done.'"

The Old Testament includes references to human and animal sacrifices—both could be demanded by God as punishment for people's sins or to keep them safe from God's wrath. Animals could also substitute for human sacrifice, as in Exodus 12 when God tells Moses that Israelite families living in Egypt should slaughter a lamb and place the blood on their doorposts so that God will skip over those houses when he goes to slaughter the first born of the Egyptians. An animal also substituted for a human in Genesis 22 when Abraham was commanded to kill his son Isaac as a measure of his faith, but at the last minute was stopped and directed to kill a ram instead. Later, with the rise of Christianity, animal sacrifice ended. Instead, Christians believe that God sacrificed his son Jesus in order to give humankind salvation; this is one of the reasons that Jesus is so often represented as a lamb in Christianity.

Today, animal sacrifice is practiced in traditional cultures around the world, and continues to play a role in Islam and Hinduism. In India, Bangladesh, and Nepal, every October brings animal sacrifices on a massive scale. In Bangladesh and some parts of India, the festival of Dussehra involves the ritual slaughter of thousands of animals at temples in honor of the Hindu goddess Sati. During the festival of Durga Puja, in order to honor the goddess Durga, animals are also slaughtered at temples throughout India and Bangladesh. In most of India, animal sacrifice is illegal and authorities and animal welfare organizations have been working to convince locals to replace the animal sacrifice, known as *bali*, with other offerings such as pumpkins, cucumbers, and other foods.

But no festival in South Asia demands more animal lives be lost than Nepal's festival of Dashain, which begins on October 15 and runs for fifteen days. Each year, hundreds of thousands of animals are sacrificed for the goddess Durga. In temples around the country, thousands of water buffaloes, pigs, goats, chickens and ducks are killed in order to please the goddess and protect against evil. So many animals are needed that they are trucked in from India and Tibet; one news report said that twenty truckloads of buffaloes were arriving daily. In 2009, more than a million animals lost their lives in this two-week period. Another Nepalese event is the month-long Gadhimai festival, which occurs every five years in November and likewise involves the sacrifice of hundreds of thousands of animals to the goddess Gadhimai, in order to end evil and bring prosperity. Gadhimai draws millions of attendees from Nepal and India, who come because animal sacrifice is legal in Nepal. In recent years, Nepalese animal rights organizations such as Animal Welfare Network Nepal (AWNN) have been attempting to stop the sacrifices and substitute new activities for the celebrations, but have thus far been unable to sway public opinion. AWNN argues that animal sacrifice is not consistent with Hindu values, and that goddesses such as Durga, who is a symbol of power and motherly love, would not want animals to be slaughtered in their name.

Proponents of animal sacrifice in South Asia note that not only are these practices cultural traditions that date back thousands of years but also that the animals live better lives than the billions of animals that are raised and killed for food every year in Western factory farms. One might also add that Westerners' distaste over practices such as this stems in part from the very public, and very bloody, way in which the animals are killed. In the United States, animals are killed in slaughterhouses which very few of us will ever see, in conditions which are, for all intents and purposes, invisible. Their suffering is thus invisible, allowing us to conveniently ignore it. The killing of animals in India, Bangladesh, and Nepal is, for sure, brutal. (Many of the animals are hacked to death and beheaded.) Comparatively speaking, though, even a few hundred thousand animals that lose their lives this way (in countries where meat consumption is quite low) is relatively minor compared to the billions of animals that lose their lives every year in the United States.

Perhaps the most controversial use of animal sacrifice has been in the United States, by practitioners of Santeria, an Afro-Caribbean religion formed by slaves in Cuba that blends African practices such as sacrifice with Catholic and native traditions. Santeria's gods are known as *orisha*; the orisha came originally from West Africa but were blended with the Catholic saints. In Santeria, animal sacrifice is a form of offering or *ebó*; other offerings include fruits, candies, and other foods, or sacrifices on the part of the practitioner, such as abstaining from alcohol. The orisha may demand that worshippers make a particular offering to them in exchange for benevolent treatment, and that offering can include animal sacrifice, especially if the worshipper has a critical need to be satisfied.

Animal sacrifice brings up a number of questions. In the United States, the debate has typically been framed between two opposing poles: freedom of religious practice and cruelty to animals. But in other cultures, the second point is not addressed at all because for many people, sacrificing an animal is not really killing an animal. Through the ritual context, the animal has been (in the eyes of believers), transmuted from a living, breathing creature into a symbol, a device that connects the sacred and the profane worlds. The animal may even, temporarily, be transformed into the god itself; by consuming the animal, ritual participants achieve communion, or oneness, with God. Another interpretation of sacrifice is that by killing the animal, the animal is freed to be with God; that is why, according to some, animals must willingly give their assent to the sacrifice. Remember that in Hinduism and Buddhism, for example, human souls can be reincarnated into animals as a spiritual punishment for a life not lived well; killing the animal is freeing the soul. In these explanations, animal sacrifice is not seen as cruel—it is necessary.

Communities of Faith and the Ethical Treatment of Animals

Should religion have something to say about the treatment of animals? Many believers think so, and look at the role that religions have historically played in helping societies to develop moral and ethical codes. For many animal lovers, religious beliefs about mercy and compassion should be extended to include animals. Theologian Andrew Linzey (1987, 1997), for example, has written extensively on the role that religious traditions play (or can play) in teaching compassion and justice regarding animals, and concludes that virtually all of the world's religions have ethical guidelines that are consistent with the humane treatment of animals.

St. Francis of Assisi, theologian Albert Schweitzer, and writer Isaac Bashevis Singer are examples of deeply religious people who took seriously their religions' commitment to compassion toward animals. The Quakers (known as the Religious Society of Friends), a seventeenth-century Christian movement, were very unusual for their time in that they felt strongly that God's creatures deserved consideration and reverence. George Fox, the founder of the Quakers, abhorred hunting and other practices, and in 1891, the Friends' Anti-Vivisection Society was founded in England in order to oppose the use of animals in science.

St. Francis, known as the patron saint of animals, was said to preach to the birds, and believed that it was the duty of man to protect nature and animals. He is the creator of the Christmas Nativity scene or crèche, and because of his beliefs that animals are God's creations just as are humans he included animals in the scene of Jesus' birth. Today, St. Francis remains an inspiration to Christians who work on behalf of animals. Many churches now hold **blessing of the animals** rituals on October 3, which is St. Francis' feast day. At these events, members of the church bring their pets to church, where the priest or monk blesses them. These blessings may have originated in pre-Christian times, but evolved with Christianity. The earliest records of such events go back not to St. Francis's feast day but to St. Anthony Abbot, whose feast day is on January 17, and who is also a patron saint of animals. Like St. Francis, St. Anthony during his lifetime was known to heal, bless, and care for animals. As early as the fifteenth century, people may have brought their farm animals to churches for blessing. Today, however, companion animals make up the majority of animals brought to blessings such as this, as their importance has increased in modern society.

The deep attachment that humans have to their animals can be seen in the various ways in which people imagine that they will share an afterlife with them. We can interpret 14,000-year-old burials of dogs to suggest that not only did humans care deeply about their dogs but that they also may have buried them with religious rituals generally used for people. For example, Egyptians entombed their mummified pet cats in their family tombs, where they were expected to be reunited in the afterlife together, and Romans buried dogs under marble headstones, but also had their companion animals killed after their own deaths, in order that their fates might be shared. With the rise of Christianity, however, animals and humans were separated after death, and Christians who wanted to be buried with their companions were not allowed to by the Church. Pet cemeteries, where pets could be buried, often in ritual fashion, emerged in the late nineteenth century, as an alternative to other ways of disposing of beloved pets.

Roman Catholic dogma says that animals do not possess an immortal soul and because of this, they will not go to heaven after death. For practicing Catholics who are animal lovers, this brings up a serious question: What happens to animals after they die? For many animal lovers whose faiths do not allow for animals to join with humans in an afterlife, they must either reconcile themselves to the fact that they will never see their beloved companion animals again, or they must come up with a new concept of an afterlife that includes animals. For many people, that afterlife is called the **Rainbow Bridge**. This refers to a meadow where pets (but not other animals) go after death, where they play together until that time that their owner dies, walks over the Bridge, and is met by his or her pets, and they all progress to heaven together. (Other versions of the Rainbow Bridge describe it as a permanent place for deceased animals to play, but they do not rejoin their owners nor do they go to heaven.) Some Christians interpret a handful of biblical verses in such a way that suggests that animals will be found in heaven. For example, Isaiah 65:25 says "the wolf and the lamb will feed together, and the lion will eat straw like the ox," which some interpret to mean that animals will go to heaven. (Others, however, say that this passage uses animals metaphorically to represent the end of conflict.) Still others have interpreted heaven to be a place in which humans and animals are welcome (or at least some animals). In most interpretations of this belief system, heaven is considered to be a place of "perfect happiness," so if one person's idea of perfect happiness includes their companion animals, they will be there, as a sort of heavenly prop. Finally, in 2009, a new business model emerged for evangelical Christians who believe in the rapture—an event whereby faithful Christians will be "caught up" to heaven to join Jesus, while the sinners will remain on earth to await Armageddon. Those Christians who have pets can now elect to pay companies to find "sinners" to care for their pets after they have been taken up in the Rapture.

Suggested Additional Readings

Chapple, Christopher Key. 1993. *Nonviolence to Animals, Earth, and Self in Asian Traditions*. Albany: State University of New York Press.

Foltz, Richard 2006. *Animals in Islamic Tradition and Muslim Cultures*. Oxford: Oneworld.

Hobgood-Oster, Laura. 2008. *Holy Dogs and Asses: Animals in the Christian Tradition*. Urbana: University of Illinois Press.

Kalechofsky, Roberta, ed. 1992. *Judaism and Animal Rights: Classical and Contemporary Responses*. Marblehead, MA: Micah Publications.

Linzey, Andrew. 1995. *Animal Theology*. Urbana: University of Illinois Press.

Sax, Boria. 1990. *The Frog King: On Legends, Fables, Fairy Tales and Anecdotes of Animals*. New York: Pace University Press.

Sax, Boria. 2001. *The Mythical Zoo: An Encyclopedia of Animals in World Myth, Legend and Literature*. Santa Barbara, CA: ABC-Clio.

Schochet, Elijah Judah. 1984. *Animal Life in Jewish Tradition: Attitudes and Relationships*. New York: Ktav.

Toperoff, Shlomo Pesach. 1995. *Animal Kingdom in Jewish Thought*. Northvale, NJ: Jason Aronson.

Waldau, Paul. 2002. *Specter of Speciesism: Buddhist and Christian Views of Animals*. American Academy of Religion Academy Series. New York: Oxford University Press.

Waldau, Paul and Kimberley C. Patton. 2006. *A Communion of Subjects: Animals in Religion, Science, and Ethics*. New York: Columbia University Press.

Webb, Stephen H. 1997. *On God and Dogs: A Christian Theology of Compassion for Animals*, New York: Oxford University Press.

Suggested Films

Gates of Heaven. DVD. Directed by Errol Morris. New York: New Yorker Films, 2004.

Holy Cow. VHS. Directed by Harry Marshall. New York: Thirteen/WNET New York, 2004.

What Do Animals and Religion Have to Do with Each Other?

LAURA HOBGOOD-OSTER
Southwestern University

Sheep gaze from the walls of the basilicas, oxen kneel as relics pass in the procession, dogs lick the hand of the person offering them a blessing, fish are resurrected after being caught and salted for consumption. Yet, still, I have been asked the question "What exactly do animals and religion have to do with each other?" so many times that my eyes glaze over and I offer my automated response. "Everything." Then I wait to see if the other primate (human) looking at me is eager for discussion or just trying to pass the time.

From the earliest days of human religious practice other-than-human animals occupied a place front and center. In his seminal essay "Why Look at Animals," John Berger states that animals "first entered the imagination as messengers and promises" not as "meat or leather or horn" (1980:4). They have inhabited the world of human religious imagination, ritual, myth, text, and community for thousands of years. But in the course of the last several decades this deeply religious relationship has been forgotten, swept aside, ignored, or, sometimes, denied.

As a scholar of the history and comparative study of religions, with a particular focus on the history of the many strands of the Christian tradition, it is amazing to me that the story of this religion is so often told empty of animals. Or, at least, empty of the overt recognition of the presence of animals. Certainly almost every child raised in the tradition hears stories from the sacred texts that are replete with animals—the Garden of Eden, Noah's Ark, Jonah and the Whale, the Nativity of Jesus. Then, however, these creatures seem to disappear as children grow and take on the supposedly more serious issues of humans.

Some have argued that Christianity is a totally human-centered religion, with the God-become-human figure of Jesus holding such a central position in the belief system that all other species can be ignored. But after years of research, digging into the stories of the tradition, analyzing the rituals and studying the imagery, that conclusion seems to be a tragic misinterpretation. Indeed, animals are everywhere in Christianity. Just open your eyes and you will see them not only lurking around every corner but also staring you directly in the eye!

Among the most interesting animals I have encountered in this research are a preaching dog, a companion lion, and a devout ox. The apostle Peter was in a long-term rhetorical battle with the heretic Simon Magus in the first century. Magus was hiding in a home in Rome when Peter arrived there, ready to take on his nemesis for another debate. They would not let Peter enter the house, so he called on the guard dog chained in front to go inside and confront the heretic.

The dog does so, preaching the Gospel on his own terms. It is an amazing scene in this early apocryphal text.

Equally amazing is the story of Saint Jerome, a doctor of the Church, early translator of the Bible, and leader in Christian monastic movements. Late in his life when he was living in a community with other monks, a lion walked into their cloister. Of course, most of the monks fled in fear. But Jerome welcomed the lion, quickly realizing that he had come for help with an injured paw. The two became constant companions for the remainder of Jerome's life. Christian iconography usually portrays the two of them together, often with the lion curled under Jerome's feet as he works on his biblical translation.

The devout ox, "il bui," is the centerpiece of a major festival that takes place in the little town of Loreto Aprutino in Italy. As the townspeople will tell you, years ago as the procession carried the bones of their patron martyr-saint from Rome to their village, a miracle occurred. People along the route knelt out of respect as the relics passed by with the exception of one farmer who was too busy plowing his fields. Then, suddenly, his draft animal, a huge ox, stopped and bowed in reverence to the martyr's bones. On the saint's official feast day, the town still recalls this miracle. A large white ox is decorated with ribbons and bells, a small girl rides on his back, and as the relics are carried in front of him, he kneels down again. Then the ox follows the procession to the church.

What does all of this mean for Christianity? Are these animals symbolic or real, or as I contend, both at the same time? Do they shift meanings in the tradition in significant ways? Do they matter?

Over and over again I come to the conclusion that indeed they do matter. Animals expand the horizon, and the ground, for religious traditions. They add wonder and mystery, complexity and beauty to religious traditions. Animals often remind humans that we are not the center of the universe, that other lives are indeed also sacred or, at least, worthy of consideration. In so doing, animals also expand the compassion footprint of religious traditions. All living beings are consequential; respect and concern for them prove central to religious teachings.

As the twenty-first century moves rapidly forward, these insights into the significance of other animals for religious and cultural traditions cannot be ignored. Humans have so altered landscapes that many animals once considered sacred are threatened with extinction. Still other animals are condemned to lives that lack any possibility of joy; one need only imagine confined animal feeding operations (factory farms) to comprehend the urgency. When my last book, The Friends We Keep: Unleashing Christianity's Compassion for Animals *(2010), was published even I was amazed at how many people were waiting for permission to* include animals in their circle of religious compassion and ethical consideration. Many thought such acts were not allowed or accepted in Christianity.

Walking into the sanctuary of the Cathedral of St. John the Divine in New York City for the annual Blessing of Animals is an overwhelming and glorious experience. Thousands of humans and animals gathered for this celebration of the life of St. Francis of Assisi—cats, dogs, turtles, lovers, families, infants, bees, and hawks. Animals first entered the imagination as messengers and promises, and they still can if one only takes the time to look for them.

16

Animals in Literature and Film

IN THE CHILDREN'S BOOK *Terrible Things: An Allegory of the Holocaust* (1989), Eve Bunting writes of "The Terrible Things" who come to the forest looking for animals with feathers, which they then take away. The other animals of the forest, the frogs and the rabbits and the squirrels, say that they do not have feathers, and in fact, that they are better than animals with feathers, and in fact, that the forest is better without the birds anyway. Next, the Terrible Things return and remove those animals with bushy tails; those remaining declared that the squirrels, which are now gone, are greedy anyway. The Terrible Things return again, and this time take all animals that swim; those that remain are glad because the frogs and the fish were unfriendly anyway. After the Terrible Things take the porcupines, they return for the white creatures; all that is left is the white rabbits, which are caught as well. The one remaining white rabbit, which hid from the Terrible Things, says, "If only we creatures had stuck together, it could have been different."

In Yann Martel's latest book *Beatrice and Virgil* (2010), the main characters are taxidermied animals named Beatrice and Virgil. Beatrice, the donkey, and Virgil, a red howler monkey, tell a story about their struggles to survive what they call "Horrors" involving the extermination of animals. Eventually, after witnessing "terrible deeds," they are both brutally killed—Virgil is beaten to death with a rifle and Beatrice is shot three times.

Both books are **allegories** about the Holocaust, told through the voices and experiences of animals. In *Terrible Things*, the Terrible Things are the

Nazis who come to remove group after group, while those who remain do nothing to stop what is happening. The lesson comes from the words of German theologian Martin Niemöller, who lamented the apathy of Germans during the Holocaust, as group after group of German citizen—Communists, trade unionists, and Jews—were purged from German society. The quote ends, "Then they came for me, and by that time no one was left to speak up." In *Beatrice and Virgil*, the terrible deeds that the animals witnessed were two Jewish women, drowning their babies in the village pond after which they drowned themselves, to escape seeing their babies killed by a group of Nazis who were chasing them.

In chapter 13, we discussed the controversies surrounding the use of Holocaust imagery to describe the suffering of animals. Both of the books discussed here use animals as a way of making sense of the Holocaust. *Terrible Things* received largely positive reviews when it came out, and has been used since then in schools to teach children about the Holocaust. *Beatrice and Virgil*, on the other hand, has gotten mixed reviews. A *New York Times* reviewer, for example, criticized Martel of trivializing the Holocaust by evoking "the extermination of animal life" and the suffering of animals (Kakutani 2010). Here again, even the hint of a comparison between the suffering of humans to the suffering of animals is seen as a trivialization of human suffering.

But why are animals so commonly used to evoke the suffering of humans? Besides *Beatrice and Virgil* and *Terrible Things, Maus: A Survivor's Tale*, the two-volume graphic novel about the Holocaust (1986, 1991) by Art Spiegelman, also tells the story of the Holocaust through animals. In these Pulitzer Prize-winning books, Jews are represented by mice (an animal that the Germans regularly used to evoke Jews), Germans are represented by cats, and other major nationalities are represented by other animals. Spiegelman has said that his use of the animal species to represent nationalities is an attempt to demonstrate how foolish it is to classify people based on characteristics such as nationality or ethnicity. But beyond Spiegelman's lampooning of racialization is the point that using animals in narratives such as this one is a useful way to depict suffering.

Animals in Literature

Why are animals used so frequently in literature? And are they always used, as in the two examples, as stand-ins for humans? Are animals ever used in literature as animals? We know that animals have played a major role in the symbolic behaviors of humans for thousands of years, through art, through religion, and through folklore and myth. As new ways of telling stories emerged, such as writing and, later, film, of course animals continued to play a major role.

Just as animals featured heavily in the myths of the great ancient civilizations, they were prominently featured in the plays and poems of the Greeks and Romans, as well as in the epic writings of India such as the Vedas and the Puranas. For instance, in the Greek tragedy *The Bacchae*, each character is likened to an animal, and the story unfolds as a contest between the hunters and the prey. *Aesop's Fables*, a collection of tales starring animals, each with a moral lesson, is probably the most famous of Greek writings involving animals. In the Indian epic poem the *Ramayana*, a number of animals play prominent roles, such as Hanuman, the monkey god, Jatayuvu, a mythical bird, and a key battle is won for Rama with the help of an army of monkeys, bears, and other animals. Homer's *Odyssey* features a dog named Argos that patiently awaits his master's return from battle; after a decade of waiting, Ulysses returns and Argos dies, having fulfilled his mission. For thousands of years, the loyalty of Argos has been a common theme in stories in Western literature involving dogs.

With the rise of Christianity in medieval Europe, much of the literature that was available was religious; it would not be until after the invention of the printing press in the fifteenth century that secular writing was more widely available. One of the earliest types of literature involving animals was the medieval bestiary, which used animals as allegorical devices to convey moral lessons to readers of the time. Each animal included in the bestiary was described in terms of its personality and moral traits; the partridge, for example, was said to steal the eggs of others, and its "immorality" was held up as a characteristic to avoid. Most Europeans at the time could not read, so bestiaries were primarily used by clergy to prepare their sermons, while illiterate church members could follow along by looking at the pictures.

Animals were used in the Age of Enlightenment as well, but at this time, with the waning influence of religion, animals were often used to parody or satire humans rather than to impart a moral lesson. For example, eighteenth-century English writer Jonathan Swift's *A Modest Proposal* (1729) is a satirical essay arguing for the breeding and sale of poor children as food in order to reduce the numbers of starving Irish. To make his point, Swift employs language generally used to refer to animals, discussing their feeding habits, their breeding, and how much they could be sold for.

In the romantic period of the nineteenth century, as the concept of conservation of nature was emerging, animals began to be treated differently in literature. Writers and poets such as William Wordsworth, Lord Byron, and John Keats wrote about wild animals and their beauty and how humans could be inspired by that wildness. With the Victorian era came the rise of natural history writing about animals, beginning with Charles Darwin. But at the same time, domesticated animals became a much more prominent feature of English and American households. With the rise of pet keeping came a new form of literature primarily aimed at children that once again used animal characters as a way to teach morality and other positive traits such as kindness.

In the twentieth and twenty-first centuries, animals continue to be major characters in literature in the West. In this firmly post-Darwinian era, where the human-animal border continues to crumble, writers have been grappling with a number of different issues through which animals can play a role. For instance, twentieth-century writers such as Franz Kafka use animality as a way to understand what it is that makes us human. In his short story "The Metamorphosis" (1915), Gregor, the protagonist of the story, wakes up one morning to find that he is now an insect. As his appearance and behavior change, his family begins to shun him and eventually he dies of starvation and an injury caused when his father threw apples at him. Another Kafka story, "A Report to the Academy" (1917), has a chimpanzee narrating a story about how he became human. The chimpanzee, Red Peter, had found himself in a cage in the cargo hold of a ship, and realized that his only way out of the cage was to be human: "There was nothing else for me to do, provided always that freedom was not to be my choice" (258). In both texts, Kafka's use of animals serves to destabilize the concept of humanity.

The twentieth century has seen a great deal of writing by women involving animals. African-American poet Maya Angelou's autobiography, *I Know Why the Caged Bird Sings* (1969), which takes its title and theme from poet Paul Lawrence Dunbar's poem "Sympathy," evokes a caged bird beating its wings fruitlessly against the bars of his cage, crying to be freed. Another African American writer, Alice Walker, wrote the short story "Am I Blue" (1989) about a horse named Blue and her feelings of loss when her companion horse, Brown (that was brought to stay with her specifically to make her pregnant), was taken away. "Am I Blue" can be read, on the one hand, as an allegory about slavery because of the way that the humans use Blue and other animals without regard for their own needs, and their lack of recognition that horses create meaningful bonds with other animals, and mourn when those bonds are broken. But "Am I Blue" does more than make animals into rhetorical devices: Walker empathizes heavily with Blue and wants the reader to recognize that animals have feelings and that those feelings need to be acknowledged and met. At the end of the story, Walker's narrator spits out a mouthful of steak that she was eating because she recognized that in eating meat she was "eating misery." But she also recognizes that when humans can treat animals in this way, they can treat humans in this way as well.

Animals in Children's Literature

Probably the area of literature that most utilizes animals is children's literature. Much of children's literature derives from folktales and fairy tales, which themselves featured animals, many talking, so it is no surprise to find that children's literature, which often evokes fantasyland settings and creatures, would also heavily feature animals.

Animals are used in children's books in one of two primary ways. They are either stand-ins for humans, representing a number of characteristics that humans have or that the author wants to teach to the reader, or they are animals—bears, foxes, or rabbits. In children's literature, they are far more commonly used as human models, and usually, as substitute children. As such, they are either presented realistically, as animals that can convey love or fear or loyalty, or they are fully anthropomorphicized, with human clothing, mannerisms, or language. Olivia the pig, the White Rabbit, the Cat in the Hat, Peter Rabbit, and Winnie the Pooh represent the latter—fully realized human characters in animal guise. The two dogs and a cat in The *Incredible Journey* (1961), who taught readers about love, perseverance, and overcoming hardship, or the title character in *Sounder* (1969), a dog that demonstrates loyalty and love, are both examples of the former—realistically portrayed animals that demonstrate important moral qualities.

Some books are a blend of the realistic portrayal of animals, with a bit of anthropomorphic portrayal as well. Richard Adams' *Watership Down* (1972), which was based on Ronald Lockley's *The Private Life of Rabbits* (1964), shows many realities of rabbits' lives, including their social structure, the way they dig warrens, mating behavior, fighting, the relationships between wild and domestic rabbits, and their daily rhythms of feeding, resting, and playing. In addition, the book is also quite explicit about the ways that rabbits die at human hands—including being shot, gassed inside the warrens, and caught in snares. Rabbits in this book are neither symbols of anything nor stand-ins for humans. But the rabbits do talk—a necessary device to give the readers insight into the minds and motivations of the characters.

Some children's book writers use animals as characters because it allows them to show diversity to children without using ethnic stereotypes. Animal characters can also do things that human characters could not because they would be dangerous or odd. Of course, animal characters do things in children's books that animals cannot do either, but the reader's disbelief is suspended because the character is not human. In fact, many animal characters in children's books behave neither like real children nor like real animals.

Animals may also play such an important role in children's literature because children seem to naturally love animals. Children of all cultures are drawn to animals from a very young age, forming attachments to them and making them central in their lives. Children also anthropomorphicize animals; scholars say that children have not reached the point that so many adults have where the animal and human worlds have become separate; animals are, to many children, playmates, parents, friends, and teachers. Because of this, children project "human" characteristics and traits onto animals—and although scientists are just now coming around to the idea that feelings such as love, anger, or jealousy are most definitely shared by humans *and* many animals, children seem to intuitively do this, using their own bodily experiences to relate to the experiences of animals. In his classic work on children's fairy tales, *The Uses of Enchantment* (1989) Bruno Bettelheim writes that the line between humans and animals is much less sharply drawn for children than adults, so the idea that animals can be children, or can turn into humans seems quite possible. In addition, children are drawn to animals which seem more like little people—bears, for example, even though fierce in real life, with their fat bodies, big heads, short limbs, and upright gait, seem like little people—and thus is born the teddy bear.

Children relate to animals, and, since at least the Victorian age, adults have understood that they could use the natural affinity between children and animals to teach children valuable social skills. Through reading about animals, children learn empathy, relationship skills, kindness, and compassion. First, children identify with animals, and from identification comes empathy. If children empathize with animals, then it is much easier for them to empathize with other humans as well, and to learn the difference between treating someone right and treating someone wrong. They also use their relationships with animals to help develop their own identities as people. Finally, children use animals as an emotional safety net—retreating to them when feeling sad or upset or scared. Because of children's connection to real animals, representations of animals are also popular with children, explaining why animals so heavily populate children's toys, children's books, and children's art.

If animals are "good for children to think," the corollary is that as children develop the skills they need to be adults, that they will (or should) no longer need animals at all. As children grow up, they are expected to

Figure 16.1. "The Tortoise and the Hare," from *Three Hundred Aesop's Fables*. (Image courtesy of Wikimedia Commons.)

shrink from animals, and if they do not, they are thought to be immature, to be hanging onto childhood. Many children's books, while emphasizing the closeness between child and animal, end with the child growing up and, sometimes, the animal's death. The most obvious example of this was found in Fred Gipson's *Old Yeller* (1956), which details the love between a boy named Travis and his dog, Yeller, but after Yeller contracts rabies, the boy must shoot him. The book ends with the replacement of Yeller by Yeller's puppy and with the gift of a horse, as Travis grows up.

According to Victorian studies scholar Tess Cosslett's *Talking Animals in British Children's Fiction, 1786–1914* (2006), children's stories featuring animals really emerged at the end of the eighteenth century, when Victorians were grappling with issues such as evolution, the treatment of animals, religion, racism, and the notion of empire. For instance, Rudyard Kipling's *The Jungle Book* (1894) is a collection of morality stories featuring talking animals, which was set in the Indian jungle. Many of the animals in the jungle are killers, but the author makes clear that the wolves, for example, never kill for pleasure, thus imparting a moral lesson. Kipling himself was born in English-ruled colonial India, and his stories are often interpreted as, in part, a celebration of British imperialism. Another popular series of animal stories that has often been interpreted as a justification for colonialism are the Babar the Elephant stories found in Jean de Brunhoff's 1931 *Histoire de Babar*. In these stories, Babar leaves the jungle and heads to the city, where he discovers the benefits of civilization and returns to the jungle to educate his fellow elephants that soon begin dressing in Western clothing like Babar. Cosslett points out the incongruities in the emergence of animal stories during the Age of Enlightenment, when Westerners were largely moving away from the religious morality stories of the past.

Ultimately, animals are useful vehicles for educating and entertaining children because of their ability to be like us and yet not like us. The make-believe world in which fantasy creatures live and animals talk to each other (and to humans) is the ideal world in which to include lessons on friendship, morality, kindness, bravery, or perseverance.

Talking Animals

Since *Aesop's Fables*, talking animals have been featured in adult and children's literature. These traditions certainly are rooted in times and places where humans and animals were seen to share worlds; they did not exist in separate spheres, so the idea of humans and animals communicating in the same language, and often with each other, did not seem so out of the ordinary. They also play a major role in children's stories because, even today, children still live in a world in which animals possess human characteristics and can be friends, teachers, or even parents.

But for at least a few hundred years, talking animals have played a major role in literature aimed at adults as well. Eighteenth- and nineteenth-century satirists used talking animals to give voice to the concerns of the poor and downtrodden. Other authors used speaking animals to express the suffering of the animals themselves. Anna Sewell's 1877 novel *Black Beauty*, for example, is an autobiography told by a horse named Black Beauty. His narrative includes the stories of a number of horses that he meets, many of whom, such as Black Beauty, suffered from the cruelty of their human owners. Sewell's intent in writing the novel was to change the treatment of horses in American society.

George Orwell's *Animal Farm* (1945), on the other hand, features a collection of talking farm animals whose uprising against the human farm owners was an allegory of the rise of Stalinism in Russia. Although the animals in *Animal Farm* are generally seen as a literary device to tell the story of human hierarchies, Orwell himself said that he originally came up with the story when he saw a boy beating a horse; he realized that men exploit animals as much as the rich exploit the poor. *Animal Farm* is then *both* about the treatment of working people under the rise of Communism and the treatment of animals by man. More recently, writer Paul Auster's 1999 novella, *Timbuktu*, told through the character of a little dog named Mr. Bones, informs the reader about the perils of homelessness (Mr. Bones' guardian, Willy Christmas, was homeless and dying) and the cruelties of humans, as well as the loyalty of a dog. Just a year later, art critic John Berger's novel, *King: A Street Story*, also shone a light on homelessness through the story of a group of squatters as told through the voice of King, one of their dogs. Both books use the canine narrator as a tool to expose the reader to the ways in which people can be so easily discarded by society. In E. B. White's *Charlotte's Web* (1952), Wilbur the pig is saved from slaughter from his friend Charlotte, a spider that convinces the farmer that Wilbur is worth saving.

It is not coincidental that, beginning in the nineteenth century, a great many of the writers who give voice to animals are women: from Anna

Sewell's *Black Beauty* to Virginia Woolf's *Flush*, the biography of a cocker spaniel named Flush who lived with poet Elizabeth Barrett Browning. Reading *Flush* tells the reader about London society in the 1930s, about female intellectuals such as Woolf and Barrett Browning, and of course about Flush himself. In many of these tales dating from the Edwardian and Victorian eras, female writers either write through the voices of animals or, in their books, befriend them and often help to free them in the process.

Animals in Film and TV

If animals are "good to think" as suggested by Lévi-Strauss in religion, folklore, and literature, especially children's literature, then they are certainly good to think in film and television as well. One of the first moving images, in fact, is of an animal. Eadweard Muybridge's 1878 series of photos depict an unnamed horse galloping; when viewed together, they show the horse in motion. With the invention of the first true motion picture camera in 1895, animals fighting, walking, running, and performing were quickly recorded onto film.

With the rise of the movie industry in the early twentieth century and television beginning in the 1950s, not only have animals been featured in movies and television shows, many have also become stars in their own right. Rin Tin Tin, Lassie, Benji, Flipper, Toto, Morris the Cat, and the Taco Bell Chihuahua all became part of the American cultural vocabulary, thanks to their performances in dozens of movies, television shows, and TV commercials. (The characters listed here were portrayed by animal actors, sometimes many. Benji, for example, was portrayed by a dog named Higgins, while at least ten different dogs portrayed Lassie from 1943 till the last movie was filmed in 2005.) The history of television and film is filled with hundreds of memorable virtual animals as well—cartoon animals such as Mickey Mouse, Bugs Bunny, Donald Duck, Dumbo, Bambi, Garfield, Roger Rabbit, or the animals from recent animated films such as those in *Madagascar, Kung Fu Panda, Ice Age, Over the Hedge,* and *Ratatouille.* It is difficult to imagine what Hollywood movies would be like without animals.

Horror films in particular have utilized animals. A popular sub-genre of horror film is the "eco-horror" film, which features nature run amok: monstrously big insects, rabbits, and other animals; killer dogs, rats, birds, and sharks; and human-animal hybrids such as werewolves or the creature from *The Fly* (1958). Wild animals such as anacondas, sharks, and pythons are often treated in horror films as monsters, and domesticated animals such as dogs, the most loyal animal of all, turn rogue and attack humans. In both cases, they are often difficult to kill or control. The 1950s and 1960s, during the height of the Cold War, saw the release of dozens of B movies featuring mutated animals wreaking havoc on humans, such as *Earth vs. the Spider* (1958), *Beginning of the End* (1957), *Night of the Lepus* (1972), and the *Godzilla* films; these may have reflected our culture's fears about the nuclear age. More recent films involving animals that attack such as *Jaws* (1975), *The Bees* (1978), *Cujo* (1983), or *Lake Placid* (1999) may have something to do with our fear that we have screwed up nature and that nature—and the animals that live in it—are "striking back." In either case, animals clearly are a useful symbol of "otherness."

Television has, in recent years, truly gone wild. Where animal actors were once seen in a variety of television shows, animals now have their own television channels. Animal Planet first went on the air in 1996 and now has dozens of shows on everything from animal hoarders to animal training to veterinarians to anticruelty investigations to dangerous animals. Their Super Bowl substitute for TV viewers who do not like football, *Puppy Bowl* (which features adorable puppies playing in a simulated football stadium, with kittens and, since 2010, bunnies and chickens at halftime), was watched by eight million people in 2008. The National Geographic Channel features a huge amount of programming dedicated to wild animals, and in 2010, they released their new channel, Nat Geo Wild, which features nothing but wild animal shows. These networks capitalize on the desire of TV viewers to watch wild and domesticated animals doing cute, dangerous, or weird things.

Even though animals are generally represented in film in much the same way they are in literature—as stand-ins for humans—in recent years, we have seen some new ways in which animals have been used. Thanks to changes in technology that make documentary filmmaking easier, and especially that make it easier to film animals in the wild, viewers have the opportunity to see wild animals in a way that we could never see them before. *Winged Migration*, Jacques Perrin and Jacques Cluzaud's 2001 film about the migration of birds, took four years to film and made viewers feel that they were flying along with the birds. In *March of the Penguins* (2005), directed by Luc Jacquet, the filmmakers filmed the Emperor penguins of Antarctica as they endured a harsh season of mating, guarding their eggs, and survival. *March of the Penguins* is a fascinating film because, although it followed the format of conventional documentaries (with its focus on groups over individuals, voice-over narration, and lack of interference on the part of the filmmakers in the lives of the animals), the survival of the animals resonated deeply with the audience, who recognized in the penguins qualities that we value in our own species. The penguins demonstrated a communal spirit, hard work, deferred gratification, and responsible parenting and sacrificed the individual for the survival of the group. Even though the animals were not anthropomorphicized, the audience learned a number of important lessons, and became emotionally invested in the survival of the birds.

Meerkat Manor was a documentary series from Animal Planet (2005–2008) that focused on a group of meerkats living in the Kalahari desert of Africa. Similar to *March of the Penguins*, the filmmakers working on *Meerkat Manor* did not interfere with the lives of the meerkats (even when they become ill, as was the case in season 3 when one of the lead characters, Flower, was bit by a snake and ultimately died of her injuries), and were able to film their lives through innovative technology such as underground burrow cameras. But unlike most documentaries, in *Meerkat Manor*, not only were the animals named, their lives were also narrated as if they were living a soap opera. Viewers were able to watch the romantic entanglements, fights, friendships, and even "gang wars" of animals whose behavior—as it is translated to us through the narration—seems awfully humanlike. We see that some of the animals are courageous, some are compassionate, and some are bullies—just like humans. And the audience became emotionally involved in the lives of the animals, expressing shock and grief at seeing some of their favorite "characters" die. Because of the success of *Meerkat Manor*, Animal Planet has created other series—such as *Orangutan Island* and *Lemur Street*—that follow the lives of wild animals in a similar narrative vein. Nat Geo Wild produces *Rebel Monkeys*, about a gang of urban monkeys in India.

Other popular films in recent years have included an animal rights message. For example, *Babe* is a 1995 film about a pig that knows that he is going to be slaughtered, and decides to develop a unique skill—herding sheep—to make himself special enough so that the farmer will save him. *Babe* was a huge success with audiences around the world, in part due to how the main character was personalized via a hairpiece and distinctive voice and behavior, and partly due to the movie's moral themes: overcoming adversity, facing challenges, treating others with respect, and seizing opportunities. But *Babe* was also heralded by animal lovers who not only rooted for Babe but also by those who hoped that the audience would realize that pigs such as Babe *are* special and should not be treated as food. However, as in *Charlotte's Web*, even though the pig is ultimately saved, and the farmer realizes that he is special, the general situation of farm animals dying for food is not dealt with. Babe (and Wilbur before him) does not challenge the idea of eating animals as food; these pigs just want to be spared of that fate themselves.

Other recent films have a similar theme—animals that are suffering from captivity or other forms of cruelty and want to escape from it—such as *Chicken Run* (2000), about a group of chickens that attempt an escape from their farm before they are turned into pot pie; *Finding Nemo* (2003), about a little fish that gets separated from his family and confined to a tank; or *Free Willy* (1993), about a whale confined in a marine mammal park that a young boy befriends and later frees. And sometimes these films, and the publicity surrounding them, have had an effect on the public. For instance, James Cromwell, the actor who portrayed the farmer in *Babe*, campaigned for People for the Ethical Treatment of Animals after the film was released, and urged kids who enjoyed *Babe* to stop eating pigs. After the release of *Free Willy*, a foundation was formed by activists who worked for years to get the real whale, Keiko, released from the marine mammal park in Mexico where he was kept.

Art historian John Berger (1980) suggests that the prevalence of animal images in modern society substitutes for a lost direct relationship with animals. This may very well be true and could account for the increasing popularity of nature films such as *Winged Migration* and *March of the Penguins* and television shows about wild animals such as *Meerkat Manor*. But what are the effects of all of this animal imagery around us?

Archaeologist and film historian Jonathan Burt (2002) says that how we see animals and how we think about them has changed thanks to the way that we see them on film. We know that many films, television shows, or advertising campaigns involving animals often result in an explosion of adoptions or purchases of the animals depicted—and, later, abandonment of those same animals. Such was the case for Dalmatians following the live-action version of Disney's *101 Dalmatians*, Chihuahuas following the Taco Bell television commercials and *Beverly Hills Chihuahua*, and Jack Russell terriers due to the popularity of Murray in the television show *Frasier*. Other films might have an effect on how we view controversial animal issues, such as increased interest in vegetarianism following the films *Babe* or *Chicken Run* or heightened concerns about the keeping of captive whales following *Free Willy*. Ironically, however, even movies with a pro-animal theme, such as *Finding Nemo*, can end up causing harm to animals. After *Finding Nemo's* release, the sales of clownfish skyrocketed, which meant that yet more fish were confined to bowls and tanks, even though the film specifically challenged the viewer to see such captivity as harmful.

Another example could be made from animal rights films. Animal rights organizations such as People for the Ethical Treatment of Animals have made dozens of films using undercover video taken from fur farms, slaughterhouses, stockyards, and animal research labs. The images of animal suffering seen in those films have played no small role in changing perceptions about the use of animals in society today. One of the best examples of this was the 26-minute PETA film *Unnecessary Fuss* (1984), which was made from footage shot in a head injury lab at the University of Pennsylvania. It includes graphic images of laboratory workers inflicting head injuries on baboons while, in some cases, laughing at the intense suffering of the animals. The name was taken from a quote by the lead researcher, Thomas Gennerelli, who said that he did not want people to know about his research because it "might stir up all sorts of unnecessary fuss" among animal lovers (Finsen and Finsen 1994:68). When the film was released, there was so much public outrage about the work being done that the lab was closed and Gennerelli's research was ended.

On the other hand, the increasing prevalence of animals in films may lead to increased awareness about the plight of animals, but it remains to be seen whether viewers will actually do something about the human impact on animals, either wild or domestic. In addition, when watching movies or television shows that have animal actors, there are other ethical issues to consider. How the animals used in films are raised and treated is only loosely regulated in the United States, and movies that are shot outside the United States do not have to obey American laws and will not get penalized if, for instance, animals are killed on set. (In the United States, the organization American Humane Association often monitors the treatment of animals on set, but not offset, and there is no law mandating that they be present.) In addition, as Randy Malamud (2007) points out, no animal chooses to be on camera. Although many domestic animals in movies or on TV today are well-loved companions whose guardians want to make them stars, many other domestic and wild animals are owned by animal trainers who keep them in kennels between productions. Some of these trainers no doubt provide exemplary conditions for their animals, and those that use exclusively domestic animals often adopt them from animal shelters. Because there is no real oversight for private trainers who train wild and domestic animals for film and television, it is difficult to know how humanely the animals are trained.

The origins and keeping of wild animals for performances is more troubling. Wild animal actors are purchased from the same dealers who supply zoos, circuses, private collectors, and game farms, and many are taken from the wild. Some trainers will acquire a single animal for use in a single production then will turn around and sell the animal back to the dealer for other uses. Other times, trainers will breed or contract with a breeder to produce multiple animals to be used for one project; once that project is over, the surplus animals must be sold. For example, the movie *Babe*, used more than nine hundred animals for the production, all of whom had to be placed after the film; even though some may have been placed into private homes as companions, others no doubt were sold to farms and eventually slaughtered for food.

One final question about the ethical implications of filming animals comes from critic Randy Malamud, who wonders whether humans should impose our gaze onto wild animals, interrupting their private lives. He asks of our interest in wildlife films and documentaries: "Does this testify to our increasing interest and concern for other animals, or does it mean we've dragged these creatures down to the level of mass entertainment?" (2010:146).

The Internet Is Made of Cats

If animals have exploded on television in recent years, the Internet has seen an eruption of animals. Cute and funny animal photos and videos are ubiquitous on the Internet, and websites featuring animals get a lot of traffic: Cute Overload averaged almost 100,000 visitors per day at the end of 2010. But unlike in so many other areas of human culture where animals are featured—such as myths, literature, or films—animals on the Internet rarely act as symbols. Instead, they act as themselves. Thanks to the popularity of video cameras and video phones, millions of people around the world can film their own animals or animals that they see in the wild, and can upload those photos to sites such as YouTube where some become instant sensations. From the earliest websites featuring animals such as Hamster Dance to hoax websites such as Bonsai Kitten to one of the most popular of all, Lolcats, animal watching on the Internet has become a popular way to waste time while working. In the last couple of years, a number of animal videos or images have turned into **Internet memes**—a concept that moves like a virus through the Internet, taking on new forms as it spreads. For example, "Dramatic Chipmunk" is a five-second video of a prairie dog turning his head around, set to music from the film *Young Frankenstein*. Although it sounds simple enough, this little video from 2007 has received ten million views, has been transformed countless times into other videos, has been the subject of countless spoofs and parodies, and was used in a television commercial for CarMax. Another popular meme is Keyboard Cat, a video of an orange tabby cat named Fatso wearing a T-shirt and appearing to play the keyboard. Besides the millions of views for the initial video, Keyboard Cat has become a popular meme whereby it is appended at the end of a video

Figure 16.2. Krusty the kitten sits in front of his own image on the monitor. (Photograph courtesy of Vicki DeMello.)

showing someone engaged in a blooper; the video is a way of "playing off" the other person, similar to gonging off a bad performer on a competition. Other popular Internet memes featuring animals include Sneezing Panda, Pedo Bear, Surprised Kitty, Bad Advice Dog, Spaghetti Cat, Cupcake Dog, and Maru. Unlike the animals featured in many other popular Internet videos, Maru, a Scottish fold cat that lives in Japan and spends much of his time playing in cardboard boxes, has become a celebrity in his own right. His YouTube channel is the eighth most popular channel in Japan and his videos average a million per episode.

Many commentators have noted that cats, in fact, are the most popular Internet animal. Joel Veitch, creator of the website RatherGood.com, made a video celebrating their popularity titled "The Internet is Made of Cats" that suggests that without cats such as Maru and Keyboard Cat, the Internet would collapse. Cats do seem to be a perfect animal for the Internet—they are not terribly trainable, so when they do something clever, it is all the more amazing. They are more aloof than dogs, and less readable, so it is fun to try to figure out what they are thinking. Cats are ubiquitous on the Internet because they are ubiquitous in our real lives—they are the most popular pet in the United States, so it should not come as a surprise that they will feature heavily in our photos and our videos. And finally, they also seem more serious than other animals, so it is funny when the camera catches them doing something stupid. That is one reason why Lolcats—photos of cats with funny, grammatically incorrect captions in a language sometimes called "catois"—are so popular. For most Lolcats, the caption is intended to convey what the cat in the photo is thinking, and many photos show cats engaging in what appear to be human behaviors. The first Lolcat was created in early 2007, by programmer Eric Nakagawa, and features a fat grey short hair cat with a pleading look on his face and the caption, "I Can Has Cheezburger?" Lolcats are celebrated because they are cats (many derive their humor from common characteristics of cats, such as ignoring humans), but also because they express the funny and dark thoughts of the people who created the images. In this sense, Lolcats are, once again, stand-ins for humans. We can relate to the sentiments expressed in Lolcat captions, because they are, after all, our own sentiments.

Suggested Additional Readings

Armstrong, Philip. 2008. *What Animals Mean in the Fiction of Modernity*. New York: Routledge.

Bousé, Derek. 2000. *Wildlife Films*. Philadelphia: University of Pennsylvania Press.

Burt, Jonathan. 2002. *Animals in Film*. London: Reaktion.

Chris, Cynthia. 2006. *Watching Wildlife*. Minneapolis: University of Minnesota Press.

Cosslett, Tess. 2006. *Talking Animals in British Children's Fiction, 1786–1914*. Aldershot, UK: Ashgate.

Fudge, Erica. 2002. *Perceiving Animals: Humans and Beasts in Early Modern English Culture*. Champaign: University of Illinois Press.

Lippit, A. M. 2000. *Electric Animal: Toward a Rhetoric of Wildlife*. Minneapolis: University of Minnesota Press.

Malamud, Randy. 2003. *Poetic Animals and Animal Souls*. New York: Palgrave MacMillan.

Mason, Jennifer. 2005. *Civilized Creatures: Urban Animals, Sentimental Culture, and American Literature, 1850–1900*. Baltimore: Johns Hopkins University Press.

Mitman, Gregg. 2000. *Reel Nature: America's Romance with Wildlife on Film*. Cambridge, MA: Harvard University Press.

Norris, Margot. 1985. *Beasts of the Modern Imagination: Darwin, Nietzsche, Kafka, Ernst and Lawrence*. Baltimore: Johns Hopkins University Press.

Paietta, Ann C. and Jean L. Kauppila. 1994. *Animals on Screen and Radio: An Annotated Sourcebook*. Metuchen, NJ: Scarecrow.

Rohman, Carrie. 2009. *Stalking the Subject: Modernism and the Animal*. New York: Columbia University Press.

Sanders Pollock, Mary and Catherine Rainwater, eds. 2005. *Figuring Animals: Essays on Animal Images in Art, Literature, Philosophy, and Popular Culture*. New York: Palgrave MacMillan.

Suggested Films

Babe. DVD. Directed by Chris Noonan. Kings Cross, Australia: Kennedy Miller Productions, 1995.

Chicken Run. DVD. Directed by Peter Lord, Nick Park. Bristol, UK: Aardman Animations, 2000.

Creature Comforts. DVD. Directed by Nick Park. Bristol, UK: Aardman Animations, 1990.

March of the Penguins. DVD. Directed by Luc Jacquet. Los Angeles: Warner Independent Films, 2005.

Websites

Bonsai Kitten: http://www.ding.net/bonsaikitten
Cute Overload: http://www.cuteoverload.com
Hamster Dance: http://www.webhamster.com
The Internet Is Made of Cats: http://www.rathergood.com/cats
Lolcats: http://icanhascheezburger.com
Maru: http://www.youtube.com/user/mugumogu

Literary Animal Encounters

PHILIP ARMSTRONG

University of Canterbury

Two travelers are marooned on two different islands. The first staves off loneliness by gathering around him a family of pets; the second is kept as a pet himself. The first shoots stray animals; the second is treated as a member of a feral species. The first domesticates goats and thereby forges a new colony, complete with its own agriculture. The second is exploited as a beast of burden, shown as a circus animal, examined as a natural-historical specimen, and condescended to by a breed of very superior horses.

Of course the first traveler is Robinson Crusoe and the second Lemuel Gulliver. If my summaries sound odd, it is because I have emphasized narrative elements that are often ignored, but are actually fundamental to both texts: namely encounters and relationships between humans and other animals.

Once you start noticing them, animals and human-animal relations appear everywhere in the literary canon. William Blake's poetry is full of exhortations to sympathy with animals as diverse as robins, wrens, and doves; with the starving dog, hunted hare, wounded skylark, misused lamb, and warhorse; with flies, spiders, moths, butterflies, and gnats. In Samuel Taylor Coleridge's Rime of the Ancient Mariner *(1798, 1817), the curse that falls on the mariner when he gratuitously shoots an albatross is exorcised only when he learns "He prayeth best, who loveth well/Both man and bird and beast." It is seldom remembered that the creature in Mary Shelley's* Frankenstein *(1818) is sutured together from animal parts as well as human ones. Herman Melville's* Moby-Dick *(1851) includes an encyclopedic survey of the ways humans have thought about whales, while progressing toward the climactic encounter between one man and one whale in particular. Virginia Woolf writes the biography of Elizabeth Barrett's spaniel in* Flush *(1933); D. H. Lawrence describes a whole bestiary of animal characters; Ernest Hemingway evokes a trophy room of animal victims.*

Contemporary literature is no less dependent on animal content. Paul Auster's Timbuktu *(2000) explores urban alienation from the perspective of a dog. Yann Martel's* Life of Pi *(2002) centers on the interdependency between a young man and a Bengal tiger adrift on a lifeboat. Margaret Atwood's* Oryx *and* Crake *(2003) contains a menagerie of genetically engineered hybrids—pigoons, rakunks, wolvogs, bobkittens, snats—designed to embody our anxieties about scientific manipulation and mass consumption of nature.*

Having located all these textual beasts, how do we read them? The default setting has conventionally been to consider their symbolic value. Animals have always and everywhere been among the most important of signifiers for human cultures: from Aesop's Fables to the bestiaries of medieval Europe, from the "totems" of animist cultures to the names of sports teams. In literature too, animals can function as metaphors for meanings that might have little to do with the animals themselves. Crusoe's husbandry of his goats allows Defoe to produce a celebratory parable about investment capitalism. Gulliver's Houyhnhnms and Yahoos are satirical embodiments of skepticism about Europe in the Age of Enlightenment. Moby-Dick is an omen of the civil war that was about to scupper the American ship of state.

Yet the animal content of texts produces a kind of excess that cannot be encapsulated by a purely symbolic interpretation. Maybe animals are too vivid, vigorous, and animated to be confined in this way. Or perhaps we are too interested in them to be entirely convinced by their reduction to metaphors. Interested in more than one sense: Animals sometimes escape our symbolic readings because in representing them we also evoke the practices, behaviors, and investments that shape our actual interactions with them.

Consideration of these interactions can modify or contradict the symbolic meaning of animals. Despite Crusoe's celebration as a paragon of self-sufficient capitalist enterprise, his agricultural economy actually depends on those goats, which Defoe can only imagine realistically because populations of livestock were in fact deliberately introduced to mid-oceanic islands by European mariners from the fifteenth century onwards. Crusoe and his goats are therefore products of what Alfred Crosby refers to, in the phrase that also provides the title of his book, as Ecological Imperialism (1986)—the same process of colonizing new terrains by means of animal agency that motivated Europeans to export pigs, sheep, cattle, rabbits, mustelids, and a host of other animals and birds. This is precisely the history mocked by Swift when he envisages English colonists degenerating into a race of feral pests, the noxious Yahoos of the last part of Gulliver's Travels. Meanwhile the Houyhnhnms, a race of rational and intelligent horses, recall the material, agricultural, and commercial dependence of eighteenth-century Europe on horsepower. But Swift reverses the slavishness of Europe's horses, as Gulliver's imagines the Houyhnhnms defeating Europeans in battle or sending missionaries to civilize them. As for Moby-Dick, chapter after chapter details the process by which nineteenth-century whalers slaughtered and processed thousands of whales into oil and spermaceti in order to lubricate the factories and light the cities of the Industrial Revolution. In all these examples, reading from animals to animal practices provides new ways into the relationships between texts and their economic, social, and historical contexts.

A third way to read animals involves exploring how human-animal narratives fulfill one of the longest-standing and most important roles of fiction: the exploration and expression of structures of feeling and sensibility. For much of the twentieth century, the evocation of emotional bonds between humans and other species seriously was—at least in "literary" (as opposed to popular) writing—regarded as a failure of taste: sentimental and puerile at best and psychologically and politically manipulative at worst. But it was not always so. For the novelists of the eighteenth century and the romantic poets who followed them, animals were fundamental to the cultural task of literature: the definition of what it meant to be human, and the refinement of ideas about sensibility and sympathy as central components of individual and civic virtue. It was not until the late nineteenth and early twentieth centuries that literary evocations of feeling for animals came to be discredited, as part of that great reaction against Victorian complacency that literary critics call modernism.

The rejection of sympathetic feelings for animals was also a product of other kinds of historical force. This was the historical moment at which science finally arrogated the supreme epistemological authority formerly held by Christianity—a triumph marked by the defeat of the anti-vivisection in the name of laboratory medicine. Meanwhile the Industrial Revolution was creating intensive agriculture, for example by introducing assembly-line slaughterhouse methods in the Chicago meatpacking district. These large social shifts required a downgrading of the cultural force of sympathy with animals, which consequently came to be considered childish, effeminate, or sentimental—a word now used pejoratively to connote emotional shallowness, cheapness, artifice, naïveté, and vapidity.

In this context, the modernist determination to eschew sentiment constitutes a bid to regain for the arts the kind of (masculinist, rigorous, intellectually muscular) credibility that was slipping away in an age of techno-scientific positivism. When modernists write about animals they are resolutely antisentimental: Hemingway and Lawrence focus, in different ways, on brutal, primitivist forms of human-animal interaction; and Woolf and Katherine Mansfield portray close emotional bonds with pets as symptoms of the effeteness of urban bourgeois life.

Yet this disdain for animal-themed sentiment remained an elite attitude. Narratives of sympathy for animals never lost their force in popular culture, and at the end of the twentieth century they also began to make a comeback in literary culture. One reason was the postmodern crisis of faith in the promises of technological modernity, along with growing concern about its treatment of nonhuman organic nature: The rise of animal advocacy and environmental consciousness were markers of this. A second reason was that postmodernism called into question the elitism of literary and artistic practice, enabling writers to rediscover the power and versatility of once-despised popular forms such as sentimental narrative.

These developments are superbly exemplified by Disgrace (2000), the Booker- and Nobel Prize-winning novel by J. M. Coetzee. The reader experiences the narrative through the point of view of its protagonist, David Lurie, whose disposition is cynical, self-interested, and dangerously inattentive to the lives of others. Lurie's refusal to consider the impact of his actions on others allows him to rationalize the rape of one of his vulnerable young students. As a result he is dismissed from his university teaching position and seeks asylum with his daughter Lucy, who lives on a smallholding far from Cape Town. Throughout the remainder of the novel, Lurie keeps failing to empathize with other human beings; he remains too bound up in his over-intellectualized view of things. Yet, humiliatingly and in spite of himself, he begins to feel compassion for animals. He assists at an animal shelter, holding unwanted dogs while they are put to death and disposing of their bodies. This experience produces one of the most emotionally intense moments in the novel:

> He had thought he would get used to it. But that is not what happens. The more killing he assists in, the more jittery he gets. One Sunday evening, driving home in Lucy's kombi, he actually has to stop at the roadside to recover himself. Tears flow down his face that he cannot stop; his hands shake.
> He does not understand what is happening to him.
> (142–143)

Lurie's abstract, rational, and dispassionate disposition is crumbling, replaced by an embodied compassion for the least privileged, least powerful, least valued of organic beings. "His whole being is gripped" by sympathy with the dying dogs: "They flatten their ears, they drop their tails, as if they too feel the disgrace of dying" (143).

Such moments exemplify a return to one of the most important and longstanding functions of literature: the exploration of complex, powerful, but difficult-to-understand states of feeling. At the same time they demonstrate the challenge posed by animals in literature, the demand that we read them more carefully, less complacently, than we might otherwise have done.

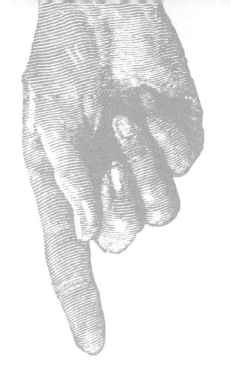

KNOWING AND
RELATING TO ANIMALS
Animal Behavior and Animal Ethics

17

Animal Behavior Studies and Ethology

HISTORY IS REPLETE WITH stories of animals performing heroic acts, either for other animals or for humans. Those stories have been repeated throughout the years in order to either humanize animals, or sometimes to teach children moral lessons. Today, in the age of YouTube, we do not just hear stories such as this; we can see them with our own eyes. During 2010, a number of such videos went viral on the Internet: from a horse that licked the wounds of another horse that had been shot with an arrow in the English countryside, keeping him alive till help came, to a dog in Chile that tried to pull another dog hit by a car from the side of a road, to a cat in Turkey that tried to massage his feline companion back to life after it had been hit by a car. In all of these cases, we see animals behaving as if they felt a strong emotion—love, loyalty, devotion—to another animal; in some cases, such as that involving the dog in Chile, an animal risked its life to try to save the life of another. Not too long ago, stories such as these would have been greeted with incredulity by animal behaviorists, who would never have granted that these animals felt anything for the other animal. Today, we respond to these tales very differently.

History of Animal Behavior Studies

Animal behavior studies first emerged as a scientific discipline in the mid-nineteenth century, but people were thinking about animal behavior long before that. We have already discussed some of the theological and philo-sophical ideas about animals dating back to the ancient Greeks. It really was not until the eighteenth century that scientists began thinking about animals naturalistically. Still in use in biology today, Carrolus Linneaus's *Systema Naturae* (1735) was the first scientific categorization system of all life, and was notable for including plants, animals, and humans—who Linneaus rec-ognized were animals too. As groundbreaking as this system was for under-standing animals, it was still based on the Greek and biblical idea that God created all of life on earth and that that life could be ranked—from lowest to highest, with man as the most perfect of all life on earth. Another drawback of the Linnaean system was the focus on species, which led to an inability to see differences in individuals. As animal behavior studies began to emerge in the nineteenth century, those scientists also focused on the normative behav-iors for a species (known as **species-typical behaviors**), and were unable or unwilling to see the differences among individuals in terms of behavior. This idea was directly responsible for the way in which people—scientists and lay-people—still say things such as "dogs do this" or "lions do that." Although it may be true that many, or even most, dogs behave in a certain way in a certain context, it is not at all true that all dogs behave that way, or that all dogs behave that way in all contexts. This is an example of what is known as **biological determinism**—the idea that animals (sometimes even humans) are hardwired to behave in certain ways, and that their biology determines their behavior. But even though scientists had been committed to the idea of species-typical behaviors, it was pet owners who first started discussing individual animals *as individuals*, and who first may have cleared the way for scientists to eventually do the same.

Charles Darwin's theory of evolution by natural selection (1859) was groundbreaking not only in its reshaping of the way that we understand human origins and the relationship between humans and other animals but also in how it shaped the field of animal behavior studies. This is because Darwin's theory not only explained the evolution of physical traits as adapta-tions to environmental stimuli, it also saw emotions, intellectual capabilities, and behaviors as having an evolutionary basis. For example, in his theory of **sexual selection**, Darwin explained the origin of sex-specific traits such as bright plumage in male birds or big tusks in male mammals as deriving from differential **reproductive strategies**. By this he referred to the fact that males, who produce far more sperm than females can ever produce eggs (not to mention live offspring), must compete with other males to assure their chances of mating, which led to males evolving showy traits that females lack. Underlying this theory is the idea that males and females do not just look different; they *behave* differently as well, and this behavior is a function of evolution.

Zoologist Charles Otis Whitman (Whitman and Riddle 1919), known as the father of the modern field of ethology, coined the term **instinct** to refer to patterns of behavior that, in the case of animals, are unconscious and triggered by certain environmental stimuli. Also known as **fixed action patterns**, they cannot be controlled by the animal itself. Whitman was succeeded by an Austrian zoologist named Konrad Lorenz (1952), who pushed animal behavior studies into the modern era. Lorenz, too, was interested in the instinctive behaviors of animals, and popularized the concept of **imprinting**, whereby some animals mimic the behaviors of the first animal that they see after birth, whether that animal is of the same species or not. A contemporary of Lorenz's (who shared the 1973 Nobel Prize with him) was Dutch ornithologist Nikolaas Tinbergen. Tinbergen (1951) was primarily interested in the causes of animal behavior, as well as in how that behavior changes as animals get older. Lorenz and Tinbergen, like others who studied the behavior of animals during this time, focused on animal behavior in the wild and did not engage in invasive experiments.

This changed with the rise of a scientific field known as **behaviorism**. Behaviorism is a psychological perspective that, in its original form, suggested that all behaviors are simple mechanical responses to external stimuli and should not be mistaken as deriving from internal thoughts or feelings. In its strictest sense, this theory suggests that animals possess no mental life whatsoever (because it cannot be directly observed)—no thoughts or feelings—just observable behavior. Behaviorism also looks at how environmental stimuli shape behavior. By introducing new stimuli, one can create new behaviors in an animal. Furthermore, animals (including humans) are essentially a blank slate on which experiences (as well as the specifics of the individual animal) shape behaviors.

One pioneer of this kind of behaviorism was psychologist B. F. Skinner, who famously trained rats in a lab to perform various activities through the use of food rewards. Skinner's work (1938) was an example of the field known as comparative psychology that aimed to compare the behaviors of animals across species. Unlike in ethology, it focused not on the evolutionary origins of animal behavior but on the environmental causes of it. Another example of this type of work was that of the Russian physician Ivan Pavlov. Pavlov is most well-known for his work (1927) demonstrating that dogs would salivate in anticipation of eating something good—a response that he was able to elicit through conditioning dogs to associate a particular sound or sensation with food (in the same way that many dogs today associate their leash with being walked, and thus get excited when the leash comes out). The term **Pavlovian** now refers to how anyone can be conditioned to respond in a particular way to a particular stimulus. Today, behavioral science and operant conditioning are often used in training animals. These are more positive methods of training than those relying on coercion or punishment. Using behaviorism, animals at zoos or laboratories, for example, can be trained to voluntarily offer a limb for blood draws Some proponents of this method of training (Sutherland 2009) see it creating a communication system between human and animal.

A more infamous example of comparative psychological studies was the Harry Harlow **maternal deprivation studies** (Griffin and Harlow 1966). Harlow, a psychologist, conducted experiments on rhesus monkeys in which baby monkeys were reared in a variety of settings in order to see how they developed. One experiment provided the babies with either a surrogate mother made of wire or one made of terrycloth; sometimes one or the other mother provided milk. The babies generally hugged the cloth mother for support, whether or not that mother was the source of milk; when deprived of the cloth mother, infants in an unfamiliar situation cried and screamed with anxiety. Even after the babies were weaned, and no longer needed the surrogate mothers for milk, they clung to the cloth mothers for comfort. Harlow's studies showed that babies with wire mothers developed psychological problems. These studies were used to counteract child-rearing theories popular in the 1950s that advised mothers to limit mother-child touching (because it was thought that it would spoil children). Other studies that Harlow conducted involved rearing baby monkeys in total isolation; as we might predict today, those monkeys developed severe psychological problems.

The two approaches—early ethology and comparative psychology—can be considered to be examples of the "nature-nurture" approach to animal behavior. Is animal behavior hardwired (that is, is it caused by nature)? Or is it caused by the environment (by nurture)?

Do animals have a mind? In other words, do they have feelings, thoughts, and intentions? Until very recently, that answer would have been "no." According to nineteenth-century psychologist C. Lloyd Morgan, "In no case may we interpret an action as the outcome of the exercise of a higher psychical faculty, if it can be interpreted as the outcome of the exercise of one which stands lower on the psychological scale" (1894:53). In other words, why interpret an animal moping around the house after his mate has died as grief when one can more easily interpret it as simply being hungry? For traditional animal behaviorists, instinct always trumps emotion. The second part of Morgan's statement, however, should be remembered. He wrote, "by no means exclude . . . higher mental processes if we already have independent evidence of their occurrence" (ibid.). In other words, even Morgan granted that if we have evidence that animals do possess higher consciousness, then we cannot discount the fact that they are using it. The question is: How does one get that evidence, if the assumption always begins with the negative—especially when animal thoughts and emotions are not that easy to observe, and thus must be inferred from observable behavior?

Animal Behavior Studies and Reductionism

As discussed in this chapter, modern ethology today owes much to the history of the field. But many practitioners today have moved far away from the **reductionism** of these early approaches, and the old school of behaviorism is largely dead. Reductionism refers to the ways in which complex phenomena can be reduced to the simplest of biological explanations. Biologist Lynda Birke (1994), for example, has written extensively about these issues and how biological determinism is at the root of much of our ill treatment of animals today. If animals are not much more than machines, as suggested by Descartes in the eighteenth century, and can only respond to environmental stimuli, as in much of behaviorism, or are hardwired to behave in certain ways, then they can certainly never be given any of the rights that we feel are owed to us as humans—the right to not be harmed, for example. She also points out that women were once defined on the basis of their biology alone—because women gave birth, for example, they were not supposed to be educated because too much learning could draw blood from the uterus and make childbearing impossible. Although we recognize today how ridiculous that sounds, it was definitely the case that women's social and political fates were often tied to their biology. Thanks to feminism, that type of biological determinism is rarely used to constrain women today, but it is definitely still used for animals that are thought of by science not as individuals but as representatives of their species.

Environmental sociologist Eileen Crist (1999) points out how the language that scientists use in discussing animals plays a role in reducing them to their biology and nothing more. She notes that language is not neutral—it plays a formative role in how animals are depicted. Crist examined animal behavior studies that were faithfully done yet radically different in the worlds of the animals that they reveal. Science, she shows, does not always mean objectivity. For instance, she shows that when we use technical language to render animal life ("attachment" instead of "love," for example), it helps conceptualize animals as objects; using ordinary language, on the other hand, reflects a regard for animals as subjects. In evaluating the work of comparative psychologists, we can also question the conditions under which the research was conducted. Animals can hardly act naturally when caged and subject to constant probing, prodding, and medication.

Anthropologist Barbara Noske (1989, 1997) writes that scientific understandings of animals are almost entirely reductionist and objectifying even though scientific understandings of humans have moved quite far away from this. The problem for Noske is that biology and ethology, both built on reductionist bases, have become *the* sciences for understanding animals; people, on the other hand, are lucky to now have cultural anthropology to define them. Animals, however, are associated fully with biological and genetic explanations that turn them into objects. The sciences of animal behavior support and reinforce this view by rejecting, or claiming as inaccessible, a subjective dimension in animal life. Whether animal behavior is understood as reinforced conditioned responses, the results of genetic hardwiring, or some other mechanism, the result is the same: animals are little more than machines and have no internal lives.

Some biologists still hold that there is an essential difference between humans (who think and feel) and animals (that, presumably, cannot). They would say that animals are governed entirely by their genes or automatically respond to their environments, with little room for individual agency or self-determination; other scientists, including many behaviorists, however, do make room for individual agency. But today, as we shall see, a whole new generation of ethologists is moving well beyond this approach. Those authors are moving away from such generalizing statements as "chimpanzees are patriarchal," "dogs are hierarchical," or "rabbits are territorial" and focusing instead on the richness and variety of attributes that may be found within an animal species. This variety, in turn, is part of what we perceive of as "personality" in an animal—and it is something people who live and work with animals may take for granted.

Scientists are often unable to see what to most of us is common sense: animals having feelings. As philosopher Jeffrey Masson (1995) writes, we should not be limited by humans as the reference point; but we need not also assume that what we humans have—emotions and a mind—animals cannot have. Scientists are reluctant to accept anecdotal evidence, even though, as Jane Goodall once said, anecdotes are careful descriptions of unusual events. Evidence based on research or experiences with animals in captivity—such as pets—is discounted because the setting is unnatural (even though the animals studied for years in laboratories certainly live in artificial environments). But that does not mean their expressions of emotion or social behavior are any less valid. Emotions and thoughts can be complex or difficult to interpret, but that does not mean they do not exist. Just because we cannot get inside the head of another human does not mean that we cannot get a good idea of how that person feels based on how they express themselves, and we certainly do not assume because we do not know what is going on in other people's heads that they do not have feelings or thoughts. The same common sense should hold for animals as well.

The Rise of Modern Ethology

Ethology, or the scientific study of animal behavior, is generally thought to have really begun with the work of Konrad Lorenz and Nikolaas Tinbergen in the 1930s. Unlike comparative psychologists, whose field emerged in order to better understand human psychology and whose research was largely conducted in laboratories, ethology (which was once considered a subfield of zoology) is focused entirely on animals. Historically, it has been conducted "in the field" with wild animals and has been much less interested in comparisons to human behavior. In addition, as we mentioned before, comparative psychology is more likely to look at environmental causes of behavior although ethology has been much more likely to focus on genetics as the cause. Today, however, the fields do overlap, even though some of these distinctions still remain today. Much of modern research—in ethology and comparative psychology—focuses on animal cognition and animal emotion, as well as the question of animal self-awareness.

Modern ethology has been heavily shaped by the work of primatologist Jane Goodall, who has been studying a population of chimpanzees in Africa since 1960 (Goodall 1971, 1986, 1990). In his introduction to Jane Goodall's *In the Shadow of Man*, biologist Stephen Jay Gould discusses Goodall's groundbreaking descriptions of the chimpanzees in Tanzania's Gombe National Park as individuals (Mike, Flo) rather than as types, as was common until that time ("alpha male" or "lactating female"). Goodall dared to name the chimps she studied, a practice unheard of in early primate research. The "alpha male" became Mike; the "lactating female" became Flo. Gould writes:

> Thus we learn that an alpha male is not always the biggest or strongest, but may win his rank by peculiar cleverness (Mike, who bluffed his way to the top by banging empty kerosene cans together) or by subtle alliances (Goblin, the present incumbent, who, although smaller than average, knows how to play that oldest game of imperialism—"divide and conquer"). Or we discover

that the main outlines of the history of the Gombe chimps during twenty-five years are not set by general principles of chimpness, but by particular events and historical peculiarities.
(1988:VI)

Goodall also recognized the importance of individual (chimpanzee) agency and historical specificity in the creation of chimp social structure. Goodall recognized that the defining feature of chimpanzee societies is not always a generalizable "chimp territorial behavior," but can be other factors, such as the splitting of one main group into two rival factions. Goodall was one of the first scientists to challenge the homogenizing and stereotyping discourses that we use to describe animals: All chimps are the same, all bats are the same.

Since that time, Goodall has written at length of her experiences (Goodall 1999, 2000), attempting to gain legitimacy and respect from her colleagues

---

BOX 17.1

JANE GOODALL

Jane Goodall was born in London in 1934 and began working with anthropologist Louis Leakey in 1958 and 1960. Leakey sent her to Tanzania to begin the first long-term study of chimpanzee behavior ever conducted. She interrupted her work in Africa to obtain a PhD in ethology in 1965 at Cambridge University, and returned to Africa where she has continued her work until the present day. Goodall's research has primarily been done in the Gombe Stream National Park where the chimpanzees there are protected by Tanzanian law—thanks to Goodall's efforts. Her studies of chimpanzee behavior challenged many long-held assumptions—for instance, she showed that chimpanzees can not only use tools but make them as well, that they self-medicate when ill, and that they engage in cooperative hunting of monkeys and other animals. Equally important, her methods influenced and changed modern ethology—until Goodall's time, no one had engaged in such long-term systematic study of wild animals, and no one had established the kind of relationships that Goodall had. Today, Goodall's work at Gombe is continued by scientists working at the Gombe Stream Research Center. Goodall spends her time traveling the world, speaking on conservation issues and advocating on behalf of chimpanzees, and overseeing the Jane Goodall Institute, which supports the research in Gombe and its youth program, Roots & Shoots, undertakes animal welfare and conservation projects around the world.

Figure 17.1. Dr. Jane Goodall visits with the Gombe chimpanzee Freud. (Photograph courtesy of Michael Neugebauer and the Jane Goodall Institute.)

who saw her work upon her return from the field as unscientific. "Objec-tive" scientists do not name the animals that they study (they number them), much less write that they have emotions, personalities, or intelligence, as Goodall did. Philosopher Jeffrey Masson discusses the fact that in scientific fields, such as Goodall's,

> [w]omen have been deemed especially prone to empathy, hence anthropo-morphic error and contamination. Long considered inferior to men precisely on the ground that they feel too much, women were thought to over-identify with the animals they studied.
> (MASSON 1995:33)

A woman's empathy toward animals, then, leads to her being less qualified to understand them!

Anthropomorphism

Until recently, if a scientist attempted to describe the behavior of an ani-mal with terms such as "sadness," "jealousy," "grief," or "joy," they would quickly be accused of that most dreaded of approaches: anthropomorphism. There are valid reasons for this attitude. No human can ever truly get inside the brain of an animal—without dissecting it—and animals cannot answer questions if we ask them how they feel, which forces us to interpret their behaviors. But the belief that animals have no intelligence, emotions, or individual personalities has certainly benefited those who exploit animals. Granting animals the ability to think, to reason, and to feel opens up a Pandora's box of issues regarding how we, as a society, should treat animals. For instance, recent research has demonstrated that not only do birds and mammals feel pain, but fish do too. Knowing that animals feel pain—even animals such as fish with their unmovable faces and seeming lack of emotion—creates some unsettling implications for people who do not want to cause suffering yet continue to, for example, eat meat or fish.

Environmental sociologist Eileen Crist goes so far as to defend anthropomorphism, which she defines as the representation of animal life in the language of the "lifeworld" (i.e., a world populated by actors whose lives are filled with action and meaning):

> In rendering action meaningful and authored, animals emerge as subjects. In turn, the portrayal of animals as subjects allows the existence of mental life to supervene with forcefulness and credibility.
> (CRIST 1999:4)

It is true that words such as love, hate, grief, and jealousy are "human" words in the sense that humans created them. But just because the words that we use to describe these emotions come from humans does not mean that we somehow own them. Even Charles Darwin, in his landmark book *The Expression of Emotions in Man and Animals* (1872), recognized that human emotions and intellectual abilities are shared with nonhuman animals, and dared use words such as "love" when describing animal behaviors and feelings. We noted already **Morgan's canon**, also known as **cognitive parsimony**, which warns not to interpret a behavior as the outcome of a higher mental capacity when it can be easily described as something simpler—such as a conditioned response. But we can also evoke Darwin with what is known as **evolutionary parsimony**, meaning that if closely related species, such as humans and chimpanzees, share behaviors, then their underlying mental processes are probably related as well.

Today, many scientists use what one might call "careful anthropomorphism," in which they try to interpret animal's mental and emotional states by using what we know of our own mental and emotional states. This then forms the bases for further study and analysis—not the endpoint.

Not everyone is on board with this move, however. Physiologist John Kennedy is highly critical of the rise of anthropomorphism in ethology; for Kennedy, "there has never been any direct evidence" that animals "are like us in having feelings and purposes and acting upon them" (Kennedy 1992:1). What would constitute evidence for Kennedy is unclear because he himself states that we only know of humans' feelings because we ourselves experience them. Because we cannot experience the feelings of animals, then we can never say that they exist. Yet it is just as unverifiable to go with Kennedy's, and the early behaviorists,' positions: that animals are just biological machines. For those who have closely lived and worked with animals, there is no question—verifiable or not—that animals share many of the same emotions that we do, although there is of course a question of whether or not we are correct in interpreting what those emotions are. But to neglect to do so would be as unthinkable as to have an anthropologist claim that the people with whom he is working do not have romantic love because marriages are arranged, courting does not occur, and he does not know how to interpret the people's songs, poetry, and gossip.

Animal Intelligence

As we mentioned before, the primary theoretical framework used to understand animal intelligence was behaviorism: One could observe the behavior of an animal, but one could not infer the motivations underlying that behavior. Starting in the 1960s, however, some scientists began to try to understand animal behavior by taking what is known about human mental processes and looking for evidence of comparable processes in other species. Beginning in the 1970s, Donald Griffin—the modern founder of **cognitive ethology**—was one of the first scientists to posit that animals are capable not only of cognition, but also of thinking, consciousness, and conscious (or deliberate) behavior. Using evidence from the field and from labs, Griffin found that animal behavior is flexible and variable, that it can change when the environment changes. (This is called **adaptive behavior**, and for a long time researchers believed only humans were capable of it.) Griffin's work (1976, 2001) shows that even what seems to be routine behavior is varied as the situation arises, and is not rigid and stereotypical, as has previously been thought.

More recently, researchers have discovered that animals are capable of tasks as complex as tool making and tool use, category formation, spatial memory, deceptive behavior, social sophistication, cognitive adaptability, symbolism, the communication of abstract feelings, and "if-then" or "purposive" thinking and behavior. For instance, researchers have documented cases in which an animal would remember where a morsel of food was placed in the past and would seek it out without being given a sensory cue. The concept that an animal might be able to remember something from the past and use that information in the present, even in the absence of a stimulus, can seem quite novel to people who believe that animals cannot think.

Animals can generalize, as can humans, but they can also compartmentalize. Ethologist John Fentress has written about a wolf named Lupey that he lived with and studied in the 1960s. Fentress discussed (2000) how Lupey could easily distinguish between people who accidentally did something wrong, and people who meant him harm. When Fentress accidentally stepped on some chicken bones that the wolf had buried, Lupey responded with growls and snarls, but quickly forgave him when he realized that Fentress had realized his mistake. On the other hand, when a caretaker hit Lupey over the head with a shovel, not only did Lupey never forgive this particular man, he also never generalized this man's offense to others; he treated the other humans as he always did, but shut the offender out of his life forever.

Researchers have also discovered that animals are able understand abstract categories. Chimpanzees, for instance, can demonstrate the relationship between two sets of objects, such as two red balls and a red and green ball, or, even more amazingly, understand that a key is to a lock as a can opener is to a can of paint, rather than as a paintbrush is to a can of paint.

All great apes make tools and use them to acquire foods; chimpanzees cooperatively hunt, using complicated strategies requiring cooperation, influence, and rank. Chimpanzees can lie, and great apes appear to have a sense of humor. They understand and can use symbols, and some may understand numbers as well. Orangutans develop new cultural practices such as using leaves as rain coverings, pillows, napkins, and even as a form of "gloves," lining their hands with them to protect them as they climb trees.

Birds such as corvids (crows and ravens) and psittacines (parrots) are social animals, have long developmental periods, and possess large forebrains—all signs of greater intelligence. Tests show that some birds can count, that they can get around detours creatively to get to food, and that they can use and make tools to get food. They can also teach each other these behaviors, use teamwork in hunting, can learn by imitation, and can improvise as the need arises. Some crows can also lie and steal, as do scrub jays. Only parrots, however, can speak in human language, and the research being done on this remarkable ability is quite astounding (and will be discussed later in this chapter).

Figure 17.2. Brownie enjoys playing tug of war with her squeaky pink bone. (Photograph courtesy of the author.)

Elephants, like humans and a few other animal species, must learn behavior as they grow up. Like birds, which also live a very long time, elephants have a lengthy developmental period during which they learn what they need to know as adults. Elephants can use tools (which they hold with their trunk), and some Asian elephants have been taught to paint realistic portraits of natural objects as well as other elephants. Elephants can perform complex tasks and can independently problem-solve; and African elephants know how to self-medicate using the same wild plants that African peoples use when they are sick. Perhaps most extraordinary are elephants' memories and sense of hearing, both of which outperform humans'. Elephants can remember places that they visited decades ago, and can hear and interpret the calls made by family members up to at least a mile away (including the screams of dying elephants). They can even recognize the urine signatures of family members decades after they have been separated from them. Similarly, sheep, a species of animal not known for their intelligence, recognize faces of acquaintances and family members and can remember them for years afterward. They can also distinguish between happy and unhappy faces, and they are calmed by being surrounded by familiar faces.

Researchers have been attempting to study the intelligence of dolphins for the last three decades, and have found that these animals may be the second smartest on the planet—next to humans. Not only do dolphins have very large brains, they also can solve a huge number of problems, can understand and respond to human language, can understand the concepts "more" and "less," and can even use a keyboard with their noses. In addition to being able to solve problems, dolphins have also demonstrated the ability to plan for the future; they can use tools (held with their mouths or wrapped around their nose) to modify their environment.

Of course, most of the familiar examples of animal intelligence are taken from the "smart" animals: apes, dolphins, corvids, psittacines, and dogs. But what about other animals? Even chickens, the world's most consumed animal and, not coincidentally, one of the most denigrated, demonstrate many varieties of intelligence. Karen Davis (1995), who has lived with and studied chickens for years, has written extensively on the intelligence of chickens. Their alarm calls, for example, differentiate between different kinds of predators (and chickens that hear the calls will respond accordingly), and their memories and ability to identify dozens of individuals are well developed. They can communicate lots of information about food sources and other important features of their environment with dozens of vocalizations; they can deceive others, and can anticipate and plan for the future.

Animal Emotions

Philosophers, scientists, and people who live with animals have had different things to say on the matter of animal emotions for thousands of years. But the core question has proven hard to answer because we can neither directly observe animal emotions (and behaviorism says that the only acceptable data can be that which is directly observed) nor obtain spoken answers to our questions.

Even so, scientists studying the emotional lives of animals have also made startling progress. Today, most ethologists grant that animals share with humans the primary human emotions of happiness, fear, anger, surprise, and sadness. And many go further and see animals as sharing with humans many secondary emotions such as regret, longing, or jealousy. One early discovery was the understanding that animals *anticipate*, and that they then demonstrate disappointment when what they anticipated does not come to pass. To animal lovers, this sounds obvious—a dog gets excited when the human puts on her running shoes, and then said dog gets dejected when said human walks out the door *alone*. One recent study demonstrated that dogs that are anxious because of being constantly left alone demonstrate pessimistic behavior; if they were humans, we would say that they see life as if it were a glass that is "half empty," as opposed to more positive dogs, that see the glass as "half full." But the fact that animals anticipate something means that they also have memory. The fact that they have memory, in turn, suggests that they can experience strong feelings—surprise, anger, and disappointment—when they are denied what they expected.

Other new discoveries include the fact that animals can feel pain—and can also carry the emotional scars associated with that pain. Anyone who has worked with animals knows they display evidence of feeling pain, of course—consider the yowl of a cat whose tail has been stepped on, the grunt (or nip) of a horse whose girth has been tightened too quickly, or the shriek of a rabbit that has been grabbed too tightly. People who work with abused animals—chimpanzees whose mothers were slaughtered in the bushmeat trade, former circus elephants, once-abused or neglected dogs, and former laboratory rabbits—can also testify to the fact that animals will continue to suffer from that emotional trauma for years afterward.

Grief is another emotion now attributed to animals, thanks especially to the studies of elephants in the wild and in captivity. Using heart-wrenching detail, Cynthia Moss (2000) has described the reactions of elephants when their relatives and friends die—from frantic denial (attempting to lift a dead elephant up to a standing position, even attempting to place food in her mouth) to partial burial of the body to a night-long vigil. Elephants even recognize the bones of other elephants, and react accordingly when finding them. Jane Goodall (1971) has written movingly of the grief of the chimpanzee Olly, who continued to care for her four-week-old baby for days after she died by carrying her, grooming her, and even attempting to nurse her before finally abandoning the body; and of Flint, who died of grief three and a half weeks after his mother Flo died. Koko, the signing gorilla, grieved deeply and publically over the loss of her kitten, All Ball. Whales, dolphins, dogs, monkeys, rabbits, and other animals also appear to grieve over the loss of their loved ones. There are also a number of anecdotal reports of animals, most from the nineteenth century, that committed suicide—often after the death of a mate and sometimes after the death of a beloved person.

What about joy, love, and friendship? Even researchers who grant animals "negative" emotions are cautious about admitting that animals have a capacity for happier feelings. But for many researchers today, as well as for those who live and work with animals, it is clear that animals do feel pleasure in all of its variations. Those who study elephants have reported on the joyous reunions that elephants have with other elephants after a long time apart. For instance, in 1999 an elephant named Shirley was relocated from her home at a zoo in Louisiana to the Elephant Sanctuary in Tennessee. There she met Jenny who, upon seeing Shirley, began to roar and beat against the gate separating them, eventually bending the metal bars in her desire to reach Shirley. When they were finally allowed near each other, they were inseparable, rarely leaving the other alone. It turns out that both elephants had once performed in the same circus—some twenty-two years before—and had never forgotten the other.

Some forms of play can be explained as simple evolutionary behavior (e.g., kittens play with yarn or chase each other to practice hunting). Other forms of play, however, are less useful in survival and cannot be so easily dismissed. Dolphins that surf, rabbits that throw toilet paper tubes into the air, animals that tease humans, monkeys that tease tigers, and parrots who creatively turn their toys upside down are not practicing survival skills, yet the animals clearly enjoy them. The babies of countless species play: piglets, lambs, goats, kittens, bunnies—all love to run, scamper, pivot, and dance. Even giant octopuses are now known to play, spraying water at targets, and rats making chirping sounds when they play, which some scientists interpret as laughter.

Some researchers now believe that animals experience even more complex emotions, such as altruism, compassion, heroism, cooperation, embarrassment, pride, and more. People who live with cats often report that if their cat, usually graceful, accidentally falls or stumbles, he or she will look around to see if anyone witnessed their clumsiness, and will immediately act as if nothing has happened; Jane Goodall has reported the same situation with chimpanzees. That sounds like just what we do if we accidentally stumble in public! Ethologist George Page (1999) says that embarrassment is an important social regulator in human communities, so it should not surprise us that it would be found in other animal species.

Many elephant observers have written of elephants joining together to help fallen comrades stand up. Elephants not only "bury" a dead friend with palm fronds and branches, they also risk their lives to defend family members, babysit for related elephants, assist ill or injured elephants, and demonstrate the symptoms of post-traumatic stress disorder after watching, for example, a related elephant being killed. Female rats will put the young of other female rats into a nest if they fall out. Baby orcas are playful and curious; their mothers often must put themselves at risk to protect their young from human-caused dangers such as moving too close to the motor of a boat. Many stories exist of dogs that wake up their humans when a fire starts, yet parrots, pigs, and even a hamster have done the same thing. Plutarch, writing in the first century, noted the depth of the relationships between chickens and their offspring:

> We daily behold hens, how they cherish their chickens, taking some of them under their spread wings, suffering others of them to run upon their backs, and taking them in again, with a voice expressing kindness and joy. When themselves are concerned, they fly from dogs and serpents; but to defend their chickens, they will venture beyond their strength and fight.
> (PLUTARCH 1874)

Finally, a new study was released in 2009 (Fraser and Bugnyar) that shows that ravens console each other after fights, and help to relieve the

Figure 17.3. "Dolphin" (Cartoon by Dan Piraro, Courtesy of http://www.bizarro.com.)

distress of their friends and kin. Biologist Marc Bekoff (2004), in fact, suggests that animals have a sense of justice, morality, and fair play, and that these concepts as they exist in humans must have evolved from animals. Not only do many animals cooperate with each other, but when an animal "cheats" by aggressively biting another animal during play, for example, or failing to notify others of a food source, then the other animals generally respond angrily, demonstrating that many animals value fair play and moral behavior. Irene Pepperberg (1999), for example, has reported that if she tricked Alex the parrot during her research, Alex would respond with what she interpreted as rage: slitted eyes, puffed feathers, opened wings, and a lowered head.

Animal Language

Anthropologists and linguists have long held that one of the defining differences between humans and nonhuman animals is the use of symbolic language. Although scholars readily acknowledge the complex systems of communication of many animals, human language uses arbitrary symbols to communicate abstract concepts to others who share the language. Most scholars of human language assert that the developments that allowed for the use of language in humans arose long after humans split off from their nearest primate relatives, and are thus not at all shared by nonhuman animals.

Animal oral communication systems are known as "**call systems**." Call systems consist of a limited number of sounds that are thought to be produced in response to specific stimuli (e.g., food or danger). In addition, animal communication uses body language, facial expressions (in mammals), scent, and other cues to communicate information to other animals. Social species such as primates, cetaceans (whales and dolphins), and many birds have extremely complex communication systems.

But can animals have *true* language? According to linguists, all human languages share a number of features that are not shared with other animals, and these include multimedia potential, cultural transmission, arbitrariness, creativity, and displacement: the ability to talk about objects or events that are remote in time and space. Only humans, it has been long thought, have languages with these qualities. However, new research in the communication systems of many nonhuman animals, known as **animal language research** (or ape language research when apes are studied), is challenging these basic ideas.

For instance, linguist Con Slobodchikoff has found that prairie dogs, rodents living in many southwest and Plains states, may have language. His research (Slobodchikoff, Perla, and Verdolin 2009) shows that prairie dog colonies have a communication system that includes nouns, verbs, and adjectives. They can tell one another what kind of predator is approaching—man, hawk, coyote, dog—and they can tell each other how fast it is moving. They can say whether a human is carrying a gun or not. They can also identify individual coyotes and tell one another which one is coming. They can tell the other prairie dogs that the approaching coyote is the one who likes to walk straight through the colony and then suddenly lunge at a prairie dog that has gotten too far away from the entrance to his burrow, or the one that likes to lie patiently by the side of a hole for an hour and wait for his dinner to appear. They can tell each other about what color clothing a person is wearing, as well as something about his size and shape. Slobodchikoff also found evidence that prairie dogs are not born knowing the calls, the way a baby is born knowing how to cry. They have to learn them. He bases this on the fact that the different prairie dog colonies around Flagstaff, Arizona, all have different dialects.

Slobodchokoff's prairie dog research not only demonstrates the complexity of prairie dog communication, but implies that prairie dogs can discuss things that are not present; in other words that their language allows for displacement. The idea that prairie dogs can discuss a tall human dressed in blue, in the absence of said human, is an extraordinary idea and challenges deeply held notions about animals' cognitive abilities. Another incredible discovery is that prairie dogs can create new words to refer to new things or creatures in their environment, in other words, that their language includes productivity.

Research with dogs, too, demonstrates previously unknown complexity with respect to language and reasoning skills. A German border collie named Rico, for example, can identify and respond to over 200 words, can pull specific play toys out of a pile when given the right command (blue snake), and can use the process of elimination to deduce that a new sound from a person must be the name of a previously unseen toy. This is known as

fast mapping, the ability to learn new words. Another border collie named Betsy, at the time of this writing, had a vocabulary of 340 words that she could recognize and respond to. She can also link photographs with the objects they represent, a skill not thought possible in dogs.

Parrots, too, have demonstrated a startling ability to use human language, and are the only animals to actually speak in human languages such as English. For instance, Alex, an African Grey parrot that died in 2007, had been trained by Irene Pepperberg (1999) to speak and comprehend English. Alex knew over a hundred words, and could identify objects by color, quantity, shape, and texture. He could carry on a conversation with Dr. Pepperberg and could clearly state his thoughts and feelings. (Pepperberg, in fact, originally taught him language because she wanted to ask him about how he saw the world.) Another well-known bird is N'Kisi, another African Grey, that works with a researcher named Aimee Morgana. N'Kisi is thought to be the only animal capable of conjugating verbs and can accurately use words to demonstrate the past and the future. He can talk about experiences that he had in the past, showing not only memory but also the ability to talk about issues that are not in the here and now. N'Kisi can express concern for others, can joke, can lie, and can pretend. At the time of this writing, Morgana has been teaching N'Kisi how to spell, and hopes that he can someday learn to read.

Other animals that have been subjects of language experiments include dolphins. For instance, Akeakamai was a bottlenose dolphin that participated in a long-term language study at a marine mammal laboratory in Hawaii until her death in 2003. She was able to understand individual words and basic sentences, and could respond appropriately. Bottlenose dolphins, such as humans, elephants, and some birds, are known as **"vocal learners,"** which means that they can imitate the complex vocalizations of others, give themselves "signature whistles," and recognize and respond to the whistles of related dolphins and friends, even when the whistles are played back to them from underwater speakers.

Perhaps the best-known research into nonhuman language capacities has been done with nonhuman primates. Researchers have been teaching primates human languages for years. Most of those have been taught sign language, and can understand signs and human speech, and respond to either human speech or signs with their own signs. The most well-known of these primates include Washoe, a chimpanzee, that eventually acquired a vocabulary of over 100 ASL signs; Lucy, the second chimp to learn ASL, that lived in a foster family until she was introduced to the wild, where she was killed by poachers; Koko, a gorilla living at a research facility called the Gorilla

Foundation in California, that regularly uses 400 ASL signs and has used 700 signs at least once; and Kanzi, a bonobo, that can communicate with researchers using a lexigram, which is a keyboard that utilizes symbols for words.

These apes have demonstrated, according to those who work with them, a wide variety of language abilities, including deception, creativity and productivity, displacement, multimedia potential, and cultural transmission. For instance, Koko often demonstrates creativity by developing new signs to communicate new thoughts. For example, nobody taught Koko the word for "ring," therefore to refer to it she combined the words "finger" and "bracelet," hence "finger-bracelet." Koko can also use language to lie. In one instance, Koko ripped a sink off of her wall and blamed her kitten, All Ball, telling Penny Patterson, the researcher who works with Koko (Patterson and Linden 1981), that All Ball did it. Koko can also tell jokes, displaying humor. Michael, a gorilla that lived with Koko for several years, attempted to give researchers a description of his mother being shot as he watched, which demonstrates displacement. Grammar is perhaps the area that is least developed by primates. But even here, there are some indications that primates can use rudimentary grammar. Roger Fouts (1997), who worked with Washoe, has pointed out that Washoe recognized the difference between "me tickle you" and "you tickle me," demonstrating a grasp of the difference between subject and object.

Critics argue that what these apes are doing does not really constitute language. Instead, some feel that the animals are very well-trained and have been conditioned to give certain signs in the expectation of certain rewards; that the apes are not really using symbolic language. In addition, no single ape appears to be able to meet every criteria of human language, although putting the skills of all of the signing apes together would meet those criteria. On the other hand, even the linguist William Stokoe, the greatest authority on **American sign language** today, has himself acknowledged the veracity of animal language studies, noting, for example, that much of Washoe's signs were spontaneous, and not made in response to human words.

Kanzi (see box 17.2) is one of the most well-known animals in ape language research today. Kanzi's ability to communicate with humans by understanding human language and responding via the use of a lexigram is astounding to those who have seen it. Perhaps the most astonishing item to come out of Savage-Rumbaugh's research with Kanzi and other bonobos at the Great Ape Trust is a remarkable document published in the *Journal of Applied Animal Welfare Science* in 2007 called "Welfare of Apes in Captive Environments: Comments on, and by, a Specific Group of Apes." Coauthored by

BOX 17.2

FAMOUS ANIMALS: KANZI

Kanzi is a bonobo (or dwarf chimpanzee) that has been studied by primatologist Sue Savage-Rumbaugh for 30 years and is well-known for his comprehension of English. Kanzi understands spoken English and uses lexigrams—symbols placed on a special keyboard that represent words—to communicate with humans. Kanzi can use more than three hundred words and can accurately respond to thousands of words used in sentences. He can point to the lexigram symbol for each word that is spoken to him, and can respond appropriately to commands involving objects and actions. Even more astonishingly, according to Savage-Rumbaugh, Kanzi vocalizes to his sister Panbanisha "words" that refer to objects. Even though he cannot speak like a human, he evidently translates English words into bonobo "language" that he can then use to communicate with others of his species. He lives with other bonobos, including his sister and his adopted mother, Matata, at Great Ape Trust, and continues to work with Savage-Rumbaugh and other researchers. He and the other bonobos at the Trust can acquire food for themselves from vending machines, prepare their food in a microwave, and watch movies on television that they have selected themselves. Kanzi has been featured in numerous documentaries on animal intelligence and animal language.

Savage-Rumbaugh and bonobos Kanzi, Panbisha, and Nyota, the article argues for better treatment of apes in captivity, and includes, based on Savage-Rumbaugh's questioning of the three bonobos, their list of what the bonobos ask for, including the ability to travel from place to place, to have fresh food of their own choosing, to maintain life long contact with family and friends, and to teach their own offspring. Part of what makes this article so extraordinary is that people—even those acting on their behalf—generally never think to actually ask animals what they want. Savage-Rumbaugh not only thought to ask, but because of her language work with these animals, was also able to elicit clear and coherent answers. (Savage-Rumbaugh's work with Kanzi has been fictionalized in writer Sara Gruen's 2010 novel *Ape House*.)

Out of all of the research done in recent years to assess and describe the intellectual and emotional capabilities of nonhuman animals, animal language research has perhaps the most wide-ranging implications. It is one thing to know that animals can think, can plan, can feel, and can suffer. Just

Figure 17.4. Kanzi and Dr. Sue Savage-Rumbaugh communicate with lexigram panels. (Photograph courtesy of the Great Ape Trust.)

knowing these basic facts can serve as the basis for reevaluating our treatment of other animals. Many people have heard about some of this research, and may even know on a gut level that animals can and do think and feel. But it is still easy to ignore. Why? Because animals have for so long been thought to have no language; they cannot tell us how they feel, so we do not have to really pay attention. But watch a YouTube video of Dr. Savage-Rumbaugh talking to Kanzi, and it is difficult to ignore the uneasy truth: Some animals can actually tell us how they feel.

The Animal Self

Dolphins and apes are two of a handful of animals that are thought to have self-awareness. In other words, they can see themselves as an object. By seeing ourselves in this way, sociologists say that we can then engage with others in a complex social environment. Sociologists have also assumed that only humans have the ability to see ourselves in this way. But how do we test this?

The best known research technique in this area is the **mirror test**, where an animal such as a chimpanzee or a dolphin is marked with paint, and then asked to look at themselves in the mirror. Unlike other animals that do not recognize that the image in the mirror is them, dolphins, great apes, elephants, and some birds can all recognize that they are looking at themselves and will look closely at the marking and often try to get the paint off of their faces. Dolphins that watch themselves in the mirror also seem to get a great deal of pleasure out of playing new tricks in front of the mirror; they appear to like to watch themselves do crazy stunts. The fact that many animals "fail" the mirror test is controversial among some researchers because it is entirely focused on vision, although many species, such as dogs, rely more heavily on other senses such as the sense of smell.

Related to the concept of self-recognition is the notion of **metacognition**. This refers to the ability of an animal to think about its own thoughts. One way that this has been tested has been to see whether some animals know, like humans know, whether they know something or not. Tests of animals' knowledge and problem-solving skills typically evaluate their ability to answer questions or solve puzzles; metacognition tests, on the other hand, evaluate whether an animal is able to respond in the negative—to say, in essence, that they do not know something; this is known as an **uncertainty response**. Monkeys, dolphins, and apes have all proven that they know when they do not know something; in addition, parrots will mumble in private when they have learned a new word but have not yet used it in front of humans.

Another example of consciousness or a sense of self is possessing a **theory of mind**, the ability to attribute mental states—beliefs, intents, desires, pretending, knowledge, etc.—to oneself and others and to understand that others have beliefs, desires, and intentions that are different from one's own. Deception, for example, requires a theory of mind, as does pointing or gazing at something in order to draw someone's attention to it. Dolphins and parrots, for example, can point with their bodies, and apes and some birds can lie. Dogs can and do follow the gaze of their owners in order to understand what that person is looking at and can interpret facial expressions and act accordingly. Primatologist Frans Plooij (2000) wrote about how he deceived a chimpanzee named Pom; she was grooming him and he wanted her to stop, so he pretended that he saw something interesting a ways off, and she followed his gaze and walked off to see what was so interesting. When she realized she had been deceived, she walked back to Plooij and slapped him across the face.

Finally, using a sociological approach, some scholars have been attempting to understand whether or not animals have a concept of self by studying the relationships between animals and humans. Through symbolic interactionism, for example, humans develop our own self in part through interacting with others. Looking at animals that interact frequently with humans, such as dogs and cats, some researchers have demonstrated that these animals do have a sense of self, precisely because they are able to understand and react to the needs of the human partner. In other words, they are able to empathize, and to take on the role of the other—a key component of selfhood. This approach also works when looking at how animals interact with other animals. They do not just behave according to instinct; they change their behaviors as the specifics of the situation demand, indicating not just behavioral flexibility but consciousness and an ability to monitor and update one's own performance.

All of this research demonstrates that many animals, through their ability to express emotions, communicate effectively with others, take on different roles, and express their own interests, have a concept of self.

If what modern researchers are finding out about nonhuman animals—that many possess self-awareness, that many can understand English and respond to it in kind, and that many share with humans complex emotions such as love, grief, hate, and jealousy—what then? If these animals share so many mental and emotional capacities with humans, what does that mean about our treatment of them? Should we continue to catch them, cage them, and display them in circuses and zoos for our pleasure? Should we continue to raise them in barbaric conditions only so that we can kill and eat them? What, if any, are our obligations to animals that may possess many of the same thoughts as we do? In chapter 18, we will address this issue.

Suggested Additional Readings

Bekoff, M. and J. A. Byers, eds. 1998. *Animal Play: Evolutionary, Comparative, and Ecological Perspectives*. Cambridge: Cambridge University Press.

Bekoff, Marc, ed. 2000. *The Smile of a Dolphin: Remarkable Accounts of Animal Emotion*. New York: Discovery Channel.

Bekoff, Marc. 2002. *Minding Animals: Awareness, Emotions and Heart*. Oxford: Oxford University Press.

Blum, Deborah. 2002. *Love at Goon Park: Harry Harlow and the Science of Affection*. Cambridge, MA: Perseus.

Byrne, R. 1995. *The Thinking Ape: Evolutionary Origins of Intelligence*. Oxford: Oxford University Press.

Crist, Eileen. 1999. *Images of Animals: Anthropomorphism and Animal Mind*. Philadelphia: Temple University Press.

Darwin, C. 1872. *The Expression of the Emotions in Man and Animals*. New York: Appleton and Company.

Dawkins, M. S. 1993. *Through Our Eyes Only? The Search for Animal Consciousness*. New York: W. H. Freeman.

De Waal, F. M. 1996. *Good Natured, The Origins of Right and Wrong in Humans and Animals*. Cambridge, MA: Harvard University Press.

Goodall, Jane. 1971. *In the Shadow of Man*. London: Collins.

Griffin, D. R. 1992. *Animal Minds*. Chicago: University of Chicago Press.

Masson, Jeffrey Moussaieff and Susan McCarthy. 1995. *When Elephants Weep: The Emotional Lives of Animals*. New York: Delta Trade Paperbacks.

Mitchell, Robert W., Nicholas S. Thompson, and H. Lyn Miles, eds. 1997. *Anthropomorphism, Anecdotes, and Animals*. Albany: State University of New York Press.

Parker, S. T., R. W Mitchell, and M. L. Boccia, eds. 1994. *Self-Awareness in Animals and Humans, Developmental Perspectives*. New York: Cambridge University Press.

Savage-Rumbaugh, S. and R. Lewin 1994. *Kanzi, the Ape at the Brink of the Human Mind*. New York: Wiley.

Sutherland, Amy. 2009. *Kicked, Bitten, and Scratched: What Shamu Taught Me About Life, Love, and Marriage*. New York: Random House.

Suggested Films

The Ape: So Human. DVD. New York: Films for the Humanities and Sciences, 1998

A Conversation with Koko. VHS. Directed by Nigel Cole. New York: Thirteen/WNET New York, 1999.

Inside the Animal Mind. DVD. Directed by Nigel Cole. Thirteen/WNET New York, 2002.

Jane Goodall: Reason for Hope. VHS. Directed by Emily Goldberg. St. Paul, MN: KTCA-TV, 1999.

Koko: A Talking Gorilla. DVD. Directed by Barbet Schroeder. Paris: Les Films du Losange, 1978.

When Animals Talk. DVD. Produced by Rob Weller. North Hollywood, CA: Weller/ Grossman Productions, 2000.

Why Dogs Smile and Chimpanzees Cry. DVD. Directed by Carol L. Fleisher. Jackson, NY: Fleischerfilm, Inc., 1999.

Websites

Ethologists for the Ethical Treatment of Animals: http://www.ethologicalethics.org

Jane Goodall Institute: http://www.janegoodall.org

Doing and Saying in Play Between Dogs and People

ROBERT W. MITCHELL
Eastern Kentucky University

Why study dog-human play? Such an obvious and commonplace event would seem out of the realm of scientific study. Yet play between dogs and people presents a fascinating psychology on the part of both participants.

My initial interest in studying dog-human play derived from my failure at studying crow communication for my doctoral research: every time the crows saw me, they flew away. The exciting thing about dogs and people playing was that they did not run away when I appeared on the scene. Watching interspecies play led me to wonder how they did it. Yes, dogs have been domesticated for eons, but I had earlier observed play interactions between rhesus monkeys and dogs, and rhesus monkeys and cats, so neither domestication nor similar phylogeny were necessary for interspecies play. I was influenced by Gregory Bateson's ideas, and the coordination of play between dogs and people seemed similar in some ways to cross-cultural exchange. In dog-human play and cross-cultural exchange, either both parties are being satisfied or the interchange will end. Bateson distinguished three types of interactions: complementary (when agent A does d, agent B does e; A does not do e, and B does not do d), symmetrical (when A does c, B does c, and vice versa), and reciprocal (when A does d, B does e, and when B does e, A does d). In Bateson's view, reciprocal interactions are likely to be maintained, whereas complementary or symmetrical interactions are not. Viewed in this way, play between dogs and people became a problem: How do dogs and people coordinate their interests to be mutually satisfying (as play seemed to be), that is, were their interactions complementary, symmetrical, and/or reciprocal?

I was also intrigued because dogs and people engaged in deception during play, often making themselves or an object appear available, only to pull back when the play partner attempted to make contact. I was coediting a book on deception in animals at the time, so wondered if the deceptions of people and dogs were similar, or if people's deceptions were more "psychologically complex" than those of dogs.

My dissertation advisor Nick Thompson and I set about to create a theory to evaluate the compatibility of what people and dogs do. Simply put, each agent engages in actions (e.g., running toward or away from the partner, moving an object toward or away from the partner, or pulling on an object that the partner is holding) that are coordinated into projects, which are the overarching goals of the players' actions. Some projects are self-keepaway (entice the partner to chase you, but avoid being contacted), object-keepaway (entice the partner to try to possess an object you possess, but do not let the partner get it), chase (attempt to contact the partner who is running away), and fakeout (move an object in and out of your control to entice the partner to attempt to obtain the object, but avoid letting the partner get it). If the projects of the dog and person are compatible, then what the dog does will satisfy what the person does, and vice versa; if incompatible, then the projects will not satisfy either partner. We called simultaneously occurring projects "routines," and predicted that most routines would be based on compatible projects. Thus, one partner playing fakeout while another played self-keepaway would be incompatible projects; one partner playing chase the object when the other was playing object-keepaway would be compatible projects. Not surprisingly, we discovered that generally pairs engaged in routines comprising compatible projects; when they did not, either one partner changed to a compatible project, or the play ended. We found this for people playing with a familiar dog, and for people playing with an unfamiliar dog. We also found that projects tended to be reciprocal: Dogs and people engaged in several of the same projects, and if dogs enacted project d to which the compatible project was e, generally people enacted e, and dogs could also enact e, to which people generally enacted d. There were some routines comprising symmetrical projects (tug o' war, for example), and some that apparently comprised complementary projects (fakeout and avoid-fakeout), but reciprocal projects were available and common. Although initially we believed that only people engaged in fakeout, and only dogs engaged in avoid-fakeout (where the player attempts to avoid responding to an apparently available object until it is clear that the player can obtain it), upon reviewing old videotapes that were not used in our original analysis, it is clear that dogs and people engage in both projects.

The deception that occurs in play derives from the fact that the overarching characterization of social play is for players to entice their partner to do x without letting the partner succeed—to control the other without being controlled. Thus, in self-keepaway the dog may entice the person to chase her, but try not to let the person catch her; and in chasing the dog, the person attempts to capture the dog in the midst of the dog doing everything it can to make that appear possible but not actually be easy. In a sense, play is born of the enjoyment of frustration! Thus, many of the projects are inherently deceptive, as appearance defies reality. Of course, none of them appear to be born of elaborate planning, though certainly some planning-in-action occurs for dog and human players.

Having videotaped the play sequences between dogs and people allowed for further examination of what was going on during play. One frequent activity on the part of the people was talking to the dog—another commonplace phenomenon. Yet again there is a problem. Why talk to dogs? Presumably dogs might know a few words, but people used a variety of words in talking to dogs, and it was saliently obvious that the words usually had no effect—saying, "Come here" *to a playing dog is not likely to result in the trained response! The whole point for the dog is to revel in being out of the person's control. So Elizabeth Edmonson and I examined what people say to dogs, to understand the purpose of the talk: Was it to control the dog (and get her attention), to have a pretend conversation, or to plan aloud (i.e., not really talk to the dog but just say what you were going to do). We derived these hypotheses from research on think-aloud protocols (where subjects say what they are thinking) and on talk to infants (similar to talk to dogs, in that infants have even less of an idea as to the meaning of talk but talk acts as an attention-getter). We discovered that people mostly talk to dogs during play to get their attention and to try (often impossibly) to control them, although sometimes people engaged in a conversation with the dog (filling in for the dog's response) and sometimes people were planning aloud. Later analyses compared talk to dogs and talk to infants during play more directly, examining frequencies of different forms of talk, as well as examining sex differences and familiarity effect. Who knew that people used baby talk more to an unfamiliar dog than to a familiar one? Sounds like a good idea if you want to make friends with a stranger.*

The next steps in my continuing analysis of dog-human play are to look into the fakeout/avoid fakeout sequences to see how much dogs and people engage in these projects, to discern the contexts in which people laugh during play, and to examine what people and dogs are doing when the people are talking. Dog-human play is a fascinating commonplace activity that provides interesting dividends when scrutinized. Even the obvious is interesting when you spend time watching.

18

The Moral Status of Animals

The alien has never once attempted to talk with me. It has
been with me, watched me, touched, handled me, for days: but
all its motions have been purposeful, not communicative. It is
evidently a solitary creature, totally self-absorbed. This would go
far to explain its cruelty.
—"MAZES" BY URSULA LEGUIN (1987)

IMAGINE THAT ONE DAY a science-fiction film came true. Planet
earth has been invaded by extremely intelligent aliens from another solar system that have come to our planet to find new food sources because they have
stripped their own planet of resources, and need to look elsewhere. These
aliens have decided to colonize earth and to use our resources for their own
food. In particular, they have chosen humans as their primary food source.
The aliens are technologically and intellectually superior to humans—that is
how they made it here in the first place. They use that superiority to justify
their wholesale exploitation of the human species, which involves, by the
way, not just hunting and killing humans but also a system of production
where humans are intensively farmed, Matrix-style, and then slaughtered.
From the perspective of the humans on earth, this is certainly a travesty. But
from the perspective of the aliens, do they have the right to raise, slaughter,
and consume humans in this way? Why or why not?

 This hypothetical situation, taken from a short story by Desmond Stewart
(1972), and the quote from the Ursula LeGuin story that opened this chapter, are good starting points to think about the question of animal rights,
and whether there is any moral justification for treating animals the way that
we now do. There is no simple answer to this question, which is designed
to make you think about the ethical choices that we do make. How do we
decide which species or individuals deserve certain kinds of treatment, but

not other kinds of treatment? If it is proper for the aliens to treat humans this way, then there must be something that distinguishes the species and gives one rights over the other. If it is not proper to treat them this way, then how do we justify doing the same thing to animals?

You might argue that the aliens have every right to do what they are doing because they are intellectually and perhaps morally superior to humans, which justifies their actions. You might even argue that it is natural—the survival of the fittest. You might argue that they have no such right because intellectual superiority is not grounds alone for the killing of other species. Or you might argue that if the farmed species—humans—can suffer, or if they are aware of what is happening to them, then it is not justifiable. Finally, you might make the argument that only certain people should be used this way—criminals, for example, or perhaps the elderly, or the mentally and physically disabled. That way only those who either deserve to be killed or perhaps only those whose lives are not "worth" as much as other people's should have to suffer. As far-fetched as this scenario may seem to you, it is a good starting point for our discussion of the moral status of animals.

History of Philosophical Debates on Animals

In the West, philosophers have been debating the moral status of animals since the ancient Greeks. The fourth-century BCE philosopher Aristotle (1943) wrote that because humans alone have the power to speak, which is indicative of rationality, humans alone have an ethical existence; nonhuman animals that lack this should serve humans. Aristotle also felt that plants, animals, and humans are all ranked on a natural hierarchy; not only are humans endowed with rationality and thus superiority over animals, but animals themselves are superior to plants because they possess consciousness. However, Aristotle felt that some humans have more rationality than others. So animals are born to serve humans and so, in their own way, are slaves.

Not all Greek philosophers shared Aristotle's views. A number of important Greek thinkers were vegetarians, including the sixth-century BCE philosopher Pythagoras, who not only abstained from eating meat but forbade his disciples from eating meat as well, saying that animals had a right to live in common with mankind, and that vegetarianism led to peaceableness (Riedweg 2005). Seven hundred years later, another Greek thinker, Plutarch, wrote extensively on the subject in his essay "Of Eating Flesh":

You ask of me then for what reason it was that Pythagoras abstained from eating flesh. I for my part do much admire in what humor, with what soul or reason, the first man with his mouth touched slaughter, and reached to his lips the flesh of a dead animal, and having set before people courses of ghastly corpses and ghosts, could give those parts the names of meat and victuals, that but a little before lowed, cried, moved, and saw; how his sight could endure the blood of slaughtered, flayed, and mangled bodies; how his smell could bear their scent; and how the very nastiness happened not to offend the taste, while it chewed the sores of others, and participated of the saps and juices of deadly wounds.
(1874:3)

But for the sake of some little mouthful of flesh, we deprive a soul of the sun and light, and of that proportion of life and time it had been born into the world to enjoy. And then we fancy that the voices it utters and screams forth to us are nothing else but uncertain inarticulate sounds and noises, and not the several deprecations, entreaties, and pleadings of each of them.
(1874:6)

Plutarch, along with other philosophers such as Pliny the Elder, shared a philosophy known as **theriophily** that holds that animals possess rationality. This viewpoint would not be taken seriously for thousands of years after the fall of the ancient Greek culture. Of all the Greek philosophers, it was Aristotle's views that influenced later thinkers about animals. Judeo-Christian thought borrowed heavily from Aristotle and absorbed the concept of the soul from Egypt; early Jewish scholars created a theology which reifies human difference from and superiority over animals. In the Book of Genesis, this position is clearly laid out: "And God said, Let us make man in our image, after our likeness: and let them have dominion over the fish of the sea, and over the fowl of the air, and over the cattle, and over all the earth, and over every creeping thing that creepeth upon the earth" (Genesis 1:26). St. Augustine, writing in the fourth and fifth centuries, also felt that humans are valued precisely because of their link to the divine, a link that is not shared by animals—or women, for that matter (Augustinus 2000). The medieval notion of the great chain of being or *scala naturae*, borrowed from Aristotle—in which God created all of life according to a hierarchy of higher and lower beings, with man just beneath God, and animals below humanity—further reinforces this view.

This ranking of human over animal was further solidified through the writings of St. Thomas Aquinas, a thirteenth-century theologian who maintained that the world is divided into persons who have reason and thus immortal souls, and nonpersons. Nonpersons are essentially things that can be used in any way to serve the interests of people (Thomas 1906). Persons are persons because they are rational and thus have intrinsic value and ought to be respected; animals, being irrational, have only instrumental value and can be used as instruments, in any way humans see fit. Aquinas wrote that the subjection of other animals to man is natural because only man can direct his own actions, and thus man owes no charity to animals.

As European philosophy evolved through the Renaissance and the Age of Enlightenment, the idea that humans possess something special that animals lack became even more important in justifying the treatment of nonhumans. Seventeenth-century philosopher René Descartes, for example, claimed (as did Aristotle) that mentality and the ability to speak were the primary characteristics separating humans from animals (Descartes 1991). Because animals are incapable of using language, Descartes considered them to be essentially machines—mindless automata that operate without higher thought or consciousness. Known as the theory of mechanism, this approach saw the whole of the natural world—with the important exception of humans, whose use of speech indicates thought—as explainable in mechanistic terms. In fact, Descartes dissected live, conscious dogs and interpreted their screams, not as pain, but as instinctive noises that a machine might make. Explaining animal behavior this way, without reference to interior states of consciousness, was a simple explanatory system that did not assume inner states that cannot be proven. He wrote that "animals are without feeling or awareness of any kind" and that they "behave as if they feel pain when they are, say, kicked or stabled" (Smith 1952:136 and 140). Not all of the great thinkers during this time shared Descartes' views. John Locke, for example, argued that animals do have feelings and can certainly feel pain, although cruelty to animals, for Locke, was wrong only because it is bad for people (Locke 1996). Another contemporary, Voltaire, went further and famously responded to Descartes by writing:

> This dog, so very superior to man in affection, is seized by some barbarian virtuosos, who nail him down to a table, and dissect him while living, the better to shew you the meseraic veins. All the same organs of sensation which are in yourself you perceive in him. Now, Machinist, what say you? answer me, has Nature created all the springs of feeling in this animal, that it may not feel? Has it nerves to be impassible? For shame! charge not Nature with such weakness and inconsistency.
> (1796:34–35)

A century after Descartes' time, Immanuel Kant (1998) wrote that rationality and autonomy were the key characteristics separating humans from other animals; because animals lack the capacity for rational moral choice, they are not moral agents and thus have no moral standing. Although animals can make choices, according to Kant, they cannot make rational choices based on carefully choosing the best course of action. Without this choice, animals have no autonomy and thus no intrinsic value, and humans owe them nothing. In addition, having autonomy gives us the capacity to engage in relationships with others and to fulfill reciprocal obligations to others. Humans are not obligated toward beings that cannot themselves have obligations. He wrote:

> [E]very rational being, exists as an end in himself and not merely as a means to be arbitrarily used by this or that will . . . Beings whose existence depends not on our will but on nature have, nevertheless, if they are not rational beings, only a relative value as means and are therefore called things. On the other hand, rational beings are called persons inasmuch as their nature already marks them out as ends in themselves.
> (KANT 1998:428)

Like John Locke, Kant did feel that cruelty to animals was wrong, but only because engaging in it is bad for people.

The eighteenth century was notable because it was a time when the first major thinkers of the modern era started to formulate a coherent theory of limited rights for animals. French philosopher Jean-Jacques Rousseau wrote that although animals may not be rational, or aware of concepts such as laws or rights, they should be protected from injury because they are sentient creatures:

> It appears, in fact, that if I am bound to do no injury to my fellow-creatures, this is less because they are rational than because they are sentient beings: and this quality, being common to men and beasts, ought to entitle the latter at least to the privilege of not being wantonly ill-treated by the former.
> (1984:14)

Rousseau's statement is an example of the difference between negative rights and positive rights. A positive right is a right to do something—to have freedom of speech, for instance, or freedom of religion, two important rights granted by the U.S. Constitution. A negative right, however, is the right to not have something happen to oneself. Rousseau notes that

Figure 18.1. Scala Naturae by Mark Dion shows the hierarchy of life. (Photograph courtesy of Mark Dion and Tanya Bonakdar Gallery, New York.)

regardless of intellectual capacity, animals should have the negative right to not be harmed.

Just a few years later, Jeremy Bentham, the founder of a school of philosophy known as utilitarianism, went even further with Rousseau's idea, and made the now-famous statement, "the question is not, Can they reason? nor Can they talk? but, Can they suffer?" Bentham was writing about the abolition of slavery in France, noting that the French finally put aside the notion that something as superficial as the color of a person's skin could mean the difference between freedom and slavery. He wrote:

> It may one day come to be recognized that the number of the legs, the villosity of the skin, or the termination of the *os sacrum* are reasons equally insufficient for abandoning a sensitive being to the same fate. What else is it that should trace the insuperable line? Is it the faculty of reason or perhaps the faculty of discourse? But a full-grown horse or dog, is beyond comparison a more rational, as well as a more conversable animal, than an infant of a day or a week or even a month, old. But suppose the case were otherwise, what would it avail? (1996:283)

Bentham's logic would later serve as the foundation for Peter Singer's work, which was revolutionary and foundational to twentieth-century animal advocacy.

Ethical Humanism and the Rights of Animals

Many of the philosophical approaches that we have just discussed can be considered together to be examples of ethical humanism. Ethical humanism is the belief that human beings are alone in deserving moral consideration, and that all humans, regardless of intelligence or ability, deserve such consideration, although animals do not. Philosopher James Rachels (1990) distinguishes between two kinds of ethical humanism, or speciesism: unqualified speciesism, which says that only those who are members of the human species deserve moral consideration, and qualified speciesism, which says that humans deserve special consideration because of something that makes them morally superior to other species. The implications of unqualified speciesism are that humans need not take animals' interests into our considerations because of their lack of moral considerability. That does not necessarily mean that, with this as a guiding ethical approach, we can still harm animals willy-nilly. There may be other reasons not to harm them—it may cause another person pain, for example, if I kill their cat, so I may choose to not kill that cat. Or living in a society in which everyone kills animals around them may lead to a very violent society, which is not good for us. But the cat's own feelings on the matter will not play a role in my decision.

As we have seen, the ethical arguments dating back to Aristotle and in favor of human power over animals all stem from speciesism—qualified or unqualified. Aristotle said that humans possess rationality that distinguishes them from animals, and gives humans power over them. The authors of the Bible used the fact that humans have a soul, and were created in God's image to separate the species and grant the one authority over the other. Thomas Aquinas used rationality and divine providence, and Descartes took language and the presence of the mind. And Kant used the capacity to make rational moral choices as the basis for who has obligations to whom, and thus who has rights.

Since Jeremy Bentham's time, philosophers have been grappling with the question: Are there really these morally significant differences that justify human exploitation of nonhuman animals, and can they be empirically verified?

To answer this question, we can look at the vast body of research that we discussed in chapter 17. Scientists have begun to challenge our assumptions about the minds of nonhuman animals, and have provided us with new questions. For instance, does a lack of language indicate a lack of thought? Ethological studies have demonstrated the complexities of animal minds, and have looked at the ability to use language in great apes, parrots, and even prairie dogs. They have also shown how emotionally complex many animal species are, and that many species of animals possess self-awareness. This research is still relatively new and, because of this, quite preliminary, but it is growing in leaps and bounds; and every year we learn more about the cognitive, intellectual, and emotional capabilities of nonhuman animals.

In terms of whether these differences justify human exploitation of animals, the issue becomes less a question of animal abilities and moves from the realm of animal behavior studies into philosophical and ethical research. Whether or not we can find a single factor that definitively distinguishes humans from nonhumans, the ethical question remains: Does that difference justify differential treatment? Certainly physical and even intellectual differences justify *some* special treatment. People who use a wheelchair cannot walk and thus are not eligible to play college football. But because a person uses a wheelchair should not keep them from having a job as a bank teller. The Americans with Disabilities Act is just one example of federal legislation ensuring that differences in skin color, physical ability, gender, or sexual orientation cannot be used as the basis for discrimination in this country. And being confined to a wheelchair certainly does not justify killing a person for food or using that person to test drugs or chemicals.

Because humans all have different physical and intellectual abilities, another problem with ethical humanism goes back to the position that *all* humans, regardless of intelligence or ability, deserve such consideration even though animals *never* do. There are human beings who are severely emotionally and intellectually disabled, who cannot do many of the things that some animals can do, and who lack some of the "morally significant" criteria, such as language or rationality. The very young and the very old also do not possess the skills and abilities that other humans do. According to qualified speciesism, it is only the possession of these qualities that gives humans special consideration, which means that humans who lack those properties—babies, comatose people, the retarded—should not be granted full-level moral consideration. For example, although most humans possess language and are thus granted special protections because they do, those humans who lack language because of disability, age, or other factors do not deserve such protection and can be eaten, tested upon, or even skinned—assuming that doing so satisfies relevant "normal" needs.

This is what is known to philosophers as the argument from **marginal cases**. The term "marginal cases" refers to humans who because of an accident of birth or disease do not possess any or all of the qualities deemed special by ethical humanists. Under current law, these "marginal humans" are granted all of the rights that other Americans have, even though they may not be able to speak, reason, or even understand their surroundings. How do we justify giving them these rights, and denying those same rights to animals that have demonstrated their capacity for reason and even language, such as Koko the signing gorilla or Kanzi, the lexigram-using bonobo? An animal rights perspective would charge that we either grant moral consideration to these animals or deny it to marginal humans.

Humanists argue that infants, the retarded, the senile, and the comatose still deserve these rights because, in the case of infants, they will one day possess the qualities of rationality and language, and in the case of the senile, they once possessed them. But what about people who either never possessed those capabilities or who never will? Humanists argue that even though *some* people do not possess language, *most* people do. In other words, humans as a species possess language, even though a few individuals do not, so the norm for the human species is language. "Marginal humans" are still humans, after all. Along the same lines, *most* apes do not use sign language, even though a few do. So the norm for great apes is to not have language, which means that even extremely intelligent apes should have no rights. Using this argument based on species norms one can rationalize the ethical humanist position—that only humans, by virtue of criteria such as language, deserve ethical consideration and all other species do not.

A related form of ethical humanism is based on the premise that animals do deserve *some* moral consideration on the basis of their sentience, but not as much as humans, for the same reasons: because they possess some quality, such as rationality or self-awareness, that elevates them above other animals. With this approach, we should not cause undo harm to animals because they are sentient: they can feel and they can suffer. But if human self-interest outweighs the interest of animals to not suffer, then humans can indeed cause animal suffering.

Peter Singer and Utilitarianism

Peter Singer is one of the most influential animal ethics philosophers of the twentieth century. His groundbreaking 1975 book, *Animal Liberation*, sets out a clear theory, based on Jeremy Bentham's work, in favor of animal liberation. Peter Singer's **utilitarian theory** (also known as **consequentialist theory**) is based on the principle of equal consideration, which means that we must give equal consideration to the interests of all creatures, and we should maximize the satisfaction of the interests of everyone—or as many as possible—who are affected by our actions.

Singer challenges the notion of ethical humanism and the idea that there is some quality or qualities that should give humans status not provided to animals. For Singer, the only quality of importance is sentience, or the ability to feel pain and pleasure, that is shared by humans and animals, although to different extents. (Even fish, according to recent studies, are now known to feel pain.) Because all sentient creatures can feel pain and experience pleasure, assuming that the level of suffering or pain is similar, Singer requires that we must give the animal's ability to feel equal weight with our own; my interests do not outweigh the interests of another creature, whose ability to feel is similar to mine. Under this approach, we can only justify killing an animal if that animal's pain and loss of life is considerably outweighed by benefits to someone else, or to a group as a whole. So for instance, just because a human likes the taste of meat is not enough to justify killing an animal for meat. One human's taste does not outweigh the animal's interest in not being killed. Even if all humans wanted to eat meat, the unimaginable suffering experienced by the billions of animals raised and slaughtered for food every year would definitely outweigh all humans' interest in eating meat. On the other hand, using Singer's theory it is not inconceivable to think of a way in which meat eating would be permissible. For instance, if the pain and suffering were removed from animal agriculture, it is conceivable that killing animals for meat might be possible under this approach. But even then, animals presumably have an interest in being alive—in enjoying life. That interest would not be outweighed by the human desire to eat meat.

Singer uses the example of marginal humans as a way to challenge ethical humanism by pointing out, that logically speaking, under qualified speciesism dogs and certainly great apes should have more rights than infants or the retarded because of their greater possession of rationality. He also shows that basing rights on certain qualities—rationality, language, autonomy, the ability to act morally, etc.—is akin to racism. Many racists believe, for instance, that their own race is morally superior to other races, that they are more intelligent, and that they are more rational. Although today we know that there are no inherited intellectual or moral differences among people of different "races" (and anthropologists and other scientists have challenged the idea that races exist empirically at all), Singer's point is that even if there were such differences, we could not justify differential treatment on the basis of them.

Tom Regan and Animal Rights

Tom Regan is another extremely influential philosopher whose 1983 book, *The Case for Animal Rights*, shaped—alongside Singer's work—animal ethics theorizing in the twentieth century and beyond. Regan differs sharply from Singer in that he feels animals have equal moral status to humans, that they should have rights, and that those rights should not be based on utilitarian principles or maximized interests.

For Regan, many animals (such as mammals and birds) have inherent value, and anything or anyone with value should automatically be accorded rights because these animals are what he calls "subjects of a life." Subjects of a life

> have beliefs and desires; perception, memory, and a sense of the future, including their own future; an emotional life together with feelings of pleasure and pain; preference- and welfare-interests; the ability to initiate action in pursuit of their desires and goals; a psychological identity over time; and an individual welfare in the sense that their experiential life fares well or ill for them, logically independently of their utility for others, and logically independently of their being the object of anyone else's interests.
> (1983:243)

For Regan, animals that are subjects of a life have interests, and it is not possible to say some interests automatically count for more than others just because of whose interests they are. What is important is that animals, such as humans, have interests. Neither humans nor animals should be eaten or experimented on, and whether or not the greater good would be met by this. Regan ultimately argues for animal rights, which should be taken as negative rights rather than positive rights. In other words, most animals presently subject to exploitation (including "marginal" humans) should have negative moral rights to noninterference rather than positive rights, such as voting rights.

Other Approaches

Philosopher Mylan Engel (2000) proposes a theory based on the idea of consistency. Engel suggests that most humans hold a number of "truths" in common, such as "Other things being equal, a world with less pain and suffering is better than a world with more pain and suffering." His theory is predicated on the idea that if a person holds this type of view, he or she could not logically support factory farming because one's own beliefs and values should reject a practice that causes so much unneeded suffering. In other words, if a harm is a harm, it is a harm no matter where or to whom or how it occurs.

Attorney Steven Wise has created a theory that hearkens back in some ways to the qualified speciesism that we discussed earlier in this chapter—that there are some qualities possessed by humans that give them special protections and special rights, and that are not shared by any (or most) animals. Wise's theory (2000, 2002) states that some animals—particularly great apes, dolphins, and parrots, because of their inherent qualities of self-awareness, intelligence, social learning, and the capacity for language—meet the criteria of legal personhood, and should be given a set of rights and protections. Wise outlines the qualities that animals must possess in order to be granted personhood. These include the ability to desire things, to act in an intentional manner, and to have a sense of self. Wise's personhood approach has faced criticism from within the animal rights community because of its privileging of certain animals over others. Most animals do not possess some of the skills that chimpanzees and bonobos possess, which makes them (according to Wise) less deserving of good treatment than others. Critics of Wise's position (such as Joan Dunayer) call it **superspecies** or **new-speciesism**: the idea that some animals are deserving of more rights than others.

Philosopher Martha Nussbaum's theory (2006), known as the capability approach is grounded in the desire, which she claims is held by animals and

BOX 18.1

ORGANIZATIONAL FOCUS: THE GREAT APE PROJECT

The Great Ape Project (GAP) is a confederation of philosophers, primatologists, and other scientists who promote the idea that great apes should be granted personhood and some basic legal rights, including the right to life, the right to freedom, and the right to not be tortured in medical experiments. This project has been spearheaded by figures such as philosophers Paola Cavalieri and Peter Singer, primatologist Jane Goodall, and attorney Steven Wise. They argue that personhood and basic human rights should be granted to great apes because of their high degree of intelligence, highly complex emotional lives, self-consciousness, and shared communication with humans. The organization has been campaigning to have the United Nations endorse what GAP is calling a Declaration on Great Apes that would provide basic human rights—of life, freedom, and to not be tortured—to these animals in all UN nations.

humans, to live a "dignified existence." For Nussbaum, what that means will be different not only for different species but also for different individuals within a species. For instance, some people have more intellectual capabilities than others, even though some have greater athletic abilities. Those different abilities mean that although we cannot all reasonably expect to achieve the same things in life, we should be granted the opportunity to achieve our greatest level of happiness and satisfaction, given our own capabilities. This holds true for animals as well. For Nussbaum, granting rights based on rationality will necessarily exclude not just animals, but some people. My dogs, for example, will never be able to go to college because of their intellectual abilities. However, they should still be allowed to flourish and to experience happiness in their own ways, and to fulfill their own capabilities—as excellent snugglers, for example, or as champion closet poopers. If my dogs were to live in a puppy mill, on the other hand, where they would be kept in wire cages without the ability to exercise or to experience affection, this would be an example of wasted capabilities, which would be a tragedy. My dogs do not need to sit on a jury or vote in order to have a dignified life; but they do need a loving home, animal and human friends, and quality food and veterinary care.

A **virtue ethics** position, as outlined by philosophers such as Cora Diamond (2001), is based on the actions that a "virtuous person" might engage in. A virtuous person would not, for example, kick a dog to death because that would demonstrate that that person's character is not virtuous at all. If the attitudes that underlie such behaviors are not compassionate or kind or virtuous, then we are not compassionate or kind. In this approach, humans and animals share membership with each other n a moral community, and members of a moral community do not kick each other to death. Rosalind Hursthouse writes:

> I began to see [my attitudes] that related to my conception of flesh-foods as unnecessary, greedy, self-indulgent, childish, my attitude to shopping and cooking in order to produce lavish dinner parties as parochial, gross, even dissolute . . . Without thinking animals had rights, I began to see both the wild ones and the ones we usually eat as having lives of their own, which they should be left to enjoy. And so I changed. My perception of the moral landscape and where I and the other animals were situated in it shifted. (2000:165–166)

A relatively new position within animal ethics theorizing comes from continental (as opposed to American) philosophy, and is based on—and challenges—the work of philosophers such as Jacques Derrida, Martin Heidegger, Gilles Deleuze, and Félix Guattari. Philosophers who work in this tradition, such as Ralph Acampora (2006) and Matthew Calarco (2008), reject the anthropocentrism of the human-animal binary, and argue instead for a new kind of relationship with animals based not on animals abilities or intelligence. Ralph Acampora, for example, borrowing from **phenomenology**, focuses on the ways in which nonhuman and human animals share the experience of living in bodies, which itself can serve as the basis for a new form of interspecies relationship based on shared understanding. These newer approaches have an advantage over the ethical approaches that grant rights or moral status based on how intelligent or capable certain animals are, and that automatically place humans—and super-intelligent animals such as primates or cetaceans—at the top of the hierarchy.

Finally, there are a number of theoretical approaches grounded in feminism that are increasingly popular today, including **ecofeminism** and the **feminist ethic of care**. Ecofeminism is a philosophy and a social movement which focuses on the links between the oppression of women and the destruction of nature. Ecofeminists examine the relationship between industrial capitalism and patriarchy as well as a host of other systems of inequality, including animal exploitation. Carol Adams and Greta Gaard are feminists who highlight the dual oppressions faced by women and animals and argue that animal rights should necessarily be a feminist position. Feminist and ecofeminist positions (Gaard 1993; Adams 1994; Adams and Donovan 1994) are structural—focusing on the power structures of society and the different forms of institutional—rather than individualistic inequality, and argue not for individual change (becoming a vegetarian, for example) but for large-scale social change.

The feminist ethic of care, on the other hand, is an individualistic approach and focuses on the relationship between humans and nonhumans. Because animals have feelings such as humans do, humans have a moral obligation to them that is not based on abstract qualities such as rights or justice, but on the idea of relationships. Proponents of this theory (see Donovan and Adams 1996, 2007) argue that other ethical approaches are too heavily based on rationality and downplay qualities associated with women such as empathy, caring and love. Theorists such as Josephine Donovan argue that we need to reinsert emotion back into discussions of animal welfare because without an emotional response to animal suffering, it is easy to see why abuse continues. As Donovan writes:

> It is also possible indeed, necessary—to ground that ethic in an emotional and spiritual conversation with nonhuman life-forms. Out of a women's relational culture of caring and attentive love, therefore, emerges the basis for a feminist ethic for the treatment of animals. We should not kill, eat, torture, and exploit animals because they do not want to be so treated, and we know that. If we listen, we can hear them.
> (1990:375)

Suggested Additional Readings

Acampora, Ralph. 2006. *Corporal Compassion: Animal Ethics and Philosophy of Body*. Pittsburgh: University of Pittsburgh Press.

Atterton, Peter and Matthew Calarco, eds. 2004. *Animal Philosophy: Essential Readings in Continental Thought*. London: Continuum.

Calarco, Matthew. 2008. *Zoographies*. New York: Columbia University Press.

Carruthers, Peter. 1992. *The Animals Issue: Moral Theory in Practice*. Cambridge: Cambridge University Press.

Francione, Gary L. 1996. *Rain Without Thunder: The Ideology of the Animal Rights Movement*. Philadelphia: Temple University Press.

Gaard, G., ed. 1993. *Ecofeminism: Women, Animals, Nature*. Philadelphia: Temple University Press.

Midgley, Mary. 1983. *Animals and Why They Matter*. Athens: University of Georgia Press.

Nussbaum, Martha C. 2006. *Frontiers of Justice: Disability, Nationality, Species Membership*. Cambridge, MA: The Belknap Press of Harvard University Press.

Rachels, James. 1990. *Created from Animals: The Moral Implications of Darwinism*. Oxford: Oxford University Press.

Regan, Tom. 2004. *The Case for Animal Rights*. Berkeley: University of California Press.

Regan, Tom and Peter Singer, eds. 1989. *Animal Rights and Human Obligations*, 2nd ed. Englewood Cliffs, NJ: Prentice-Hall.

Sapontzis, Steve F. 1987. *Morals, Reason, and Animals*. Philadelphia: Temple University Press.

Singer, Peter. 2006. *In Defense of Animals: The Second Wave*. Malden, MA: Blackwell.

Sunstein, Cass R. and Martha Nussbaum, eds. 2004. *Animal Rights: Current Debates and New Directions*. New York: Oxford University Press.

Wise, Steven M. 2000. *Rattling the Cage: Toward Legal Rights for Animals*. Cambridge, MA: Perseus.

Suggested Films

Earthlings. DVD. Directed by Shaun Monson. Burbank, CA: Nation Earth Films, 2005.

In Defense of Animals: A Portrait of Peter Singer. VHS. Directed by Julie Akeret. Oley, PA: Bullfrog Films, 1989.

Jane Goodall: Reason for Hope. VHS. Directed by Emily Goldberg. St. Paul, MN: KTCA-TV, 1999.

The Morality of Awareness

KATHIE JENNI
University of Redlands

My work explores the intersection of animal ethics and moral psychology: the study of the inner dimensions of moral life such as emotion, perception, imagination, and motivation. My concern is the morality of awareness: *the ethics of managing our consciousness of suffering. Educated persons know that nonhuman animals are abused and exploited in horrific ways, but we are remarkably skilled at keeping that knowledge vague or suppressed, at keeping it out of mind as we go about our daily lives. Self-deception, avoidance, willful ignorance, and forgetfulness are powerful forces that threaten the awareness of suffering that can move us to take action to help. In light of these forces, my writing explores ethical questions about our attention to animal suffering at human hands. What degree of awareness is appropriate and morally required? How can we remain robustly aware of suffering without succumbing to depression and despair? How should those who care about animals "manage" our psyches to move ourselves to action? How can we achieve a realistic sense of what the world is like, maintain our mental health, and enhance our capacities to make the world a better place? What does respect for animals require of us as we seek to make their abuse visible and public without dishonoring them?*

Inattention *is a critically important, but often-neglected phenomenon. In dramatic cases of evil such as genocide, as well as everyday tolerance of hidden horrors, people regularly act (or fail to act) in ways that their moral values clearly condemn. Why is it that we often do what, at some level, we believe that we should not? Why do we often fail to do what we clearly should do, in light of our moral commitments? Weakness of will and self-deception are important explanations, and both have received exhaustive and insightful treatment by philosophers and psychologists. What remains less examined is the simple phenomenon of inattention: failure to attend to morally important aspects of our lives. In "Vices of Inattention" (2003), I examine moral dimensions of inattention: what makes it problematic, what vices it reflects, what obligations we have to overcome it, and how we might try to do that. I argue that inattention obscures our responsibilities to prevent harm, erodes our own autonomy, manifests a lack of virtues such as courage and compassion, and undermines integrity. For these reasons, we have moral obligations of attentiveness. We should attend (at least) to apparent violations of our moral values in which we are personally implicated, which we have some power to affect, and to which we have been directed by clues that something is amiss. We need to be aware enough of suffering to form personal policies aligned with our values (such as a policy of not buying factory-farmed products), but not so acutely aware of it that we cannot continue to function. Maintaining this kind of awareness is a matter of holding our understanding of suffering at arm's length: being conscious of it without fully confronting its horror much of the time. In this way we can maintain consciousness of moral atrocities, and be mindful of them in deliberation and conduct, without going mad or falling into despair.*

Related questions arise from my experience teaching animal ethics: in particular, from the dramatic responses of students to visual presentations of animal suffering. In reading about the atrocious conditions in which animals live and die for meat production, students respond with authentic concern, but rarely with passion or action. But upon seeing animals' treatment on factory farms via documentary videos, they experience moral epiphanies. The issue takes on urgent and vital importance; the suffering that they witness arouses deep empathy and moral outrage; many resolve to do something to help, either by boycotting factory-farmed products or by joining activist efforts to end the practice altogether. This powerful response to the visual is heartening and frightening. It is heartening because it shows the latent empathy of people who can otherwise seem callous or indifferent. But it is frightening in that without visual prompts, it seems that empathic concern can remain dormant indefinitely. Moreover, although some students make life changes when they learn about factory farming, most of them allow the images—along with the empathy, outrage, and moral convictions that the images arouse—to fade away with time. In "The Power of the Visual" (2005), I explore the moral implications of the power of the visual to move us morally, and our tendency to forget. I explore the nature of the moral insights provided by the visual, the problematic nature of our disparate responses to facts that are seen and unseen, and our responsibilities to enliven our imaginations when images are absent.

Some images of animal abuse are so graphic, the treatment they capture so degrading and cruel, that they approach the pornographic. (The ones that moved me to consider this were internet videos of Chinese workers skinning raccoon dogs alive for fur.) What is the most responsible approach to such records of horror? Is it more respectful to animal victims to watch them, or to look away? In "Bearing Witness to Animal Suffering" (2009), I explore the notion of bearing witness to animal suffering as a manifestation of respect. I begin by asking why it is important to bear witness to human atrocities such as the Holocaust. Some rationales are forward-looking and consequentialist: We bear witness to stir moral motivation and preventive action. But there are also backward-looking and expressive reasons: We keep the memory of atrocity alive to show respect for the dead, to express our solidarity and grief, to honor survivors, to affirm the moral value of both the lost and the saved. Nonhuman victims of atrocity differ importantly from human ones, of course: The animal dead did not value being remembered, and animal survivors (probably) do not share a collective consciousness of horror. Yet memorial obligations persist. Bearing witness to human-animal violence affirms the moral status of animal victims and expresses respect through attention to their suffering. As with human-to-human atrocities, bearing witness to animals' mistreatment is part of constitutive justice—something we owe to animals that suffered and died as victims of brutal wrongdoing.

Yet bearing witness raises moral dilemmas, so that it matters greatly how we do so. One problem is that visual testimony does not find its way only to compassionate audiences, but also to voyeurs who find pleasure in observing animal torture. If we bear witness to pay respect to victims, this is the last outcome that we could desire. Toward the end of finding respectful ways to remember, I explore the importance of bearing witness in private and in social contexts, of who attends to animal suffering, and of how and through what media we do so. I conclude that how to bear witness is a matter of moral judgment and discernment that those who would honor the animal dead must take on.

Humans vastly underestimate our power to enhance or obscure our own awareness of suffering. I hope through my work to reveal how this is so, and so to pursue better moral perception—not only for its own sake, but always and primarily for the practical benefit of saving animals from torment.

19

The Animal Protection Movement

Twenty-six years after my coronation various animals were
declared to be protected—parrots, mainas, aruna, ruddy geese,
wild ducks, nandimukhas, gelatas, bats, queen ants, terrapins,
boneless fish . . . tortoises, porcupines, squirrels, deer, bulls,
okapinda, wild asses, wild pigeons, domestic pigeons, and all
four-footed creatures that are neither useful nor edible. Those
nanny goats, ewes, and sows which are with young or giving
milk to their young are protected, and so are young ones less
than six months old. Cocks are not to be caponized, husks
hiding living beings are not to be burnt and forests are not to be
burnt either without reason or to kill creatures. One animal is
not to be fed to another.
—"EDICTS OF ASHOKA," THE FIFTH PILLAR

ASHOKA WAS THE EMPEROR of India in the third century BCE.
He converted to Buddhism as emperor and was responsible for the spread of
Buddhism out of India into East Asia. During his reign, the world's first animal
protection laws were passed, including a ban on sacrifice. After Ashoka's death,
however, his feelings toward animals disappeared into history, and the world
would not see new animal protection laws until almost 2,000 years had passed.

Conserving Nature

The animal rights movement is a relatively recent social movement, with
roots in the eighteenth century. But before there could be an animal rights
movement, there had to be a shift in social thinking about nature, to which
animals were largely thought to belong.

According to historians such as Keith Thomas (1983), prior to the six-teenth century, most Europeans saw nature—and the animals that inhabited it—as something frightening, dangerous, and out of control. It was only when people began to feel like they had some sort of control over nature that they could view it more affectionately. It was at that time that animals began to make an appearance in children's fairy tales, for example, as frightening representations of the wild, but also as talking creatures whose stories could impart lessons to children. By about the nineteenth century, a new idea—that nature was something worth conserving—began to appear in the minds of Europeans and Americans.

Of course, many non-Western cultures, as we have discussed in chapters 2 and 5, had very different understandings of nature and animals, and animals' relationships to humans and culture. In those cultures, conservation as an idea was not necessary because many indigenous peoples did not share with Europeans the idea that nature was a force to be subdued and humans were created differently from, and superior to animals. Of course, it would be simplistic to argue that only Western nations have harmed the environment or animals. Native Americans probably caused the extinction of large num-bers of megafauna that once lived in North America, and China has been responsible for a huge amount of environmental devastation. But whether from culture or economic growth, it is not hard to see that Western Europe and the United States themselves have been responsible for much of the destruction of the natural world.

The nineteenth century saw the beginnings of the conservation movement in North America, as artists such as Albert Bierstadt and Frederic Edwin Church began painting landscapes of the American West, and writers such as Henry David Thoreau began writing about nature, inciting the curiosity and interest of urban Americans. John Muir, for example, was inspired by the nature writings of Ralph Waldo Emerson, whose description of Yosemite Valley led Muir to visit, build a cabin there, and ultimately, to help conserve the area. Muir became an outspoken activist for the preservation of Amer-ica's wild spots, helping to protect the areas that would ultimately become Yosemite National Park and Sequoia National Park, and cofounding the Sierra Club, the nation's preeminent environmental organization.

In the early twentieth century, perhaps the most important conservation-ist in terms of his impact on public policy was President Theodore Roosevelt. Ironically, Roosevelt was a big-game hunter, who enjoyed traveling around the world and killing large animals whose bodies he preserved as trophies. But as a hunter, Roosevelt realized that without government regulation, the animals that he so enjoyed killing would disappear forever. As president,

Roosevelt fought some of the most disturbing trends in hunting such as the rise of commercial hunting and the extinction of species at the turn of the twentieth century.

During his tenure as president, Roosevelt helped create laws that restricted the amount of animals hunters could kill, established the U.S. Forest Service in 1905 to manage government forestlands, and used the Antiquities Act of 1906 to create dozens of national parks and wildlife refuges. The movement that he and others began later became the American conservation movement, but its early incarnations had nothing to do with preserving nature or wild animals for their own intrinsic value. The purpose was to preserve wild animals and their habitats for the use of Americans, who at that time were largely hunters. It was to ensure that the public owned the land and the wildlife on it, and that neither business interests nor the wealthy few could destroy them. Ironically, it was primarily the wealthy American hunters who made up the early conservation movement. They, like Roosevelt, were concerned about the loss of wildlife and wild lands for their recreational desires, although working-class hunters viewed wildlife as an important economic resource for fur, skin, or feathers. Even Roosevelt, horrified by the loss of the bison of the northern plains, was originally drawn to the territory in order to kill a bison. He eventually added one of the few remaining animals to his own trophy collection.

Roosevelt was a conservationist, conserving the wilderness and the wild animals that lived there for future use. On the other hand, John Muir was a preservationist, advocating the preservation of nature for its own sake. It is with Muir's philosophy that animal rights as a movement is more closely allied. But the animal rights movement ultimately had its own beginnings in Europe.

The Movement's Precursors

The animal rights movement proper began in the nineteenth century in Europe, but the first animal protection laws were passed much earlier— in the seventeenth century. Thomas Wentworth, Lord Deputy of Ireland, passed legislation in 1635 prohibiting the pulling of horses' tails and pulling wool off of live sheep. Just a few years later, in 1641, Nathaniel Ward, a Puritan minister who wrote the colony of Massachusetts' first set of laws in 1641, included in those laws language prohibiting "tyranny or crueltie towards any bruite creature which are usuallie kept for man's use." In particular, *The Body of Liberties* mandated that cattle must be rested and given food and water when being driven from place to place.

BOX 19.1

FIRST ANTICRUELTY LAW, 1635

None shall plow or work horses by the tail. Whereas in many places of this kingdome, there hath been a long time used a barbarous custome of ploughing, harrowing, drawing and working with horses, mares, gledings, garrans and colts, by the taile, whereby (besides the cruelty used to the beasts) the breed of horses is much impaired in this kingdome, to the great prejudice thereof. *Barbarity of the custom, a prejudice to the breed of horses.* Whereas also divers have and yet do use the like barbarous custome of pulling of the wooll yearly from living sheep instead of clipping or shearing of them; be it therefore enacted by the Kings's most excellent Majesty, and the lords spirituall and temporall, and the commons in this present Parliament assembled, that no person or persons whatsoever, shall after one yeare next ensuing the end of this present Parliament, plough, harrow, draw or worke with any horse, gelding, mare, garran or colt, by the taile, nor shall cause, procure of suffer any other to plough up or harrow his ground, or to draw any other carriages with his horses, mares, geldings, garrans or colts, or any of them, by the taile; and that no person or persons whatsover, shall, after the end of this present Parliament, pull the wool of any living sheep, or cause or procure to be pulled, instead of shearing or clipping of them; and if any shall doe contrarie to this act, and the intention thereof, that the justices of assize at the generall assizes to be holden before them, and the justices of peace at their quarter-sessions, shall have power by this act to enquire of, heare and determine all and every offence and offences done contrary to this present act, and to punish the offendors which shall do contrary to the same, by fine and imprisonment, as they in their discretion shall think fit.

Prior to this time, the English Parliament had passed a number of acts that placed limits on a number of practices involving animals, but none were anticruelty laws. Instead, they either limited certain activities to certain times (such as excluding cockfighting on Sundays) or certain places. For example, in 1488, Parliament passed an act prohibiting animal slaughter within the walls of London because of the blood, odors, and other foul things emanating from the practice. Bearbaiting, cockfighting, and bullbaiting were all subject to a number of acts and ordinances starting in the thirteenth century, many of which were aimed at controlling the unruly activities of the poor. But it was not until the seventeenth century that the first law prohibiting "animal cruelty" was passed.

From the seventeenth century until the nineteenth century, there was not a single animal protection law passed in either England or the United States (or elsewhere). In fact, the question of animal rights was so out of place for thinkers at that time that in 1792, after the publication of feminist Mary Wollstonecraft's book, *A Vindication of the Rights of Women with Strictures on Political and Moral Subjects*, English philosopher Thomas Taylor published the satirical work, *A Vindication of the Rights of Brutes*. Taylor thought the idea of rights for women so preposterous he argued that if women should have them, why should not animals have them too?

However, the late eighteenth century did see the emergence of a large body of philosophical writings on the question of animals—writings that would impact the formation of the animal rights movement in the twentieth century. Barrister Jeremy Bentham, for example, father of the Utilitarian school of philosophy, is the most influential figure of the period, crafting the first major challenge to the prevailing notion that animals are brutes with no souls, no reason, and no rights. Instead, Bentham argued that what is important in evaluating animals from a moral perspective is not reason, but the ability to suffer (Bentham 1781).

Other late-eighteenth-century thinkers shared Bentham's sentiments. A number of lawmakers in England and the United States wrote of their desire to see the passage of anticruelty laws. Many shared artist William Hogarth's sentiment that cruelty to animals could lead to cruelty toward people, and supported legislation that would ban the former. Physician (and signer of the Declaration of Independence) Benjamin Rush, for example, wrote that he was

> so satisfied of the truth of a connection between morals and humanity to brutes that shall find it difficult to restrain my idolatry for that legislature that shall first establish a system of laws, to defend them from outrage and oppression.
> (RUSH 1812)

Other writers opposed cruelty to animals for the animals' sake alone. Eighteenth-century English poet William Cowper's feelings toward three hares with whom he lived led him to take a public stand against blood sports. In a piece he wrote for *Gentlemen's Magazine*, Cowper wrote:

> You will not wonder, Sir, that my intimate acquaintance with these specimens of the kind has taught me to hold the sportsman's amusements in abhorrence; he little knows what amiable creatures he persecutes, of what gratitude they

are capable, how cheerful they are in spirits, what enjoyments they have of life, and that impressed as they seem with a peculiar dread of man, it is only because man gives them peculiar cause for it.
(1784:414)

Quakers and other members of other faiths also took stands against animal cruelty. The Quakers famously opposed hunting, and Reverend Charles Daubeny wrote "A Sermon on Cruelty to Dumb Animals" in 1799 which took the position that as God's creatures, they deserved equal treatment with man, and that those who abuse them will suffer in the hereafter, if not in this lifetime. Another preacher, Reverend Humphrey Primatt, wrote *A Dissertation on the Duty of Mercy and Sin of Cruelty to Brute Animals* in 1776. He pleaded:

> See that no brute of any kind . . . whether intrusted to thy care, or coming in thy way, suffer thy neglect or abuse. Let no views of profit, no compliance with custom, and no fear of ridicule of the world, ever tempt thee to the least act of cruelty or injustice to any creature whatsoever. But let this be your invariable rule, everywhere, and at all times, to do unto others as, in their condition, you would be done unto.
> (MERZ PEREZ AND HEIDE 2004:8)

One of the first major works on animal cruelty was John Oswald's *The Cry of Nature or An Appeal to Mercy and Justice on Behalf of the Persecuted Animals* (1791). Oswald was a political revolutionary who served in the British Army in India, and was inspired by Hinduism to become a vegetarian. He advocated compassion for humans and animals, and suggested that "there would come a day when the growing sentiment of peace and good-will toward men will also embrace, in a wide circle of benevolence, the lower orders of life" (1791). Just a few years later, Thomas Young published *An Essay on Humanity to Animals* (1798). Young recognized the links between the suffering of animals and the suffering of marginal peoples such as prisoners, slaves, and the poor, and advocated for the abolition of the slave trade as well as humane treatment of animals. And finally, the Englishman John Lawrence wrote *A Philosophical and Practical Treatise on Horses and on the Moral Duties of Man towards the Brute Creation*, that asked, "Can there be one kind of justice for men, and another for brutes?" (1796).

Also in the second half of the eighteenth century, we see a number of children's stories appear that encourage children, for the first time in print, to treat animals kindly. Dorothy Kilner's *The Life and Perambulation of a Mouse*

(1783), Sara Trimmer's *Fabulous Histories* (1786), Mary Wollstonecraft's *Original Stories from Real Life* (1788), Thomas Day's *History of Sandford and Merton* (1789), and John Aikin and Arma Barbauld's *Evenings at Home* (1795) are all examples of this trend. Here, children are encouraged to treat animals with kindness so that they may also learn to treat other social inferiors with kindness, such as slaves and servants.

The Animal Rights Movement: The First Wave

The animal rights movement really began in the nineteenth century with two separate movements: the prevention of cruelty movement and the antivivisectionist movement. These two distinct but overlapping movements can be together referred to as the first wave of the animal rights movement. This movement also overlapped extensively with the first wave of the women's rights movement—the suffrage movement—as well as the abolitionist movement. In fact, many of the same people were working to overthrow slavery, to give women the vote, and to end the suffering of animals, demonstrating again the link that we discussed in chapter 13 between the suffering of animals and the suffering of people.

The nineteenth century saw the first organized efforts to prevent animal cruelty in England, and later the United States. Activists attempted to get a bill passed in Parliament banning bullbaiting in 1800 but, after months of debate, it did not pass; another attempt was made in 1802 that was also unsuccessful. A more general bill, for "preventing wanton and malicious cruelty to animals" was proposed by Lord Erskine in 1809. Although it passed the House of Lords, it failed to pass the House of Commons, either in 1809 or 1810 when he proposed it again. It would not be till 1822 that Richard Martin introduced and saw passed the nation's first anticruelty law in 200 years. This act prevented the "cruel and improper Treatment of Horses, Mares, Geldings, Mules, Asses, Cows, Heifers, Steers, Oxen, Sheep, and other Cattle," and prohibited the cruel beating or ill treatment of these animals, subject to a minimum ten shilling fine or imprisonment up to three months in jail. Richard Martin, the Irish member of Parliament who ushered the 1822 legislation (commonly known as "Martin's Act") through the House of Commons, realized that the police would not have the manpower or will to enforce the bill. In 1824, the Reverend Arthur Broome stepped up and founded the Royal Society for the Prevention of Cruelty to Animals (RSPCA), which took as its mission the protection of animals and the enforcement of Martin's Act. There was only one other animal protection organization ever founded in England prior to that time (the Society for the Suppression and Prevention of Wanton Cruelty to Animals in 1809), but the organization did not last long. The RSPCA became the world's longest-lived animal protection organization, protecting cats from being skinned and dogs from being used as draft animals and alleviating the misery of draft horses. The organization also tried to ban dogfighting (which, along with cockfighting, was finally prohibited in 1835), bullfighting, and bullbaiting, and worked with police officers to educate them on the importance of Martin's Act.

As we discussed in chapter 9, experimentation on animals became a standard part of medical research starting in the seventeenth century, with wild animals, farm animals, and domestic cats and dogs being used as subjects. By the nineteenth century, knowledge about these experiments had made many members of the public uneasy. Even before this time, a number of important European thinkers such as Alexander Pope and Samuel Johnson had been critical of animal experimentation. But it was not until the nineteenth century that anti-vivisectionists organized themselves, forming the first anti-vivisection societies in England and leading to the first investigation of animal experimentation in England by a royal commission in 1875.

From the beginning, leading figures from other social justice movements of the period were involved in the fight to end animal experimentation. The Society for the Protection of Animals Liable to Vivisection, for example, was cofounded by women's rights advocate Frances Power Cobbe. Other groups from the period included the Society for the Abolition of Vivisection, the International Association, and the London Anti-Vivisection Society. The work of these organizations led to the passage of the world's first anti-vivisection legislation in 1876, the Cruelty to Animals Act. The law did not abolish animal experimentation, but instead mandated that physicians be licensed by the Home Secretary and that facilities conducting experimentation be inspected. Anti-vivisectionists, however, wanted more; they attempted to get Parliament to ban vivisection entirely, but were unsuccessful.

In the United States, the anticruelty movement was strongly influenced by what was happening in England at the time, and can be largely traced to the work of one man, Henry Bergh. Bergh was a New York City gentleman who traveled extensively, and witnessed the abuse of animals around the world. He visited London in 1865, and met with the then-president of the RSPCA, a meeting which led to Bergh's founding the American Society for the Prevention of Cruelty to Animals (ASPCA) upon his return to the United States the following year. Bergh found, as did his contemporaries in England, that the animals who suffered the most at that time were the large work animals—especially horses. They were being beaten and worked almost to death, and starved once they could work no more. Also as in England, there were a number of popular pastimes that involved cruelty to animals, such as bearbeating, dogfighting, and cockfighting. Within months of the founding of the ASPCA, New York passed a major anticruelty law that banned the cruel treatment of animals and included overdriving and overloading horses, cockfighting, bullbaiting, dogfighting, lack of food or water for animals, and abandoning them. In addition, the state gave the ASPCA (and Bergh himself) the power to enforce them. Bergh was known as "The Great Meddler" for involving himself in what had once been the private practices of Americans; now, when cart drivers beat their horses, Bergh would intervene, letting the drivers know that these practices were illegal. New York's laws became models for others to be passed around the country. During this time, the federal government passed its very first animal protection law, the Twenty-Eight-Hour Law of 1873. This law mandated that animals being shipped by rail be provided with rest, food, and water after 28 hours of transit. (Interestingly, it was only in 2005 that, thanks to the work of animal rights organizations, the USDA recognized that the dominant mode of transit for farm animals today is truck, not train, and began enforcing the law in that context.)

In addition, the ASPCA was the model for other SPCAs and humane societies that sprung up around the country starting in the late 1860; groups in Pennsylvania, Massachusetts, and San Francisco were among the first. By 1890, thirty-one states had such organizations. Originally concerned with enforcing anticruelty laws, many groups soon began running animal shelters following a model developed in Philadelphia. (Unlike England, which passed anticruelty laws that took effect in the entire country, the United States only has state-wide legislation.) The United States actually did see a number of states pass anticruelty laws prior to Bergh's time. New York passed the nation's first such law in 1828, followed by Massachusetts in 1835, and Wisconsin and Connecticut in 1838. But those laws had little chance for enforcement until the formation of the state ASPCAs, of which New York's was the first in 1866.

Ironically, it took one of the animal anticruelty laws that Bergh helped pass to help children suffering abuse in the United States. Prior to 1873, children were like animals—they were the property of the parents, and had no rights of their own or protection under the law. Bergh was introduced to a foster child named Mary Ellen Wilson who was being starved and beaten by her foster mother. Because no American law could intervene in the "private" activities between parents and children, the ASPCA brought the case before a court and convinced the court that the child was an animal, and thus deserved protection under the law of 1867. Bergh then received so many pleas on behalf of children that in 1875 he founded the American Society for the Prevention of Cruelty to Children to advocate on their behalf. Two years later, the American Humane Association was founded, in order to protect children and animals.

As in England, the other major arm of the growing animal protection movement in the United States was found in the anti-vivisection movement. The first organization in the United States dedicated to the abolition of vivisection was the American Anti-Vivisection Society (AAVS), founded in 1883 by Caroline Earle White. Others quickly followed: the New England Anti-Vivisection Society in 1895, the Vivisection Reform Society in 1903, and the New York Anti-Vivisection Society in 1908. But unlike in England, the U.S. anti-vivisection movement resulted in no laws at all—the American medical establishment, which opposed any regulations or restrictions at all on vivisection, was too powerful. In fact, the first law regulating animal experimentation in the United States would not appear until 1966 with the passage of the Laboratory Animal Welfare Act (later shortened to the Animal Welfare Act). Still, this movement played an important role in promoting the view that animals were God's creatures that should not be tortured.

As in England, many of the leading American anti-vivisectionists were drawn from other social justice movements. Caroline Earle White, for example, was a Quaker whose attorney father represented free blacks in his legal practice, and she was actively involved in the abolitionist movement. This was also an example of a type of nineteenth-century pursuit known as "social feminism," in which women such as White worked to expand the role of women in public institutions by the creation of separate institutions. For example, twenty years before the founding of the AAVS, she was the leading force in organizing the Pennsylvania Society for the Prevention of Cruelty to Animals, but because of her gender, she was not allowed to serve on the board.

By the 1920s, the anti-vivisection movement in England and the United States had lost much of its steam, but the humane movement was gaining power in both countries. With industrialization and the invention of the automobile, the use of horses as beasts of burden declined, so the movement's focus shifted from large animals such as horses and cattle to small animals such as cats and dogs. At the same time, pet keeping was increasing in importance in both countries, and issues relating to dogs and cats became ever more important to the public.

The Animal Rights Movement: The Second Wave

If the nineteenth-century animal protection movement was centered around opposition to vivisection, on the one hand, and protection of large working animals, on the other hand, the movement of the twentieth century began with concerns about companion animals. The United States in particular was the focus of vast economic and demographic changes, with fewer and fewer people working on farms or living in rural communities, and more people living in cities and, increasingly, suburbs—far away from animal agriculture. As Americans grew more distant from wild and farmed animals, companion animals became more important, a trend that really picked up after the end of World War II and has continued, unabated, ever since. The expansion of pet keeping led in part to the growth of humane attitudes toward animals and greater concern toward companion animals in particular. From the state SPCAs founded in the late nineteenth century, more and more communities began forming local humane societies to protect animals. The mid-twentieth century also saw the founding of a handful of national humane organizations, including The American Welfare Institute in 1950, The Humane Society of the United States, which split, in 1954, from the American Humane Association, and The Society for Animal Protective Legislation in 1955. These organizations, and others like them, were responsible for every major animal protection law in this country, including the Humane Methods of Slaughter Act (1958), the Laboratory Animal Welfare Act (1966), the Endangered Species Act (1969), the Wild Horse and Burro Protection Act (1970), and the Marine Mammal Protection Act (1972). It was in this era that Americans began to see specific kinds of treatments against certain kinds of animals as a social problem, in the ways that the feminist and civil rights movement identified sexism and racism as social problems.

Even though activists had been working to protect animals in England and the United States since the early nineteenth century, it really was not until the 1970s that what most people think of as the animal rights movement was born. This movement, or the second wave of animal rights, arose with the publication of philosopher Peter Singer's groundbreaking book, *Animal Liberation*, in 1975. Singer's work and, in the 1980s, the work of Tom Regan provided the philosophical underpinnings for the modern animal rights movement and gave rise to the current generation of animal activists. The 1970s were a particularly fertile time for the emergence of an animal rights philosophy, because it was also at this time that we saw the maturation of the civil rights movement as well as the rise of the women's movement. In fact, the 1970s was a time during which a number of social movements arose, including self-help, New Age, women's spirituality, the men's movement, and the ecology movement. At this time, there was a turning inward among many in the United States such that much of the social activism of the earlier period shifted toward an emphasis on personal transformation. According to many observers, this was a moment of profound importance—a revolution of human consciousness that affected how people thought about religion, society, others, and the self. The animal rights movement, then, even though ostensibly focused on an external "other" (i.e., animals) is also heavily self-referential, attracting men and women who engage in a great deal of introspection as well.

The Modern Animal Rights Movement

Today, the animal rights movement is an international movement that can be roughly divided into three areas: philosophical debate regarding the status of animals, legislative work, and direct action. Although the nineteenth century had been about humane treatment of working animals plus the abolition of (or restrictions on) vivisection, and the early twentieth century was about the humane treatment of companion animals, the modern animal rights movement encompasses a variety of issues and concerns, from the end of the use of animals in circuses to the promotion of a nonanimal diet to the opposition to the wearing of fur or leather.

Even though those who define themselves primarily as animal protection advocates, or animal welfare advocates, seek to protect animals from abuse and harm, the animal rights movement seeks an end to the status of animals as property, and an end to the use of animals in research, and for food, clothing, and entertainment. This conflict—between animal rights activists or abolitionists and animal welfare advocates, as well as between those (sometimes called pragmatists) who argue for the special treatment of certain species such as apes and dolphins—has been present for years in the movement.

Except for the Humane Society of the United States (HSUS), founded in 1954 and today a leader in the modern animal rights movement, all of the other American organizations at the center of this movement were founded since the 1970s. This includes the Animal Legal Defense Fund established in 1979 and a leader in working to gain animals legal victories, People for the Ethical Treatment of Animals (PETA) in 1980, the Farm Animal Reform Movement in 1981, Farm Sanctuary in 1986, and Compassion over Killing in 1995.

The 1970s, 1980s, and 1990s were the years of protests, demonstrations, public outreach, and direct action. Animal rights activists, either alone or with the support of major organizations, publically protested the use of animals in medicine, in agriculture, and in entertainment, and sought to educate the public about their treatment, generally with the assistance of the media. Organizations such as PETA and HSUS provided pamphlets, stickers, and flyers to distribute, and events such as **Fur-Free Friday** (held the day after Thanksgiving to oppose the use of fur in fashion) became public symbols of the movement. Shadow groups such as the **Animal Liberation Front** (not really a group so much as a loose confederation of individuals with shared goals) performed direct action, breaking into fur farms, animal labs, and factory farms, freeing animals, and taking undercover footage of animal suffering. Because of actions such as this, the animal rights movement has sometimes been compared to the **prolife movement**. Both movements not only protect "those without a voice" but also use similar methods, which include picketing locations where animals or fetuses are harmed and even the homes of physicians who are thought to cause that harm, engaging in **nonviolent civil disobedience**, and, within the militant wings of the movements, engaging in property damage. The difference, however, is that although militant anti-abortionists have actually killed people, no animal rights activist, no matter how extreme, has ever harmed a person during a protest or raid. The question about methods—direct action versus legislation—has been simmering in the animal rights movement since the 1970s, with each side claiming more victories.

Protests, demonstrations, and direct action—even though still methods employed by many in the movement—are rarely used by the major organizations that spend their resources on lobbying and legislation. On the other hand, undercover videos are still widely used by organizations big and small to direct public attention to the many abuses of animals in U.S. industries today—especially the agricultural and entertainment industries.

What makes a successful animal rights campaign? According to sociologist Rachel Einwohner (1999a), some campaigns are less successful if the practice—hunting, say, or biomedical research—is seen to be necessary by those who engage in it. Attending a circus, on the other hand, or wearing fur, is not seen as a necessary activity, so convincing people to forego attending a circus or wearing a fur coat is much easier than asking a hunter to stop hunting. Einwohner argues that animal rights activists should consider their target audience when engaging in protests, and whether or not it is feasible for that audience to be persuaded of their arguments. In general, animal rights organizations are fighting a battle on multiple fronts: They are fighting against tradition, the profit motive of the corporations who depend on animals for their profits, human greed and vanity, the intractability of the legal system, and the invisibility of much of animal oppression. Sociologist Bonnie Berry (1997) suggests as well that as human rights recede in a society, which she argues occurred in the United States during the George W. Bush administration from 2000 to 2008, activists will have a much harder time convincing the public to embrace animal rights.

Animal rights activists, as well as animal welfare activists and organizations, scored a number of important victories during the last few decades of the twentieth century. One of the movement's biggest accomplishments was educating the public about the unnecessary suffering of animals in product testing, which has resulted in a huge number of major companies having stopped using animal testing for their cosmetic and household products. In general, nonanimal methods have radically increased in product testing and, to a lesser extent, in medical research. Another major accomplishment has been the rising awareness of the cruelties inherent in factory farming, and the resulting growth in vegetarianism and veganism. Finally, each state's anti-cruelty statutes, many of which were originally passed in the late nineteenth century, have been considerably strengthened in recent years, thanks to the work of animal rights organizations.

In fact, it is this last issue—the involvement of the animal rights movement in legislation—that is perhaps the most important area in which the movement is involved today. Although just a generation ago, animal rights activists were thought to be college students noisily holding protests in front of McDonald's or Kentucky Fried Chicken, today they are just as likely to be working quietly behind the scenes, as lawyers, publicists, or lobbyists, changing public attitudes and influencing legislation. In the past few years, the movement has passed major legislation, in the United States and elsewhere in the world. Even though the United States still has only three federal laws related to animal welfare—the Animal Welfare Act, the Twenty-Eight Hour Law, and the Humane Methods of Slaughter Act—these laws do not address cruelty. This is why each state has their own cruelty statutes, many of which, unfortunately, exclude farm animals from any sort of protection. Other nations, however, have laws that protect animals. For example, all of the nations comprising the European Union must abide by the European Convention for the Protection of Pet Animals, which protects companion animals from pain, suffering, and abandonment. Similar laws exist—the European Convention for the Protection of Animals for Farming Purposes, the European Convention for the Protection of Animals during International Transport, and the European Convention for the Protection of Animals for Slaughter—that cover farmed animals. All of these laws were passed in the past twenty-five years, and all have more comprehensive protection than any federal law in the United States. In addition, the European Union has either banned or been phasing out many of the most egregious farm animal practices, such as **battery cages** for egg-laying hens, **gestation crates** for pregnant pigs, and **veal crates** for veal calves—all practices that are still legal in the United States.

In recent years, some states—thanks to the work of organizations such as HSUS—have passed laws protecting farm animals. From 1990 to 2008, state legislatures introduced thirty-two animal protection measures, twenty-two of which passed. Florida and Arizona, for example, passed laws outlawing the use of gestation crates (in 2002 and 2006, respectively), and in 2008, California passed the nation's strictest anti-farm animal cruelty law when it passed Proposition 2. This legislation outlawed gestation crates and veal crates, and also mandated that chickens be kept in cages large enough for them to turn around, lie down, and fully extend their wings. The animal rights movement of the 1980s would have been unable to marshal the public support to get these initiatives onto state ballots, much less obtain the majority of the states' votes. But today, as the animal rights movement has matured, these types of victories seem like only the beginning.

Animal rights organizations currently have a number of major legislative aims, all of which involve federal legislation. This includes amending the Humane Methods of Slaughter Act to include poultry (and rabbits), amending the Animal Welfare Act to include rodents and birds, protecting **downed animals** from slaughter, and extending California's Proposition 2 protections to all farmed animals in the country. The HSUS, for example, has created a political action committee known as Humane USA, whose sole purpose is to lobby federal politicians and candidates for animal-friendly legislation.

Some organizations have their sights set on even greater goals. One campaign, for example, would have animals (or certain animals, such as great apes and cetaceans) gain legal standing as "persons," which would safeguard their liberty and protect them from suffering, regardless of the patchwork of laws that exist today. Attorney Steven Wise founded the Nonhuman Rights Project that aims to gain legal **personhood** status for great apes. Wise's goal is to find an imprisoned chimpanzee or other great ape on whose behalf he could file a lawsuit based on the **writ of habeas corpus**, which was once used to free slaves in the British Empire. Another goal is to establish a Federal Animal Protection Commission, similar to the Commission on Civil Rights that Congress established in 1957, to investigate violations of animal protection laws and promote better laws.

BOX 19.2

ORGANIZATIONAL FOCUS: PETA

PETA is one of the world's largest animal rights organizations, with an annual budget of over $30 million and approximately two million members worldwide. Founded in 1980 by Ingrid Newkirk and Alex Pacheco, PETA became well-known for their fight to save the Silver Spring monkeys (discussed in chapter 9). Their campaigns cover factory farming, animal experimentation, companion animal cruelty, fur, and the use of animals in entertainment.

They are among the most successful and the most controversial of all animal protection organizations. Their staff engages in undercover operations that result in highly public video releases. Their advertisements and other public campaigns are thought provoking and much-criticized, including their use of naked celebrities in ads and sometimes controversial spokespersons such as Pamela Anderson and Alec Baldwin. Many of their campaigns are intended to shock the public, such as when PETA compared the use of animals in agriculture to Jews dying in the Holocaust, or when they parodied the dairy industry's "Got Milk" campaign with a "Got Beer" advertising campaign aimed at encouraging college students to drink beer instead of milk.

Controversy notwithstanding, PETA's campaigns have been extremely successful. Thanks to their efforts, McDonald's and Wendy's have modified the way that some of the animals used in their food are raised, a host of major designers have stopped using fur in their clothing, and a number of major cosmetics corporations have stopped animal testing.

Demographics: Who Becomes an Animal Rights Activist?

What makes a person become an animal rights activist? By using Kellert's animal attitudes scale, we can say that animal rights activists are those who have a moralistic attitude toward animals, defined as "the primary concern for the right and wrong treatment of animals, with strong opposition to exploitation or cruelty toward animals" (1980:89). More than any other type of person, those who gravitate toward this movement score the highest in moralism. As we noted in chapter 11, according to scholars there are three broad sets of factors related to attitudes about animals: social position (class, age, gender, education, income, employment, ethnicity), environmental attitudes, and current animal-related experiences and practices. In addition, as

Figure 19.1. Mercy for Animals activists protest the cruelty associated with modern egg production. (Photograph courtesy of Mercy for Animals.)

we have discussed elsewhere in this text, attitudes differ according to the species of animal (with pandas, for example, generating more positive attitudes than snakes or rats) and the role that that species plays in human culture.

Statistics show that animal activists are more likely to be white, female, urban, middle class, and young, and to have jobs that do not depend on animal use (Plous 1998; Lowe and Ginsberg 2002). In recent years, animal rights support may be becoming more diffuse in society at large, in particular among younger, less educated, and less well-to-do individuals (Jerolmack 2003). They are also more likely to have had pets as a child, and to have pets as an adult. They are more likely to score higher on empathy on personality tests, and to have positive attitudes toward the environment. According to a 1984 survey of readers of *Animals Agenda*, 65 percent of respondents classified themselves as atheists or agnostics, demonstrating that lack of religious affiliation is correlated with animal rights advocacy. Sociologist David Nibert (1994) has also found that support for animal rights is correlated to other political positions. For instance, support for animal rights is related to gun control (high), acceptance of violence (low), and acceptance of diversity for people (high), although opposition to animal rights is related to gun access (high), opposition to abortion (high), racism and homophobia (high), and acceptance of violence (high). As with many of these correlations, it is difficult to tell the causal relationship among the variables, however. But certainly, political conservatism tends to be adversely correlated to animal rights activism. Religious affiliation, too, is correlated to support for animal rights (DeLeeuw et al. 2007), with creationists and other religious fundamentalists being negatively correlated with support for animal rights.

Other scholars suggest that a disposition toward an optimistic worldview, coupled with an idealistic moral absolutism, is an important component in the psychological makeup of those engaged in social activism such as animal rights. Psychologist Hal Herzog has a different take on why some people become animal rights activists (Herzog and Golden 2009). Even though most research on what people think about animal use has concentrated on positive emotions such as empathy, Herzog examines the role that a negative emotion—visceral disgust—plays in animal activism. His study finds that animal activists are more prone to disgust than those who are neither animal activists nor vegetarians. Interestingly, Herzog also demonstrated that many (45 percent, based on his study) animal activists are not vegetarians, demonstrating that although vegetarianism and animal activism are strongly correlated, being an animal activist does not necessarily mean that one will be a vegetarian, and being a vegetarian does not necessarily mean that one will be an animal activist. In addition, Herzog's study did not specify what *kind* of animal activist these respondents were—animal welfare, animal rights, etc. On the other hand, in a study by Brooke Dixon Preylo and Hiroko Arikawa (2008), male vegetarians definitely demonstrate more empathy toward other animals than do nonvegetarians, showing a correlation between vegetarianism and animal activism, at least for men. This study found that whether one was a vegetarian for moral or other reasons (such as health), the correlation between a vegetarian diet and empathy toward other animals was consistent.

Sociologist Lyle Munroe (2005) found that activists got involved in the movement through three main avenues. They had intellectual, emotional, and practical motivations. Intellectual motivations could have included hearing a talk by an animal rights activist or reading a book by a leading philosopher in the movement; emotional reasons, which are the most numerous, could range from seeing images of animals being slaughtered to the love of a companion animal; and practical reasons include actual experiences that activists had, such as rescuing an animal. Welsey Jamison, Caspar Wenk, and James Parker (2000) see getting involved in animal rights as akin to religious conversion, with one or more formative experiences being the catalysts for conversion. Once converted, new activists join a community of "believers" that reinforces their beliefs and provides a support system for them. Ultimately, these sociologists argue that animal rights activism becomes a functional religion, offering a belief system, set of rituals, and community of shared believers.

In terms of age, younger activists are more prominent among those who engage in demonstrations and direct action, and older people tend to make up much of the leadership of the movement. As we mentioned before, in terms of ethnicity and class, the movement is overwhelmingly white and middle class. Women make up about 75 percent of activists and about 70 percent of donors to animal protection organizations. On the other hand, men make up the majority of activists who are engaged in direct action, and also make up the majority of the leadership in the animal welfare and animal rights side of the movement. The gender breakdown in the animal protection movement can be correlated to a broad difference in two types of work: what Lyle Munroe (2005) calls "caring work," primarily engaged in by women, and intellectual work, primarily engaged in by men.

A number of scholars have addressed the question of gender—why are so many more women involved in animal welfare and animal rights? Some point to the many parallels between the exploitation of animals and the exploitation of women. For example, men who abuse women often abuse animals, and vice versa. As feminist scholar Carol Adams (1991) has pointed out, many of the worst abuses in animal agriculture occur toward female animals such as egg-laying hens and dairy cattle. Women and animals are often associated symbolically; both may be constrained by their bodies, and both are linguistically linked through degrading terms such as those we discussed in chapter 14. Women are also the target of two major animal rights campaigns: the anti-fur and the anti-product-testing movement. Those products (fur and cosmetics) are largely consumed by women. They are also the main subject of one of the most long-running and controversial advertising campaigns in the movement: PETA's "I'd rather go naked than wear fur" campaign, which features photos of naked women—generally celebrities— and the iconic tagline. Many feminists, within and outside of the movement, find these ads (and other PETA ads that feature naked or scantily clad women) objectifying and degrading.

In addition, women are socialized to be more empathetic than men, and are the caretakers of the species; thus, it makes sense that they would be more drawn to caring for animals. And it is not just women who score higher on empathy than men in tests. When surveyed on sex role orientation, men whose scores rank them closer to "feminine" also score higher on

Figure 19.2. Anita Carswell at a weekly anti-fur demonstration. (Photograph courtesy of Anita Carswell.)

empathy and pro-animal attitudes (Herzog, Betchart, and Pittman 1991). And finally, historically, women were heavily involved in the abolition movement and the suffragist movement; many of those same women saw the links among the different forms of oppression and were also involved in animal protection. Unfortunately, the major involvement of women in the movement, combined with the movement's strong emotional appeal, make the movement easy to criticize from the outside—it is "too emotional," and thus not important enough to take seriously.

Sociologist Lyle Munroe sums up the motivations of those who are involved in the animal rights movement by paraphrasing Sigmund Freud: "people need love and work to give their lives meaning" (2005:1). In the context of the animal rights movement, that means that people use their love of animals to drive their commitment to the movement, and they put that caring to work in their activism, through making personal changes such as going vegetarian or vegan, and through political and social advocacy. That this movement has made significant legal and social changes is without a doubt; how these changes will continue to play themselves out in the human-animal relationship of the future is the question.

Place of the Movement in Contemporary Society

The animal rights movement is both more mainstream, yet also more politically threatening, than ever before. The organizations that were founded in the 1970s and 1980s are maturing and have increasingly focused their efforts on legislation rather than demonstrations. Veterinarians, lawyers, physicians, and other professionals now play a major role in the movement, and it has been bolstered in part by the spread of human-animal studies across college campuses around the world. Although the study of human-animal relations began in philosophy departments, it has now expanded into disciplines in the social sciences, the humanities, and even the natural sciences, and hundreds of law schools now offer courses in animal law. Many of the scholars who consider these issues from an academic perspective are also involved in practical applications of their theories, typical of other academic disciplines that are grounded in social justice movements (namely ethnic, women's, and environmental studies). Even though animal rights advocates may cheer this development, others feel that it can inhibit the growth of the field because many scientists will be threatened by the presence of animal rights activism in their universities.

In the past decade, dozens of new laws have been passed at the state and local level giving animals additional layers of protection. In 1999, Humane USA, the nation's first political action committee devoted to electing animal-friendly candidates to state and federal office, was formed. Although the federal government has been slow to take up animal legislation, in 2009 animal-friendly congressional members formed the bipartisan Congressional Animal Protection Caucus, a group of over 80 Representatives who work to pass animal welfare legislation. Through the actions of Humane USA, the Congressional Animal Protection Caucus (and its predecessor, the Friends of Animals Caucus), and animal rights organizations, this country has seen a number of new laws passed in recent years. In 2011, animal activists' goals included bills banning horse slaughter, the aerial killing of wolves, the trade in bear parts, canned hunting, invasive research on chimpanzees, puppy mills, and Internet hunting. They also would like to see Congress pass a tax credit for spaying and neutering pets, a law mandating that companies that make fur-trimmed garments label their products accurately, and a law protecting wild horses from slaughter and implementing sanctuaries and birth control as a humane means of control.

Even with all of this professional activity, many still consider animal rights to be a fringe movement at best, and a "terrorist" enterprise at worst. In fact, in 2006, Congress passed the Animal Enterprise Terrorism Act, which "prohibits any person from using force, violence, or threats for the purpose of damaging or interfering with the operations of an animal enterprise" and increases the penalties for property crimes (or bodily harm, even though no person has ever been harmed by animal rights activism) motivated by animal rights. It was passed thanks to extensive lobbying by the biomedical industry, which had been frequent targets of ALF raids, but the new law protects any enterprises that use or sell animals or animal products. Animal rights activists—most of whom do not engage in direct action, much less terrorism, or illegal activities such as raids—saw the passage of this bill as a way to stifle activists' constitutional rights of protest. In addition, many activists are worried that **whistleblowers** and those taking undercover videos of animal operations will be afraid to undertake these activities, or any other type of civil disobedience, in the future. It is worth noting that no other industry in the United States has ever received such protections against public protests and other forms of activism. However, those businesses that depend on animal use and exploitation, such as biomedicine and agribusiness, have been able to marshal their considerable resources in order to secure such a law.

Suggested Additional Readings

Beers, Diane. 2006. *For the Prevention of Cruelty: The History and Legacy of Animal Rights Activism in the United States.* Athens, OH: Swallow.

Finsen, Lawrence and Susan Finsen. 1994. *The Animal Rights Movement in America.* New York: Twayne.

Francione, Gary L. 1996. *Rain Without Thunder: The Ideology of the Animal Rights Movement.* Philadelphia: Temple University Press.

Garner, Robert, ed. 1996. *Animal Rights: The Changing Debate.* New York: New York University Press.

Guillermo, Kathy Snow. 1993. *Monkey Business: The Disturbing Case That Launched the Animal Rights Movement.* Washington, DC: National Press Books.

Guither, Harold D. 1998. *Animal Rights: History and Scope of a Radical Social Movement.* Carbondale: Southern Illinois University Press.

Munro, Lyle. 2005. *Confronting Cruelty: Moral Orthodoxy and the Challenge of the Animal Rights Movement (Human-Animal Studies).* Leiden, the Netherlands: Brill Academic Publishers.

Silverstein, Helena. 1996. *Unleashing Rights: Law, Meaning, and the Animal Rights Movement.* Ann Arbor: University of Michigan Press.

Wand, Kelly. 2003. *The Animal Rights Movement.* San Diego: Greenhaven Press.

Suggested Films

Behind the Mask: The Story of the People Who Risk Everything to Save Animals. DVD. Directed by Shannon Keith. Los Angeles: Uncaged Films, 2006.

Peaceable Kingdom: The Journey Home. VHS. Directed by Jenny Stein. Ithaca, NY: Tribe of Heart, 2009.

The Witness. DVD. Directed by Jenny Stein. Ithaca, NY: Tribe of Heart, 2000.

Websites

American Anti-Vivisection Society: http://www.aavs.org

American Humane Association: http://www.americanhumane.org

American Society for the Prevention of Cruelty to Animals: http://www.aspca.org

Defenders of Wildlife: http://www.defenders.org

Great Ape Project: http://www.greatapeproject.org

Humane Research Council: http://www.humanespot.org

Humane Society of the United States: http://www.humanesociety.org

Mercy for Animals: http://www.mercyforanimals.org

The Nature Conservancy: http://www.nature.org

People for the Ethical Treatment of Animals: http://www.peta.org

Physicians Committee for Responsible Medicine: http://www.pcrm.org

World Wildlife Fund: http://www.wwf.org

20

The Future of the Human-Animal Relationship

Man is the only creature that consumes without producing. He does not give milk; he does not lay eggs; he is too weak to pull the plough; he cannot run fast enough to catch rabbits. Yet he is lord of all the animals. He sets them to work, he gives back to them the bare minimum that will prevent them from starving, and the rest he keeps for himself.
(ORWELL 1946:7)

WHERE WILL OUR RELATIONSHIP with animals go in the twenty-first century? Even though we no longer need meat, fur, or leather to survive, most people seem unwilling or unable to shed their dependence on the products of animal agriculture. Even more intractable may be our connection to our companion animals. According to a 2002 American Animal Hospital Association pet owners' survey, 73 percent of Americans have signed a greeting card from their dog, 86 percent include pets in holiday celebrations, 46 percent plan all or most of their free time around their animals, 58 percent include pets in family portraits, and almost half have more photos of their pets than their partners. Given this level of commitment, it is difficult to imagine most pet lovers being willing to live without animals.

In addition, virtual animals are playing ever greater roles in our lives—through the television shows and movies that we watch, the video games that we play, the books that we read, and through the Internet websites that we visit. For many people, a workday is not complete without a little bit of time wasted on a website such as I Can Has Cheezburger? or Cute Overload.

It appears that humans need and want to stay connected to animals. Since the disappearance from human society of wild animals and even the farm animals on whom we still depend, some scholars see the presence and importance of pets as a testimony to an enduring and necessary link between human and nature that provides, albeit incompletely, in ecologist Paul Shepard's words, "a glimmer of that animal ambience, sacredness, otherness" (1996:141).

Sociologist Adrian Franklin (1999) positions the changing attitudes towards animals in the postmodern era as deriving from a number of cultural and economic changes in society. In particular, he suggests that ontological insecurity (the feeling that one's physical and social environment is unpredictable), risk reflexivity (the feeling that humans and animals are in danger because of the damage that humans have done to the world), and misanthropy (a feeling of dislike or distaste toward and about humans) have all shaped our evolving relationships with animals. Thanks to these three attitudes, humans have been drawing closer to animals than ever before.

In addition, the information that the public is receiving from the field of ethnology, discussed in chapter 17, has also strongly influenced the changing role of animals in society today. As we learn more and more about the abilities and feelings of animals (They can use human language! They bury their dead!), the more many people realize that the border between animal and human (animal) is porous and shifting, and may not even mean much at all. Certainly, we feel less able today to justify many of the ways in which we treat at least some animals, based on how much we now know about their intellectual and emotional capabilities, and how close many animals really are to us.

Today, the relationship between human and domestic animal continues to evolve. Companion animals are becoming ever more drawn into human lives, providing love and companionship but also, perhaps, filling a more complicated need for humans to connect with other species. On the other hand, agricultural animals are becoming increasingly distanced from us, shut away in factory farms, and slaughtered in secret. But farmed animal and pet are moving physically further away from us as well, as they are genetically engineered and even cloned to ever more exacting human specifications, losing much of their "animal-ness" in the process.

Another trend today is the rise in humane sentiments among the population—in the United States and elsewhere in the world. More people are contributing to animal welfare organizations than ever before, and more people have become personally involved in animal issues—fostering animals, cutting out meat from their diet, protesting circuses, or writing letters to the editor regarding animal issues in their hometowns. But this rise in animal activism is coupled with, and caused in part by, the rise in exploitation of some animals. This is perhaps the hardest thing to get a handle on with respect to the human-animal relationship today: how we are creating deeper relationships with animals, bringing them closer to us and making them ever more important parts of our lives yet at the same time tolerating ever-increasing levels of cruelty, confinement, and suffering in agriculture primarily but also in biomedicine. One response to these two poles—concern and intimacy on the one hand and distancing and invisibility on the other—is the increasing importance of the animal protection movement in the United States and around the world.

In just the past two decades, legislative efforts to change the role of animals in our lives have radically increased. European countries, in particular, have passed a number of major laws that protect livestock while they are being raised, during transport, and during slaughter, and that protect laboratory animals from some of the worst abuses. Although the United States lags behind European efforts, the past twenty years have seen a number of important laws passed in state legislatures protecting animals. Most of those laws protect only companion animals, providing greater penalties for (companion) animal cruelty, for example, but a handful protect wild animals and even livestock as well. The future will no doubt bring more extensive laws protecting animals, and the personhood movement that we discussed in chapter 19 is gaining steam.

Sometimes it takes having the extraordinary occur to expose the cultural sentiments about a subject. For instance, 2005's Hurricane Katrina exposed a number of contradictory aspects of the human-animal relationship as it exists in contemporary America. It demonstrated the great love for their pets that many people have; many of those people stayed behind after mandatory evacuations were announced and at great risk to themselves because they did not want to leave their pets behind. It also showed the altruism of hundreds of volunteers who came to the Gulf Coast to help rescue the animals left behind; thanks to the work of these selfless people, at least 10,000 animals were saved. But it also exposed many other, less positive sentiments, such as the government's lack of concern for pets in crafting evacuation policies. Even though many people were forced by government officials to leave their pets, others apparently did so voluntarily, leaving tens of thousands of dependent dogs, cats, and other animals to starve or drown. And of the 10,000–15,000 animals that were rescued after Katrina, only 15–20 percent were ever reunited with their families; why did so many families fail to reclaim their animals? And animals living in industrial farming conditions fared much worse, with hundreds of thousands of chickens, cows, goats, and other animals left behind to die; very few people shed a tear over that loss of life. The good news is that after this disaster exposed so many cracks in our animal caretaking system, President Bush signed the Pets Evacuation and Transportation Act in October 2006, which mandates that state and local authorities include animals in evacuation plans. A dozen states have since passed their own such laws. If anything, the example of Hurricane Katrina exposes our cultural ambivalence toward animals, an ambivalence that appears to be growing in the twenty-first century.

Bibliography

Abdill, Margaret N. and Denise Juppe, eds. 1997. *Pets in Therapy*. Ravensdale, WA: Idyll Arbor.

Acampora, Ralph. 2005. "Zoos and Eyes: Contesting Captivity and Seeking Successor Practices." *Society & Animals* 13(1).69–88.

Acampora, Ralph. 2006. *Corporal Compassion: Animal Ethics and Philosophy of Body*. Pittsburgh: University of Pittsburgh Press.

Acampora, Ralph, ed. 2010. *Metamorphoses of the Zoo: Animal Encounter after Noah*. Lanham, MD: Lexington.

Adams, Frost, L. 1991. "Pets and Lovers: The Human—Companion Animal Bond in Contemporary Literary Prose." *Journal of Popular Culture* 25(1):39–55.

Adams, Carol and Josephine Donovan, eds. 1994. *Animals and Women: Feminist Theoretical Explorations*. Durham, NC: Duke University Press.

Adams, Carol J. 1991. *The Sexual Politics of Meat: A Feminist—Vegetarian Critical Theory*. New York: Continuum.

Adams, Carol J., ed. 1993. *Ecofeminism and the Sacred*. New York: Continuum.

Adams, Carol J. 1994. *Neither Man nor Beast: Feminism and the Defense of Animals*. New York: Continuum.

Adams, Carol J. 2004. *Pornography of Meat*. New York: Continuum.

Adams, Frost, L. 1991. "Pets and Lovers: The Human—Companion Animal Bond in Contemporary Literary Prose." *Journal of Popular Culture* 25(1):39–55.

Aftandilian, David, ed. 2007. *What Are the Animals To Us? Approaches from Science, Religion, Folklore, Literature, and Art*. Knoxville: University of Tennessee Press.

Aikin, John, Anna Laetitia (Aikin) Barbauld, and Dalziel Brothers. 1879. *Evenings at Home*. London: Frederick Warne.

Akhtar, Salman and Vamik D. Volkan, eds. 2003. *Cultural Zoo: Animals in the Human Mind and its Sublimations*. Madison, CT: International Universities.

Alford, V. 1978. *The Hobby Horse and Other Animal Masks*. London: Merlin Press.

Alger, Janet and Steve Alger. 2003. *Cat Culture: The Social World of a Cat Shelter*. Philadelphia: Temple University Press.

Allen, Barbara. 2009. *Pigeon*. London: Reaktion.

Allen, C., and M. Bekoff. 1999. *Species of Mind: The Philosophy and Biology of Cognitive Ethology*. Cambridge, MA: MIT Press.

Allen, Jessica M., Diane Hammon Kellegrew, and Deborah Jaffe. 2000. "The Experience of Pet Ownership as a Meaningful Occupation." *Canadian Journal of Occupational Therapy* 57(4):271–278.

Allen, Karen Miller. 1985. *The Human—Animal Bond: An Annotated Bibliography*. Metuchen, NJ: Scarecrow.

Allen, Mary. 1983. *Animals in American Literature*. Urbana: University of Illinois Press.

Anderson, J. K. 1985. *Hunting in the Ancient World*. Berkeley: University of California Press.

Anderson, Patricia K. 2003. "A Bird in the House: An Anthropological Perspective on Companion Parrots." *Society & Animals* 11(4):393–418.

Anderson, R. K., B. L. Hart, and L. A. Hart, eds. 1984. *The Pet Connection: Its Influence on Our Health and Quality of Life*. Minneapolis: University of Minnesota Press.

Anderson, R. S., ed. 1984. *Pet Animals and Society*. London: Bailliere Tindall.

Anderson, Virginia DeJohn. 2004. *Creatures of Empire: How Domestic Animals Transformed Early America*. New York: Oxford University Press.

Animal Studies Group. 2006. *Killing Animals*. Champaign: University of Illinois Press.

Anthony, Lawrence and Graham Spence. 2009. *The Elephant Whisperer: My Life with the Herd in the African Wild*. New York: Thomas Dunne.

Appleby, Michael C. 1999. *What Should We Do About Animal Welfare?* Oxford: Wiley-Blackwell Science.

Appleby, Michael C. and Barry O. Hughes, eds. 1997. *Animal Welfare*. Wallingford, England: CAB International.

Aquinas, Saint Thomas and Berardus Bonjoannes. 1906. *Compendium of the Summa Theologica of St. Aquinas: Pars Prima*. London: Thomas Baker

Archetti, E. P. 1997. *Guinea Pigs: Food, Symbol and Conflict of Knowledge in Ecuador*. Translated by V. Napolitano and P. Worsley. Oxford: Berg.

Aristotle and Benjamin Jowett. 1943. *Aristotle's Politics*. New York: Modern Library.

Arkow, Phil. 2004. *Pet Therapy: A Study and Resource Guide for the Use of Companion Animals in Selected Therapies*, 9th ed. Stratford, NJ: Privately printed.

Arluke, Arnold. 2002. "Health Implications of Animal Hoarding: Hoarding of Animals Research Consortium." *Health and Social Work* 27(2):125–136.

Arluke, Arnold. 2006. *Brute Force: Animal Police and the Challenge of Cruelty*. Lafayette, IN: Purdue University Press.

Arluke, Arnold. 2008. "Hope and Conflict in the Social World of Animal Sheltering," in *Between the Species: Readings in Human-Animal Relations*, ed. Arnold Arluke and Clinton Sanders. Boston, MA: Pearson Education, pp. 270–280.

Arluke, Arnold and Carter Luke. 1997. "Physical Cruelty towards Animals in Massachusetts, 1975–1996." *Society & Animals* 5(3):195–204.

Arluke, Arnold and Clinton Sanders. 1996. *Regarding Animals*. Philadelphia: Temple University Press.

Arluke, Arnold and Clinton Sanders, eds. 2009. *Between the Species: A Reader in Human—Animal Relationships*. Boston, MA: Pearson Education.

Armbruster, Karla and Kathleen R. Wallace, eds. 2001. *Beyond Nature Writing: Expanding the Boundaries of Ecocriticsm*. Charlottesville: University of Virginia Press.

Armstrong, Philip. 2008. *What Animals Mean in the Fiction of Modernity*. New York: Routledge.

Armstrong, Susan and Richard Botzler. 2008. *The Animal Ethics Reader*. London: Continuum.

Arnold, Albert J., ed. 1996. *Monsters, Tricksters, and Sacred Cows: Animal Tales and American Identities*. Charlottesville: University of Virginia Press.

Ascione, Frank. 1997. "Humane Education Research: Evaluating Efforts to Encourage Children's Kindness and Caring Toward Animals." *Genetic, Social and General Psychology Monographs* 123(1):57–77.

Ascione, Frank. 2005. *Children and Animals: Exploring the Roots of Kindness and Cruelty*. West Lafayette, IN: Purdue University Press.

Ascione, Frank. 2008. *The International Handbook of Animal Abuse and Cruelty: Theory, Research and Application*. West Lafayette, IN: Purdue University Press.

Ascione, Frank and Phil Arkow, eds. 1999. *Child Abuse, Domestic Violence, and Animal Abuse: Linking the Circles of Compassion for Prevention and Intervention*. West Lafayette, IN: Purdue University Press.

Ascione, Frank, Claudia Weber and David S. Wood. 1997. "The Abuse of Animals and Domestic Violence: A National Survey of Shelters for Women Who Are Battered." *Society & Animals* 5(3):205–218.

Atterton, Peter and Matthew Calarco, eds. 2004. *Animal Philosophy: Essential Readings in Continental Thought*. London: Continuum.

Augustinus, Aurelius. 2000. *The Works of St. Augustine: A Translation for the 21st Century*. Translated by John E. Rotelle, and Edmund Hill. Part I: Books. Volume 13. New York: New City.

Baker, Steve. 1993. *Picturing the Beast: Animals, Identity and Representation*. Manchester, UK: Manchester University Press.

Baker, Steve. 2000. *The Postmodern Animal*. London: Reaktion.

Balcombe, Jonathan. 2000. *The Use of Animals in Higher Education*. Washington, DC: Humane Society.

Balcombe, Jonathan P. 2006. *Pleasurable Kingdom: Animals and the Nature of Feeling Good*. New York: MacMillan.

Baldick, Julian. 2000. *Animals and Shaman: Ancient Religions of Central Asia*. New York: New York University Press.

Bancel, Nicolas, Pascal Blanchard, Gilles Boëtsch, Eric Deroo, Sandrine Lemaire, and Charles Forsdick, eds. 2009. *Human Zoos: From the Hottentot Venus to Reality Shows*. Liverpool: University of Liverpool Press.

Baraty, Erica and Elisabeth Hardouin-Fugier. 2004. *Zoo: A History of Zoological Gardens in the West*. London: Reaktion.

Baron, David. 2004. *The Beast in the Garden: A Modern Parable of Man and Nature*. New York: Norton.

Beck, Alan and Aaron Katcher. 1996. *Between Pets and People: The Importance of Animal Companionship*. West Lafayette, IN: Purdue University Press.

Becker, Marty and Danelle Morton. 2002. *The Healing Power of Pets: Harnessing the Amazing Ability of Pets to Make and Keep People Happy and Healthy*. New York: Hyperion.

Beers, Diane. 2006. *For the Prevention of Cruelty: The History and Legacy of Animal Rights Activism in the United States.* Athens, OH: Swallow.

Beetz, Andrea and Anthony Podberscek, eds. 2009. *Bestiality and Zoophila: Sexual Relations with Animals.* West Lafayette, IN: Purdue University Press.

Beirne, Piers. 1994. "The Law is an Ass: Reading E.P. Evans' 'The Medieval Prosecution and Capital Punishment of Animals.'" *Society & Animals* 2(1):27–46.

Beirne, Piers. 2009. *Confronting Animal Abuse—Law, Criminology, and Human—Animal Relationships.* New York: Rowman and Littlefield.

Bekoff, Marc. 2000. *Strolling with Our Kin: Speaking for and Respecting Voiceless Animals.* New York: Lantern.

Bekoff, Marc, ed. 2000. *The Smile of a Dolphin: Remarkable Accounts of Animal Emotions.* New York: Discovery.

Bekoff, Marc. 2002. *Minding Animals: Awareness, Emotions and Heart.* Oxford: Oxford University Press.

Bekoff, Marc. 2004. "Wild Justice and Fair Play: Cooperation, Forgiveness and Morality in Animals." *Biology and Philosophy* 19(4):489–520.

Bekoff, Marc. 2005. *Animal Passions and Beastly Virtues: Reflections on Redecorating Nature.* Philadelphia: Temple University Press.

Bekoff, Marc. 2007. *The Emotional Lives of Animals: A Leading Scientist Explores Aimal Joy, Sorrow, and Empathy—and Why They Matter.* Novato, CA: New World Library.

Bekoff, Marc, ed. 2007. *Encyclopedia of Human–Animal Relationships.* Westport, CT: Greenwood.

Bekoff, Marc, ed. 2009. *Encyclopedia of Animal Rights and Animal Welfare,* 2nd ed. Westport, CT: Greenwood.

Bekoff, Marc, C. Allen, and G. Burghardt, eds., 2002. *The Cognitive Animal.* Cambridge, MA: MIT Press.

Bekoff, Marc and J. A. Byers, eds. 1998. *Animal Play: Evolutionary, Comparative, and Ecological Perspectives* Cambridge: Cambridge University Press.

Bekoff, Marc, and D. Jamieson, eds. 1996. *Readings in Animal Cognition.* Cambridge, MA: MIT Press.

Bell, Michael. 2004. *An Invitation to Environmental Sociology.* Thousand Oaks, CA: Pine Forge.

Benson, G. John and Bernard E. Rollin, eds. 2004. *The Well-Being of Farm Animals.* Oxford: Wiley-Blackwell.

Bentham, J. 1781/1982. *An Introduction to the Principles of Morals and Legislation,* ed. J. H. Burns and H. L. A. Hart. London: Methuen.

Benton, J. 1992. *The Medieval Menagerie: Animals in the Art of the Middle Ages.* London: Abbeville.

Berger, John. 1971. "Animal World." *New Society* 18:1042–43.

Berger, John. 1977. "Animals as Metaphor." *New Society* 39:504.

Berger, John. 1977. "Vanishing Animals." *New Society* 39:664–665.

Berger, John. 1977. "Why Zoos Disappoint." *New Society* 40:122–123.

Berger, John. 1980. *About Looking.* New York: Pantheon.

Berry, Bonnie. 1997. "Human and Non-Human Animal Rights and Oppression: An Evolution toward Equality." *Creative Sociology.* 25:155–160.

Best, Steve and Anthony Nocella. 2004. *Terrorists or Freedom Fighters: Reflections on the Liberation of Animals.* New York: Lantern.

Bieder, Robert E. 2005. *Bear*. London: Reaktion.

Birke, Lynda. 1994. *Feminism, Animals, and Science: The Naming of the Shrew*. Buckingham: Open University Press.

Birke, Lynda. 2003. "Who—or What—Are the Rats (and Mice) in the Laboratory?" *Society & Animals* 11(3):207–224.

Birke, Lynda and R. Hubbard, eds. 1995. *Reinventing Biology: Respect for Life and the Creation of Knowledge*. Bloomington: Indiana University Press.

Birke, Lynda and Mike Michael. 1998. "The Heart of the Matter: Animal Bodies, Ethics, and Species Boundaries." *Society & Animals* 6(1):245–261.

Birke, Lynda and Luciani Parisi. 1999. "Animals Becoming." In *Animal Others: On Ethics, Ontology and Animal Life*, ed. Peter Steves. Albany: State University of New York Press, pp. 55–74.

Bissonette, John A. and P. Krausman, eds. 1995. *Integrating People and Wildlife for a Sustainable Future*. Bethesda, MD: The Wildlife Society.

Blum, Deborah. 2002. *Love at Goon Park: Harry Harlow and the science of affection*. Cambridge, MA: Perseus.

Boakes, R. A. 1984. *From Darwin to Behaviorism: Psychology and the Minds of Animals*. Cambridge: Cambridge University Press.

Boat, Barbara W. and Juliette C. Knight. 2000. "Experiences and Needs of Adult Protective Services Case Managers When Assisting Clients Who Have Companion Animals." *Journal of Elder Abuse & Neglect* 12(3&4):145–155.

Boehrer, Bruce, ed. 2007. *A Cultural History of Animals in the Renaissance*. Oxford: Berg.

Boehrer, Bruce Thomas. 2002. *Shakespeare Among the Animals: Nature and Society in the Drama of Early Modern England*. Early Modern Cultural Studies. New York: Palgrave.

Bostock, John and Henry Thomas Riley, eds. 1890. *The Natural History of Pliny*, Volume 2. London: G. Bell and Sons.

Bostock, S. 1993. *Zoos and Animal Rights: The Ethics of Keeping Animals*. London: Routledge.

Bourdillion, M. F. C. and M. Fortes, eds. 1980. *Sacrifice*. London: Academic Press.

Bouse, Derek. 2000. *Wildlife Films*. Philadelphia: University of Pennsylvania Press.

Bradley, Janis. 2007. *Dog Bites: Problems and Solutions*. Policy paper. Ann Arbor, MI: Animals and Society Institute.

Bradshaw, Gay. 2009. *Elephants on the Edge: What Animals Teach Us About Humanity*. New Haven, CT: Yale University Press.

Bright, M. 1984. *Animal Language*. London: BBC.

Brightman, Robert. 1993. *Grateful Prey: Rock Cree Human-Animal Relationships*. Berkeley: University of California Press.

Brody, H. 2001. *The Other Side of Eden*. London: Faber and Faber.

Bryld, Mette and Nina Lykke. 2000. *Cosmodolphins: Feminist Cultural Studies of Technology, Animals, and the Sacred*. New York: Zed.

Budiansky, Stephen. 1997. *The Covenant of the Wild*. London: Phoenix.

Budiansky, Stephen. 1999. *If a Lion Could Talk: How Animals Think*. London: Phoenix.

Budiansky, Stephen. 2000. *The Truth About Dogs*. New York: Viking.

Buettinger, Craig. 1997. "Women and Antivivisection in Late Nineteenth-Century America." *Journal of Social History* 30(Summer):857–872.

Bulbeck, Chilla. 2005. *Facing the Wild: Ecotourism, Conservation, and Animal Encounters*. London: Earthscan.

Bulliet, Richard. 2005. *Hunters, Herders, and Hamburgers: The Past and Future of Human–Animal Relationships*. New York: Columbia University Press.

Burghardt, Gordon M. and Harold A. Herzog, Jr. 1980. "Beyond Conspecifics: Is Brer Rabbit Our Brother?" *Bioscience* 30(11):763–768.

Burghardt, Gordon M. and Harold A. Herzog, Jr., eds. 1989. *Perceptions of Animals in American Culture*. Washington, DC: Smithsonian Institution.

Burgon, H. 2003. "Case Studies of Adults Receiving Horse-Riding Therapy." *Anthrozoös* 16(3):263–276.

Burke, Edmund and Jonathan C. D. Clark. 2001. *Reflections on the Revolution in France*. Stanford, CA: Stanford University Press.

Burkhardt, Richard. W., Jr. 2005. *Patterns of Behavior: Konrad Lorenz, Niko Tinbergen, and the Founding of Ethology*. Chicago: University of Chicago Press.

Burns, J. H. and H. L. A. Hart, eds. 1996. *The Collected Works of Jeremy Bentham*. Oxford: Oxford University Press.

Burt, Jonathan. 2002. *Animals in Film*. London: Reaktion.

Burt, Jonathan. 2006. *Rat*. London: Reaktion.

Byrne, R. 1995. *The Thinking Ape: Evolutionary Origins of Intelligence*. Oxford: Oxford University Press.

Byrne, R., and Whiten, A. 1988. *Machiavellian Intelligence: Social Expertise and the Evolution of Intellect in Monkeys, Apes, and Humans*. Oxford University Press.

Caesar, Terry. 2009. *Speaking of Animals—Essays on Dogs and Others*. Leiden, the Netherlands: Brill Academic.

Calarco, Matthew. 2008. *Zoographies*. New York: Columbia University Press.

Candland, D. 1993. *Feral Children and Clever Children: Reflections on Human Nature*. Oxford: Oxford University Press.

Caporael, L. R. and C. M. Heyes, 1997. "Why Anthropomorphize? Folk Psychology and Other Stories" in *Anthropomorphism, Anecdotes, and Animals*. Robert W. Mitchell, Nicholas S. Thompson, and H. Lyn Miles, eds. Albany: State University of New York Press.

Carbone, Larry. 2004. *What Animals Want: Expertise and Advocacy in Laboratory Animal Welfare Policy*. New York: Oxford University Press.

Carlisle-Frank, Pamela and Tom Flanagan. 2006. *Silent Victims: Recognizing and Stopping Abuse of the Family Pet*. Lanham, MD: University Press of America.

Carlisle-Frank, Pamela and Joshua M. Frank. 2006. "Owners, Guardians and Owner-Guardians: Differing Relationships with Pets." *Anthrozoös* 19(3):225–242.

Carlson, Laurie Winn. 2001. *Cattle: An Informal Social History*. Chicago: Ivan R. Dee.

Carnell, Simon. 2009. *Hare*. London: Reaktion.

Carruthers, Peter. 1992. *The Animals Issue: Moral Theory in Practice*. Cambridge: Cambridge University Press.

Carter, Paul. 2005. *Parrot*. London: Reaktion.

Cartmill, Matt. 1993. *A View to a Death in the Morning: Hunting and Nature through History*. Cambridge, MA: Harvard University Press.

Cassidy, Rebecca. 2002. *The Sport of Kings: Kinship, Class and Thoroughbred Breeding in Newmarket*. Cambridge: Cambridge University Press.

Cassidy, Rebecca and Molly Mullen. 2007. *Where the Wild Things Are Now: Domestication Reconsidered*. New York: Berg.

Castricano, Jodey. 2008. *Animal Subjects: An Ethical Reader in a Posthuman World*. Waterloo, ON: Wilfred Laurier University Press.

Catechism of the Catholic Church. 1994. London: Geoffrey Chapman.

Cavalieri, Paola. 2009. *The Death of the Animal: A Dialogue*. New York: Columbia University Press.

Cavalieri, Paola and Peter Singer, eds. 1993. *The Great Ape Project: Equality Beyond Humanity*. New York: St. Martin's.

Cavell, Stanley, Cora Diamond, John McDowell, Ian Hacking, and Cary Wolfe. 2009. *Philosophy and Animal Life*. New York: Columbia University Press.

Chandler, Cynthia. 2005. *Animal-Assisted Therapy in Counseling*. New York: Routledge.

Chapple, Christopher Key. 1993. *Nonviolence to Animals, Earth, and Self in Asian Traditions*. Albany: State University of New York Press.

Chaudhuri, Una. 2003. "Animal Geographies: Zooësis and the Space of Modern Drama." *Modern Drama* (46) 4:646–662.

Chaudhuri, Una. 2007. "Animal Rites: Performing Beyond the Human." *Critical Theory and Performance*, rev. ed., ed. Joseph Roach and Janelle Reinelt. Ann Arbor: University of Michigan Press.

Cherfas, J. 1989. *The Hunting of the Whale*. Harmondsworth UK: Penguin.

Chris, Cynthia. 2006. *Watching Wildlife*. Minneapolis: University of Minnesota Press.

Clark, S. and S. Lyster. 1997. *Animals and Their Moral Standing*. London: Routledge.

Clarke, P., ed. 1990. *Political Theory and Animal Rights*. London: Pluto.

Clarke, S. R. L. 1977. *The Moral Status of Animals*, Oxford: Oxford University Press.

Clubb, R., M. Rowcliffe, K. U. Mar, P. Lee, C. Moss, and G. J. Mason. 2008. "Compromised Survivorship in Zoo Elephants." *Science* 322:1949.

Clutton-Brock, Juliet. 1981. *Domesticated Animals from Early Times* Austin: University of Texas Press.

Clutton-Brock, Juliet, ed. 1989. *The Walking Larder: Patterns of Domestication, Pastoralism, and Predation*. London: Hyman.

Clutton-Brock, Juliet. 1999. *A Natural History of Domesticated Mammals*, 2nd ed. Cambridge: Cambridge University Press.

Coetzee, J. M. 1980. *Waiting for the Barbarians*. Harmondsworth, UK: Penguin.

Coetzee, J. M. 1999. *The Lives of Animals*. Princeton, NJ: Princeton University Press.

Coetzee, J. M. 2003. *Elizabeth Costello*. New York: Viking.

Cohen, Carl and Tom Regan. 2001. *The Animal Rights Debate*. Lanham, MD: Rowman & Littlefield.

Cohen, E. 1994. "Animals in Medieval Perceptions: The Image of the Ubiquitous Other," In *Animals and Human Society: Changing Perspectives*, ed. Aubrey Manning and James Serpell. London: Routledge, pp. 59–80.

Connor, J. 1997. "Cruel Knives? Vivisection and Biomedical Research in Victorian English Canada." *Canadian Bulletin of Medical History*, 14:37–64.

Copeland, Marion W. 2003. *Cockroach*. London: Reaktion.

Coppinger, Raymond and Laura Coppinger. 2001. *Dogs: A Startling New Understanding of Canine Origin, Behavior, and Evolution*. New York: Scribner.

Corbey, Raymond. 2005. *The Metaphysics of Apes: Negotiating the Animal–Human Boundary*. New York: Cambridge University Press.

Coren, Stanley. 2003. *The Pawprints of History: Dogs and the Course of Human Events*. New York: Free Press.

Cornwall, I. 1968. *Prehistoric Animals and Their Hunters*. London: Faber and Faber.

Cosslett, Tess. 2006. *Talking Animals in British Children's Fiction, 1786–1914*. Aldershot, UK: Ashgate.

Cowper, William. 1784. *The Gentleman's Magazine*, Vol. 54.

Cox, Christopher and Nato Thompson, eds. 2005. *Becoming Animal: Contemporary Art in the Animal Kingdom*. Cambridge: MIT Press.

Creager, Angela, and Jordan William, ed. 2005. *The Animal/Human Boundary: Historical Perspectives*. Rochester, NY: University of Rochester Press.

Crist, Eileen. 1999. *Images of Animals: Anthropomorphism and Animal Mind*. Philadelphia: Temple University Press.

Croke, Vicki. 1997. *The Modern Ark: The Story of Zoos, Past, Present and Future*. New York: Scribner.

Cronon, William. 1992. *Nature's Metropolis: Chicago and the Great West*. New York: W.W. Norton.

Cronon, William. 1996. *Uncommon Ground: Rethinking the Human Place in Nature*. New York: W.W. Norton.

Culhane, S. 1986. *Talking Animals and Other People*. New York: St. Martin's.

Curnutt, Jordan. 2001. *Animals and the Law*. Santa Barbara, CA: ABC-CLIO.

Curtis, L. P. 1997. *Apes and Angels: The Irishman in Victorian Caricature*. Rev. ed. Washington, DC: Smithsonian Institution.

Daly, B. and L. Morton. 2003. "Children with Pets Do Not Show Higher Empathy: A Challenge to Current Views." *Anthrozoös* 16(4):298–314.

Daly, B. and L. Morton. 2006. "An Investigation of Human Animal Interactions and Empathy as Related to Pet Preference, Ownership, Attachment and Attitudes in Children." *Anthrozoös* 19(2):113–127.

Daly, B. and L. Morton. 2009. "Empathetic Differences in Adults as a Function of Childhood and Adult Pet Ownership and Pet Type." *Anthrozoös* 22(4):371–382.

Darnton, Robert. 1984/2001. *The Great Cat Massacre and Other Episodes in French Cultural History*. New York: Basic.

Darwin, Charles. 1859/1985. *On the Origin of Species*. New York: Penguin.

Darwin, Charles. 1871/1981. *The Descent of Man*. Princeton, NJ: Princeton University Press.

Darwin, Charles. 1965. *The Expression of Emotions in Man and Animals*. Chicago: University of Chicago Press.

Daston, Lorraine and Gregg Mitman, eds. 2005. *Thinking with Animals: New Perspectives on Anthropomorphism*. New York: Columbia University Press.

Davis, H. and Dianne Balfour, eds. 1992. *The Inevitable Bond: Examining Scientist-Animal Interactions*. Cambridge: Cambridge University Press.

Davis, Karen. 1995. "Thinking Like a Chicken: Farm Animals and the Feminine Connection" in *Animals and Women: Feminist Theoretical Explorations*, ed. Carol J. Adams and Josephine Donovan. Durham, NC: Duke University Press, pp. 192–212.

Davis, Karen. 2001. *More than a Meal: The Turkey in History, Myth, Ritual and Reality*. New York: Lantern.

Davis, Karen. 2005. *The Holocaust and the Henmaid's Tale: A Case for Comparing Atrocities*. New York: Lantern.

Davis, Susan and Margo DeMello. 2003. *Stories Rabbits Tell: A Natural and Cultural History of a Misunderstood Creature*. New York: Lantern.

Davis, Susan G. 1997. *Spectacular Nature: Corporate Culture and the Sea World Experience*. Berkeley: University of California Press.

Dawkins, Marian. 1980. *Animal Suffering: The Science of Animal Welfare*. London: Chapman and Hall.

Dawkins, Marian. 1993. *Through Our Eyes Only? The Search for Animal Consciousness*. Oxford: Freeman.

Day, Thomas. 2010. *The History of Sandford and Merton*. Charleston, SC: Nabu Press.

Decker, Daniel J. et al. 2001. *Human Dimensions of Wildlife Management in North America*. Bethesda, MD: The Wildlife Society.

DeGrazia, David. 1996. *Taking Animals Seriously: Mental Life and Moral Status*. Cambridge: Cambridge University Press.

Dekkers, M. 1994. *Dearest Pet*. London: Virago.

Deleeuw, Jamie L., Luke W. Galen, Cassandra Aebersold, and Victoria Stanton. 2007. "Support for Animal Rights as a Function of Belief in Evolution, Religious Fundamentalism, and Religious Denomination." *Society & Animals* 15(4):353–363.

DeMello, Margo. 2007. "The Present and Future of Animal Domestication," in *A Cultural History of Animals: Volume 6, The Modern Age*, ed. Randy Malamud. Oxford: Berg, pp. 67–94.

DeMello, Margo. 2010. "Becoming Rabbit: Living with and Knowing Rabbits." *Spring A Journal of Archetype and Culture*, 83(Spring):237–252.

DeMello, Margo, ed. 2010. *Teaching the Animal: Human-Animal Studies across the Disciplines*. New York: Lantern.

DeMello, Margo. 2011. "Blurring the Divide: Human and Animal Body Modification," in *A Companion to the Anthropology of Bodies/Embodiments*, ed. Fran Mascia-Lees. Hoboken, NJ: Wiley-Blackwell, pp. 338–352.

Dent, A. 1976. *Animals in Art*. London: Phaidon.

Derr, Mark. 2004. *Dog's Best Friend: Annals of the Dog-Human Relationship*. Chicago: University of Chicago Press.

Derrida, Jacques. 2008. *The Animal That Therefore I Am: More to Follow*. New York: Fordham University Press.

Descartes, René. 1991. *The Philosophical Writings of Descartes*. 3 vols. Translated by John Cottingham, Robert Stoothoff, and Dugald Murdoch. Cambridge: Cambridge University Press.

Descola, P., and G. Pálsson, eds. 1996. *Nature and Society: Anthropological Perspectives*. London: Routledge.

De Silva, A. et al., eds. 1965. *Man and Animal*. London: Educational Productions.

Desmond, Jane. 1999. *Staging Tourism: Bodies on Display from Waikiki to Sea World*. Chicago: University Chicago Press.

De Waal, Frans. 1983. *Chimpanzee Politics: Power and Sex among Apes*. London: Unwin.

De Waal, F. M. 1996. *Good Natured, The Origins of Right and Wrong in Humans and Animals*. Cambridge, MA: Harvard University Press.

De Waal, Frans. 1998. *Bonobo: The Forgotten Ape*. Berkeley: University of California Press.

De Waal, Frans. 2001. *The Ape and the Sushi Master: Cultural Reflections by a Primatologist*. New York: Basic.

Diamond, Cora. 2001. *The Realistic Spirit*. Cambridge, MA: MIT Press.

Diamond, Jared. 1997. *Guns, Germs, and Steel: The Fates of Human Societies*. New York: Norton.

Dillard, Jennifer. 2008. "A Slaughterhouse Nightmare: Psychological Harm Suffered by Slaughterhouse Employees and the Possibility of Redress through Legal Reform." *Georgetown Journal on Poverty Law & Policy* 15(2):391–408.

Dion, Mark and Alexis Rockman, eds. 1996. *Concrete Jungle: A Pop Media Investigation of Death and Survival in Urban Ecosystems*. New York: Juno.

Dizard, Jan. 1994. *Going Wild: Hunting, Animal Rights and the Contested Meaning of Nature*. Amherst: University of Massachusetts Press.

Dizard, Jan E. 2003. *Mortal Stakes: Hunters and Hunting in Contemporary America*. Boston: University of Massachusetts Press.

Dobbs, David. 2000. *The Great Gulf: Fishermen, Scientists, and the Struggle to Revive the World's Greatest Fishery*. Washington DC: Island.

Dolan, Kevin. 1999. *Ethics, Animals and Science*. Ames: Iowa State University Press.

Dolines, F., ed. 1999. *Attitudes to Animals: Views on Animal Welfare*. Cambridge: Cambridge University Press.

Dombrowski, Daniel. 1997. *Babies and Beasts: The Argument from Marginal Cases*. Urbana: University of Illinois Press.

Donaldson, Jean. 2001. *Dogs Are from Neptune*. Montreal: Lasar Multimedia.

Donovan, Josephine. 1990. "Animal Rights and Feminist Theory." *Signs*, 15(2):370–375.

Donovan, Josephine and Carol Adams, eds. 1996. *Beyond Animal Rights: A Feminist Caring Ethic for the Treatment of Animals*. New York: Continuum.

Donovan, Josephine and Carol J. Adams, eds. 2007. *The Feminist Care Tradition in Animal Ethics: A Reader*. New York: Columbia University Press.

Douglass, C. B. 1997. *Bulls, Bullfighting, and Spanish Identities*. Tucson: University Arizona Press.

Douglas, M. 1957. Animals in Lele Religious Symbolism. *Africa* 27(1):46–58.

Douglas, M. 1970. *Natural Symbols*. New York: Vintage.

Douglas, M. 1990. "The Pangolin Revisited: A New Approach to Animal Symbolism," In *Signifying Animals: Human Meaning in the Natural World*, ed. Roy Willis. London: Routledge, 25–36.

Douglas, Mary. 1970. *Natural Symbols*. Harmondsworth, UK: Penguin.

Douglas, Mary. 1975. *Purity and Danger*. London: Routledge.

Dunayer, Joan. 2001. *Animal Equality: Language and Liberation*. Derwood, MD: Ryce.

Dunayer, Joan. 2004. *Speciesism*. Derwood, MD: Ryce.

Dundes, Alan, ed. 1994. *The Cockfight: A Casebook*. Madison: University of Wisconsin Press.

Dupré, John. 2002. *Humans and Other Animals*. Oxford: Clarendon.

Eaton, J. 1995. *The Circle of Creation: Animals in the Light of the Bible*. London: SCM.

Ebenstein, Helene and Jennifer Wortham. 2001. "The Value of Pets in Geriatric Practice: A Program Example." *Journal of Gerontological Social Work* 35(2):99–115.

Einarsson, N. 1993. "All Animals are Equal But Some are Cetaceans: Conservation and Culture Conflict" in *Environmentalism: The View from Anthropology*, ed. Kay Milton. London: Routledge, pp. 73–84.

Einwohner, Rachel L. 1999a. "Gender, Class, and Social Movement Outcomes: Identity and Effectiveness in Two Animal Rights Campaigns." *Gender and Society* 13(1):56–76.

Einwohner, Rachel L. 1999b. "Practices, Opportunity, and Protest Effectiveness: Illustrations from Four Animal Rights Campaigns." *Social Problems* 46(2):169–186.

Einwohner, Rachel L. 2002. "Motivational Framing and Efficacy Maintenance: Animal Rights Activists; Use of Four Fortifying Strategies." *Sociological Quarterly* 43(4):509–526.

Eisnitz, Gail. 2007. *Slaughterhouse: The Shocking Story of Greed, Neglect, and Inhumane Treatment Inside the U.S. Meat Industry*. Amherst, NY: Prometheus.

Elder, Glen, Jennifer Wolch, and Jody Emel. 1998. "Race, Place, and the Human-Animal Divide." *Society & Animals* 6(2):183–202.

Ellis, Colter. 2009. "The Gendered Process of Cattle (Re)Production." Paper presented at the Animals: Past, Present, and Future conference at Michigan State University, East Lansing, MI.

Ellis, Colter and Leslie Irvine. 2010. "Reproducing Dominion: Emotional Apprenticeship in The 4-H Youth Livestock Program." *Society & Animals* 18(1):21–39.

Ellis, Richard. 2005. *Tiger Bone and Rhino Horn: The Destruction of Wildlife for Traditional Chinese Medicine*. Washington DC: Island.

Emberley, J. V. 1997. *The Cultural Politics of Fur*. Ithaca, NY: Cornell University Press.

Emel, Jody. 1995. "'Are You Man Enough, Big and Bad Enough?' An Ecofeminist Analysis of Wolf Eradication in the United States." *Society and Space: Environment and Planning*. 13:707–734.

Emel, Jody and Julie Urbanik. 2010. "Animal Geographies: Exploring the Spaces and Places of Human-Animal Encounters," in *Teaching the Animal: Human Animal Studies across the Disciplines*, ed. Margo DeMello. New York: Lantern, pp. 202–217.

Engel, Mylan, Jr., and Kathie Jenni. 2010. "Examined Lives: Teaching Human-Animal Studies in Philosophy," in *Teaching the Animal: Human Animal Studies across the Disciplines*, ed. Margo DeMello. New York: Lantern, pp. 60–102.

Evans, D. 1992. *A History of Nature Conservation in Britain*. London: Routledge.

Evans, E. P. 1906. *The Criminal Prosecution and Capital Punishment of Animals*. London: William Heinemann.

Evans, Rhonda, DeAnn K. Gauthier, and Craig J. Forsyth. 1998. "Dogfighting: Symbolic Expression and Validation of Masculinity." *Sex Roles* 39(11/12):825–838.

Evans-Pritchard, Edward. 1940. *The Nuer: A Description of the Modes of Livelihood and Political Institutions of a Nilotic People*. Oxford: Oxford University Press.

Faver, Catherine A. and Alonzo M. Cavazos, Jr. 2007. "Animal Abuse and Domestic Violence: A View from the Order." *Journal of Emotional Abuse* 7(3):59–81.

Faver, Catherine A. and Elizabeth B. Strand. 2003. "Domestic Violence and Animal Cruelty: Untangling the Web of Abuse." *Journal of Social Work Education* 39:237–253.

Faver, Catherine A. and Elizabeth B. Strand. 2007. "Fear, Guilt, and Grief: Harm to Pets and the Emotional Abuse of Women." *Journal of Emotional Abuse* 7(1):51–70.

Favre, David. 2003. *Animals: Welfare, Interests, and Rights*. East Lansing: Animal Law and History Web Center, Michigan State University–Detroit College of Law.

Fentress, John C. "Lessons of the Heart," In *The Smile of a Dolphin: Remarkable Accounts of Animal Emotions*, ed. Marc Bekoff. New York: Discovery, pp. 68–71.

Ferguson, Moira. 1998. *Animal Advocacy and Englishwomen, 1780–1900: Patriots, Nation, and Empire*. Ann Arbor: University of Michigan.

Fiddes, N. 1991. *Meat: A Natural Symbol*. London: Routledge

Fine, Aubrey, ed. 2000. *Handbook on Animal-Assisted Therapy: Theoretical Foundations and Guidelines for Practice*. San Diego: Academic.

Fine, Gary A. and Lazaros Christoforides. 1991. "Dirty Birds, Filthy Immigrants, and the English Sparrow War: Metaphorical Linkage in Constructing Social Problems." *Symbolic Interaction* 14(4):375–393.

Finsen, Lawrence and Susan Finsen. 1994. *The Animal Rights Movement in America*. New York: Twayne.

Fitzgerald, Amy J., Linda Kalof, and Thomas Dietz. 2009. "Slaughterhouses and Increased Crime Rates: An Empirical Analysis of the Spillover from 'the Jungle' into the Surrounding Community." *Organization & Environment* 22(2):158–184.

Flannery, Tim. 1994. *The Future Eaters: An Ecological History of the Australian Lands and People*. Chatswood, NSW: Reed.

Flynn, Clifton P. 2000. "Woman's Best Friend: Pet Abuse and the Role of Companion Animals in the Lives of Battered Women." *Violence Against Women* 6:162–177.

Flynn, Clifton P. 2001. "Acknowledging the 'Zoological Connection': A Sociological Analysis of Animal Cruelty." *Society & Animals* 9(1):71–87.

Flynn, Clifton P. 2002. "Hunting and Illegal Violence Against Humans and Other Animals: Exploring the Relationship." *Society & Animals* 10(2):137–154.

Flynn, Clifton, ed. 2008. *Social Creatures: A Human and Animal Studies Reader*. New York: Lantern.

Fogle, Bruce. 1983. *Pets and Their People*. New York: Viking.

Foltz, Richard 2006. *Animals in Islamic Tradition and Muslim Cultures*. Oxford: Oneworld.

Foucault, Michel. 1998. *The History of Sexuality, Volume1: The Will to Knowledge*. London: Penguin.

Fouts, Roger. 1997. *Next of Kin: What Chimpanzees Taught Me About Who We Are*. New York: William Morrow.

Francione, Gary L. 1995. *Animals, Property, and the Law*. Philadelphia: Temple University Press.

Francione, Gary L. 1996. *Rain without Thunder: The Ideology of the Animal Rights Movement*. Philadelphia: Temple University Press.

Francione, Gary L. 2000. *Introduction to Animal Rights: Your Child or the Dog?* Philadelphia: Temple University Press.

Francione, Gary L. 2008. *Animals as Persons: Essays on the Abolition of Animal Exploitation*. New York: Columbia University Press.

Franklin, Adrian. 1999. *Animals and Modern Cultures: A Sociology of Human–Animal Relations in Modernity*. London: Sage.

Franklin, Julian H. 2005. *Animal Rights and Moral Philosophy*. New York: Columbia University Press.

Franklin, Sarah. 2007. *Dolly Mixtures: The Remaking of Genealogy*. Durham, NC: Duke University Press.

Fraser, David. 2008. *Understanding Animal Welfare: The Science in Its Cultural Context*. Oxford: Wiley-Blackwell.

Fraser, David, Daniel Weary and Marina A.G. von Keyserlingk. 2010. "Two Interdisciplinary Courses on the Use and Welfare of Animals," in *Teaching the Animal: Human Animal Studies across the Disciplines*, ed. Margo DeMello. New York: Lantern, pp. 341–365.

Fraser, Orlaith and Thomas Bugnyar. 2009. "Do Ravens Show Consolation? Responses to Distressed Others." *PLoS ONE* 5(5):8.

Freeman, Carol. 2010. *Paper Tiger: A Visual History of the Thylacine*. Leiden, the Netherlands: Brill.

French, R. D. 1975. *Antivivisection and Medical Science in Victorian Society*. Princeton, NJ: Princeton University Press.

French, R. K. 1999. *Dissection and Vivisection in the European Renaissance*. Brookfield, VT: Ashgate.

Frey, R. G. 1980. *Interests and Rights: The Case against Animals*. Oxford: Clarendon.

Frey, R. G. 1983. *Rights, Killing, and Suffering: Moral Vegetarianism and Applied Ethics*. Oxford: Basil Blackwell.

Freyfogle, Eric T. and Dale D. Goble. 2008. *Wildlife Law: A Primer*. Washington, DC: Island.

Frommer, Stephanie S. and Arnold Arluke. 1999. "Loving Them to Death: Blame-Displacing Strategies of Animal Shelter Workers and Surrenderers." *Society & Animals* 7(1):1–16.

Fudge, Erica. 2002. *Animal*. London: Reaktion.

Fudge, Erica. 2002. *Perceiving Animals: Humans and Beasts in Early Modern English Culture*. Champaign: University of Illinois Press.

Fudge, Erica. 2006. *Brutal Reasoning: Animals, Rationality, and Humanity in Early Modern England*. Ithaca, NY: Cornell University Press.

Fudge, Erica, Ruth Gilbert, and Susan Wiseman, eds. 1999. *At the Borders of the Human: Beasts, Bodies and Natural Philosophy in the Early Modern Period*. New York: St. Martin's.

Fukuda, K. 1997. "Different Views of Animals and Cruelty to Animals: Cases in Fox-Hunting and Pet-Keeping in Britain." *Anthropology Today* 13(5):2–6.

Fuller, R., ed. 1981. *Fellow Mortals: An Anthology of Animal Verse*. Plymouth, UK: Macdonald and Evans.

Gaard, Greta, ed. 1993. *Ecofeminism: Women, Animals, Nature*. Philadelphia: Temple University Press.

Gaard, Greta. 1993. "Living Interconnections with Animals and Nature," in *Ecofeminism: Women, Animals, Nature*, ed. Greta Gaard. Philadelphia: Temple University Press, pp. 647–653.

Galdikas, Biruté. 1995. *Reflections of Eden: My Years with the Orangutans of Borneo*. New York: Little, Brown and Company.

Garber, Marjorie. 1997. *Dog Love*. New York: Touchstone.

Garner, Robert. 2005. *The Political Theory of Animal Rights (Perspectives on Democratization)*. Manchester, UK: Manchester University Press.

Gates, P. 1997. *Animal Communication*. Cambridge: Cambridge University Press.

Gee, Nancy, Elise Crist and Daniel Carr. 2010. "Preschool Children Require Fewer Instructional Prompts to Perform a Memory Task in the Presence of a Dog." *Anthrozoös* 23(2):173–184.

Geertz, C. 1994. "Deep Play: Notes on the Balinese Cockfight," in *The Cockfight: A Casebook*, ed. Alan Dundes. Madison: University of Wisconsin Press, 94–132.

George, Kathryn Paxton. 2000. *Animal, Vegetable, or Woman? A Feminist Critique of Ethical Vegetarianism*. Albany: State University of New York Press.

George, W. 1962. *Animal Geography*. London: Heinmann.

George, W. 1969. *Animals and Maps*. London: Secker & Warburg.

Gigliotti, Carol, ed. 2009. *Leonardo's Choice: Genetic Technologies and Animals*. Dordrecht, the Netherlands: Springer.

Gill, Jerry H. 1997. *If a Chimpanzee Could Talk and Other Reflections on Language Acquisition*. Tucson: University of Arizona Press.

Glosecki, Stephen O. 1996. "Movable Beasts: The Manifold Implication of Early Germanic Animal Imagery," in *Animals in the Middle Ages: A Book of Essays*, ed. N. C. Flores. New York: Garland, 3–23.

Gluck, J. P. and S. R. Kubacki. 1991. "Animals in Biomedical Research: The Undermining Effect of the Rhetoric of the Besieged." *Ethics and Behavior* 1(3):157–173.

Gobster, Paul H. and R. Bruce Hull, eds. 2000. *Restoring Nature: Perspectives from the Social Sciences and Humanities*. Washington DC: Island.

Goedeke, Theresa. 2004. "In the Eye of the Beholder: Changing Social Perceptions of the Florida Manatee." *Society & Animals* 12(2):99–116.

Goedeke, Theresa. 2010. "Putting Society Back in the Wild: 'Wildlife & Society' Curriculum as a Tool for Teaching Ecology," in *Teaching the Animal: Human Animal Studies across the Disciplines*, ed. Margo DeMello. New York: Lantern, pp. 366–388.

Goedeke, Theresa L. 2005. "Devils, Angels or Animals: The Social Construction of Otters in Conflict Over Management," in *Mad About Wildlife,* eds. Ann Herda-Rapp and Theresa L. Goedeke. Boston: Brill, pp. 25–50.

Goodall, Jane. 1971. *In the Shadow of Man*. Boston: Houghton Mifflin.

Goodall, Jane. 1986. *The Chimpanzees of Gombe: Patterns of Behavior*. Boston: The Belknap Press of the Harvard University Press.

Goodall, Jane. 1990. *Through a Window: 30 Years Observing the Gombe Chimpanzees*. Boston: Houghton Mifflin.

Goodall, Jane and Phillip Berman. 1999. *Reason For Hope: A Spiritual Journey*. New York: Warner.

Goodall, Jane. 2000. *40 Years At Gombe*. New York: Stewart, Tabori, and Chang.

Goodall, Jane and Marc Bekoff. 2002. *The Ten Trusts: What We Must Do to Care for the Animals We Love*. San Francisco: Harper.

Gould, J. L., and C. G. Gould. 1994. *The Animal Mind*. New York: Scientific American Library, W. H. Freeman.

Grandin, Temple. 2005. *Animals in Translation: Using the Mysteries of Autism to Decode Animal Behavior*. New York: Scribner.

Greek, C. Ray and Jean Swingle Greek. 2000. *Sacred Cows and Golden Geese: The Human Costs of Experiments on Animals*. New York: Continuum.

Grier, Katherine. 2006. *Pets in America: A History*. Durham: University of North Carolina Press.

Griffin, Donald. 1976. *The Question of Animal Awareness: Evolutionary Continuity of Mental Experience*. New York: Rockefeller University Press.

Griffin, Donald. 2001. *Animal Minds: Beyond Cognition to Consciousness*. Chicago: University of Chicago Press.

Griffin, Gary A. and Harry F. Harlow. 1966. "Effects of Three Months of Total Social Deprivation on Social Adjustment and Learning in the Rhesus Monkey." *Child Development* 37(3):533–547.

Griffith, Marcie, Jennifer Wolch, and Unna Lassiter. 2002. "Animal Practices and the Racialization of Filipinas in Los Angeles." *Society & Animals* 10(3):221–248.

Groves, Julian McAllister. 1997. *Hearts and Minds: The Controversy over Laboratory Animals*. Philadelphia: Temple University Press.

Gruen, Lori. 1990. "Gendered Knowledge? Examining Influences on Scientific and Ethological Inquiries," in *Interpretation and Explanation in the Study of Animal Behavior*, eds. Marc Bekoff and Dale Jamieson. Boulder, CO: Westview, pp. 56–73.

Gruen, Lori. 1993. "Dismantling Oppression: An Analysis of the Connection between Women and Animals," in *Ecofeminism: Women, Animals, Nature*, ed. Greta Gaard. Philadelphia: Temple University Press, pp. 60–90.

Gruen, Lori. 2009. "The Moral Status of Animals," in *The Stanford Encyclopedia of Philosophy*, ed. Edward N. Zalta, Stanford: Stanford University. http://plato.stanford.edu/archives/fall2010/entries/moral-animal. Accessed April 2, 2012.

Gruen, Lori and Kari Weil. 2010. "Teaching Difference: Sex, Gender, Species," in *Teaching the Animal: Human Animal Studies across the Disciplines*, ed. Margo DeMello. New York: Lantern, pp. 127–143.

Guggenheim, S. 1994. "Cock or Bull: Cockfighting, Social Structure, and Political Commentary in the Phillipines," in *The Cockfight: A Casebook*, ed. Alan Dundes. Madison: University of Wisconsin Press, 133–173.

Guillermo, Kathy Snow. 1993. *Monkey Business: The Disturbing Case that Launched the Animal Rights Movement*. Washington, DC: National.

Guither, Harold D. 1998. *Animal Rights: History and Scope of a Radical Social Movement*. Carbondale: Southern Illinois University Press.

Hahn, D. 2003. *The Tower Menagerie*. London: Simon and Schuster.

Hall, Molly J., Anthony Ng, Robert J. Ursano, Harry Holloway, Carol Fullerton, and Jacob Casper. 2004. "Psychological Impact of the Animal–Human Bond in Disaster Preparedness and Responses." *Journal of Psychiatric Practice* 10(6):368–374.

Ham, Jennifer and Matthew Senior. 1997. *Animal Acts: Configuring the Human in Western History*. New York: Routledge.

Hancocks, D. 2001. *A Different Nature: The Paradoxical World of Zoos and Their Uncertain Future*. Berkeley: University of California Press.

Hanselman, Jan L. 2001. "Coping Skills Interventions with Adolescents in Anger Management Using Animals in Therapy." *Journal of Child and Adolescent Group Therapy* 11(4):159–195.

Hansen, K. M., C. J. Messinger, M. M. Baun, et al. 1999. "Companion Animals Alleviating Distress in Children." *Anthrozoös* 12(3):142–148.

Hanson, Elizabeth. 2002. *Animal Attractions: Nature on Display in American Zoos*. Princeton, NJ: Princeton University Press.

Haraway, Donna. 1989. *Primate Visions: Gender, Race, and Nature in the World of Modern Science*. New York: Routledge, Chapman & Hall.

Haraway, Donna. 1991. *Simians, Cyborgs, and Women: The Reinvention of Nature*. London: Routledge.

Haraway, Donna. 1991. "Situated Knowledges: The Science Question in Feminism and the Privilege of Partial Perspective," in *Simians, Cyborgs and Women*, ed. Donna Haraway. London: Free Association, pp. 575–599.

Haraway, Donna. 1997. *Modest_Witness@Second_Millenium. FemaleMan_Meets_Oncomouse: Feminism and Technoscience*. New York: Routledge.

Haraway, Donna. 2003. *The Companion Species Manifesto: Dogs, People, and Significant Otherness*. Chicago, IL: Prickly Paradigm.

Haraway, Donna. 2007. *When Species Meet*. Minneapolis: University of Minnesota Press.

Harbolt, Tami and Tamara H. Ward. 2001. "Teaming Incarcerated Youth with Shelter Dogs for a Second Chance." *Society & Animals* 9(2):177–182.

Harbolt-Bosco, Tami. 2003. *Bridging the Bond: The Cultural Construction of the Shelter Pet*. West Lafayette, IN: Purdue University Press.

Hardin, Rebecca. Forthcoming. *The Heart of Parkness: Trophies, Tours, and Transvalued Wildlife in the Western Congo Basin*.

Hargrove, Eugene, ed. 1992. *The Animal Rights/Environmental Ethics Debate: The Environmental Perspective*. Albany: State University of New York Press.

Harris, Marvin. 1966. "The Cultural Ecology of India's Sacred Cattle." *Current Anthropology* 7:51–66.

Harris, Marvin. 1974. *Cows, Pigs, Wars and Witches*. New York: Vintage.

Harris, Marvin. 1983. "Some Factors Influencing Selection and Naming of Pets." *Psychological Reports* 53(3, Pt. 2):1163–70.

Harris, Marvin. 1985. *Good to Eat: Riddles of Food and Culture*. London: Allen and Unwin.

Harris, Marvin. 1987. *The Sacred Cow and the Abominable Pig: Riddles of Food and Culture:* New York: Touchstone.

Hart, L. 1995. "The Role of Pets in Enhancing Human Well-Being: Effects for Older People," in *The Waltham Book of Human-Animal Interaction: Benefits and Responsibilities of Pet Ownership*, ed., I. Robinson. Exeter: Pergamon, pp. 19–31.

Hawkins, Ronnie Z. 1998. "Ecofeminism and Nonhumans: Continuity, Difference, Dualism and Domination." *Hypatia* 13(1):158–197.

Hawley, Fred. 1993. "The Moral and Conceptual Universe of Cockfighters: Symbolism and Rationalization." *Society & Animals* 1(2):159–168.

Hearne, Vicki. 1987. *Adam's Task: Calling Animals by Name*. New York: Alfred A. Knopf.

Hearne, Vicki. 1994. *Animal Happiness*. New York: HarperCollins.

Hearne, Vicki. 1995. "A Taxonomy of Knowing Animals: Captive, Free-Ranging, and at Liberty." *Social Research* 62(3):441–456.

Hearne, Vicki. 2007. *Adam's Task*. New York: Skyhorse Publishing.

Hediger, H. 1970. *Man and Animal in the Zoo*. London: Routledge and Kegan Paul.

Heidegger, Martin. 1971. "A Dialogue on Language (between a Japanese and an Inquirer)," in *On the Way to Language*. Translated by Peter D. Hertz. New York: Harper & Row.

Heimlich, Katherine. 2001. "Animal-Assisted Therapy and the Severely Disabled Child: A Quantitative Study." *Journal of Rehabilitation* 67(4):48–54.

Henninger-Voss, Mary, ed. 2002. *Animals in Human Histories: The Mirror of Nature and Culture*. Rochester: University of Rochester Press.

Henry, Bill. 2006. "Empathy, Home Environment, and Attitudes Towards Animals in Relation to Animal Abuse." *Anthrozoös* 19(1):17–34.

Henry, Bill and Cheryl E. Sanders. 2007. "Bullying and Animal Abuse: Is There A Connection?" *Society & Animals* 15(2):107–126.

Herda-Rapp, Ann and Theresa L. Goedeke, eds. 2005. *Mad about Wildlife: Looking at Social Conflict over Wildlife*. Leiden, the Netherlands: Brill Academic Publishers.

Hergovich, Andreas, Bardia Monshi, Gabriele Semmler, and Verena Zieglmayer. 2002. "The Effects of the Presence of a Dog in the Classroom." *Anthrozoös* 15(1):37–50.

Herzog, Harold, Jr., 2010. "Are Humans the Only Animals that Keep Pets?" Blog on *Psychology Today* website. http://www.psychologytoday.com/blog/animals-and-us/201006/are-humans-the-only-animals-keep-pets.

Herzog, Harold, Jr. 2010. *Some We Love, Some We Hate, Some We Eat: Why It's So Hard to Think Straight About Animals*. New York: Harper Collins.

Herzog, Harold, Jr., and G. Burghardt. 1988. "Attitudes toward Animals: Origins and Diversity," in *Animals and People Sharing the World*, ed. Andrew Rowan. Hanover, NH: University Press of New England, pp. 75–94.

Herzog, Harold, Jr., and Lauren Golden. 2009. "Moral Emotions and Social Activism: The Case of Animal Rights." *Journal of Social Issues* 65(3):485–498.

Herzog, Harold A., Jr., 1988. "Cockfighting and Violence in the South," in *The Encyclopedia of Southern Culture*, ed. W. Ferris. Chapel Hill: University of North Carolina Press, pp. 1477–1478.

Herzog, Harold A., Jr., 1993. "The Movement is My Life: The Psychology of Animal Rights Activism." *Journal of Social Issues* 49(1):103–120.

Herzog, Harold A., Jr., Nancy S. Betchart, and Robert B. Pittman. 1991. "Gender, Sex Role Orientation, and Attitudes Toward Animals." *Anthrozoös* 4(3):184–191.

Herzog, Harold A., Jr., Tamara Vore, and John C. New. 1988. "Conversations with Veterinary Students: Attitudes, Ethics, and Values." *Anthrozoös* 2(3):181–188.

Hicks, C. 1993. *Animals in Early Medieval Art*. Edinburgh: Edinburgh University Press.

Hinde, R. 1970. *Animal Behavior: A Synthesis of Ethology and Comparative Psychology*. London: McGraw-Hill.

Hines, Linda M. 2003. "Historical Perspectives on the Human-Animal Bond." *American Behavioral Scientist* 47(1):7–15

Hoage, R. J. and William A. Deiss, eds. 1996. *New Worlds, New Animals: From Menagerie to Zoological Park in the Nineteenth Century*. Baltimore: The Johns Hopkins University Press.

Hobgood-Oster, Laura. 2008. *Holy Dogs and Asses: Animals in the Christian Tradition*. Urbana: University of Illinois Press.

Hobgood-Oster, Laura. 2010. *The Friends We Keep: Unleashing Christianity's Compassion for Animals*. Waco, TX: Baylor University Press.

Hogan, Linda, Deena Metzger, and Brenda Peterson, eds. 1998. *Intimate Nature: The Bond Between Women and Animals*. New York: Ballantine.

Holmberg, Tora. 2009. *Investigating Human/Animal Relations in Science, Culture and Work*. Uppsala, Sweden: Uppsala University Press.

Houston, Pam, ed. 1995. *Women on Hunting*. Hopewell, NJ: Ecco.

Houston, W. 1993. *Purity and Monotheism: Clean and Unclean Animals in Biblical Law*. Sheffield: JSOT.

Howell, S. 1996. "Nature in Culture or Culture in Nature? Chewong Ideas of 'Humans' and Other Species," in *Nature and Society: Anthropological Perspectives*, eds. P. Descola and G. Pálsson. London: Routledge, 127–144.

Hubert, H. and Marcel Mauss. 1981. *Sacrifice: Its Nature and Functions*. Chicago: Midway Reprints.

Hughes, T. 1995. *Collected Animal Poems*. London: Faber and Faber.

The Humane Society of the United States (HSUS), First Strike Campaign. 2003. *2003 Report of Animal Cruelty Cases*. Washington, DC: HSUS.

Hume, C. W. 1957. *The Status of Animals in the Christian Religion*. London: Universities Federation for Animal Welfare.

Hursthouse, Rosalind. 2000. *Ethics, Humans and Other Animals: An Introduction with Readings*. New York: Routledge.

Hyers, Lauri. 2006. "Myths Used to Legitimize the Exploitation of Animals: An Application of Social Dominance Theory." *Anthrozoös* 19(3):194–210.

Ingold, Tim. 1980. *Hunters, Pastoralists and Ranchers*. Cambridge: Cambridge University Press.

Ingold, Tim, ed. 1988. *What is an Animal?* London: Routledge.

Ingold, Tim. 1994. "From Trust to Domination: An Alternative History of Human-Animal Relations," in *Animals and Human Society*, eds. Aubrey Manning and James Serpell . New York: Routledge, pp. 61–76.

Ingold, Tim. 2001. "Animals and Modern Cultures: A Sociology of Human-Animal Relations in Modernity." *Society & Animals* 9(2):183–188.

Irvine, Leslie. 2001. "The Power of Play." *Anthrozoös* 14(3):151–160.

Irvine, Leslie. 2002. "Animal Problems/People Skills: Emotional Interactional Strategies in Human Education." *Society & Animals* 10(1):63–91.

Irvine, Leslie. 2004. *If You Tame Me: Understanding Our Connections with Animals*. Philadelphia: Temple University Press.

Irvine, Leslie. 2009. *Filling the Ark: Animal Welfare in Disasters*. Philadelphia: Temple University Press.

Isenberg, Andrew C. 2000. *The Destruction of the Bison: An Environmental History, 1750–1920*. New York: Cambridge University Press.

Jacobs, Joseph. 1894. *The Fables of Aesop*. New York: Shocken.

Jalongo, Mary Renck, ed. 2004. *The World's Children and Their Companion Animals: Developmental and Educational Significance of the Child/Pet Bond*. Olney, MD: Association for Childhood Education International.

Jamieson, Dale. 2002. *Morality's Progress: Essays on Humans, Other Animals, and the Rest of Nature*. New York: Oxford University Press.

Jamison, W. V., C. Wenk, and J. V. Parker. 2000. "Every Sparrow that Falls: Understanding Animal Rights Activism as Functional Religion." *Society & Animals* 8(3): 305–330.

Janega, James. 2007. "That Researcher in the Ape House? She Was Studying You." *Chicago Tribune*, April 26, 2007.

Jaschinski, Britta. 1996. *Zoo*. London: Phaidon.

Jaschinski, Britta. 2003. *Wild Things*. London: Phaidon.

Jasper, James M. 1996. "The American Animal Rights Movement," in *Animal Rights: The Changing Debate*, ed. Robert Garner. New York: New York University Press.

Jasper, James M. and Dorothy Nelkin. 1992. *The Animal Rights Crusade: The Growth of a Moral Protest*. New York: Free Press.

Jepson, Jill. 2008. "A Linguistic Analysis of Discourse on the Killing of Nonhuman Animals." *Society & Animals* 16(2):127–148.

Jerolmack, Colin. 2003. "Tracing the Profile of Animal Rights Supporters: A Preliminary Investigation." *Society & Animals* 11(3):245–258.

Jerolmack, Colin. 2007. "Animal Archeology: Domestic Pigeons and the Nature-Culture Dialectic." *Qualitative Sociology Review* 3(1):74–95.

Jory, Brian and Mary-Lou Randour. 1998. *The Anicare Model of Treatment for Animal Abuse*. Ann Arbor, MI: Animals and Society Institute.

Joseph, Cheryl. 2010. "Teaching Human-Animal Studies in Sociology," in *Teaching the Animal: Human Animal Studies across the Disciplines*, ed. Margo DeMello. New York: Lantern, pp. 299–340.

Joy, Melanie. 2002. "Toward a Non-Speciesist Psychoethic." *Society & Animals* 10(4): 457–458.

Joy, Melanie. 2009. *Why We Love Dogs, Eat Pigs, and Wear Cows: An Introduction to Carnism*. Newburyport, MA: Conari.

Kahn, Peter H., Jr., and Stephen R. Kellert, eds. 2002. *Children and Nature: Psychological, Sociocultural, and Evolutionary Investigations*. Cambridge, MA: MIT Press.

Kalechofsky, Roberta, ed. 1992. *Judaism and Animal Rights: Classical and Contemporary Responses*. Marblehead, MA: Micah Publications.

Kalechofsky, Roberta. 2003. *Animal Suffering and the Holocaust: The Problem with Comparisons*. Marblehead, MA: Micah Publications.

Kalland, A. 1993. "Management by Totemization: Whale Symbolism and the Anti-Whaling Campaign." *Arctic* 46(2):124–33.

Kalof, Linda. 2007. *Looking at Animals in Human History*. London: Berg.

Kalof, Linda and Amy Fitzgerald, eds. 2007. *The Animals Reader: The Essential Classic and Contemporary Writings*. New York: Berg.

Kalof, Linda and Brigitte Resl, eds. 2007. *A Cultural History of Animals*. New York: Berg.

Kant, I., 1785. *The Groundwork for the Metaphysics of Morals*. Translated by Mary J. Gregor. Cambridge: Cambridge University Press, 1998.

Kappeler, Susanne. 1995. "Speciesism, Racism, Nationalism . . . Or the Power of Scientific Subjectivity," in *Animals and Women: Feminist Theoretical Explorations*, ed. Carol J. Adams and Josephine Donovan. Durham, NC: Duke University Press, pp. 320–352.

Kass, Philip H., John C. New Jr., Janet M. Scarlett, and Mo D. Salman. 2001. "Understanding Animal Companion Surplus in the United States: Relinquishment of Nonadoptables to Animal Shelters for Euthanasia." *Journal of Applied Animal Welfare Science* 4(4):237–248.

Kean, H. 1999. *Animal Rights: Political and Social Change in Britain Since 1800*. London: Reaktion.

Kean, H. 2003. "An Exploration of the Sculptures of Greyfriars Bobby, Edinburgh, Scotland, and the Brown Dog, Battersea, South London, England." *Society & Animals* 11(4):353–373.

Kean, Hilda. 2001. "Imagining Rabbits and Squirrels in the English Countryside." *Society & Animals* 9(2):163–175.

Kehoe, Monica. 1991. "Loneliness and the Aging Homosexual: Is Pet Therapy an Answer?" *Journal of Homosexuality* 20(3):137–142.

Kellert, Stephen and J. K. Berry. 1985. *A Bibliography of Human/Animal Relations*. Lanham, MD: University Press of America.

Kellert, Stephen and J. K. Berry. 1987. "Attitudes, knowledge, and behaviors toward wildlife as affected by gender." *Wildlife Society Bulletin* 15:363–371.

Kellert, Stephen and Edward Wilson, ed. 1993. *The Biophilia Hypothesis*. Washington, DC: Island.

Kellert, Stephen R. 1980. "American attitudes toward and knowledge of animals: An update". *International Journal for the Study of Animal Problems* 1(2):87–119.

Kellert, Stephen R. 1985. "Historical trends in perceptions and uses of animals in 20th century America." *Environmental Review* 9:34–53.

Kellert, Stephen R. 1989. "Perceptions of animals in America." In *Perceptions of Animals in American Culture*, ed. R.J. Hoage. Washington, DC: Smithsonian Institution.

Kellert, Stephen R. 1994. "Attitudes, knowledge and behavior toward wildlife among the industrial superpowers." In *Animals and Human Society,* ed. Aubrey Manning and James Serpell. New York: Routledge.

Kemmerer, Lisa. 2006. *In Search of Consistency: Ethics and Animals.* Leiden, the Netherlands: Brill Academic Publishers.

Kemmerer, Lisa. 2006. "Verbal activism: 'Anymal.'" *Society & Animals* 14(1):9–15.

Kennedy, J. S. 1992. *The New Anthropomorphism.* Cambridge: University of Cambridge Press.

Kete, Kathleen. 1995. *The Beast in the Boudoir: Petkeeping in Nineteenth-Century Paris.* Berkeley: University of California Press.

Kheel, Marti. 1985. "Speaking the Unspeakable: Sexism in the Animal Rights Movement." *Feminists for Animal Rights Newsletter* 2(Summer/Fall):1–7.

Kheel, Marti. 1995. "License to Kill: An Ecofeminist Critique of Hunters' Discourse," In *Animals and Women: Feminist Theoretical Explorations,* eds. Carol J. Adams and Josephine Donovan. Durham, NC: Duke University Press, pp. 85–125.

Kheel, Marti. 2004. "The History of Vegetarianism," in *The Encyclopedia of World Environmental History*, eds. Shepard Krech III, C. Merchant, and J. R. McNeil. New York: Routledge, pp. 1273–1278.

Kheel, Marti. 2004. "Vegetarianism and Ecofeminism: Toppling Patriarchy with a Fork," in *Food for Thought: The Debate Over Eating Meat*, ed. Steve F. Sapontzis. Amherst, NY: Prometheus, pp. 327–341.

Kheel, Marti. 2006. "The Killing Game: An Ecofeminist Critique of Hunting." *Journal of the Philosophy of Sport* 23(1):30–44.

Kheel, Marti. 2008. *Nature Ethics: An Ecofeminist Perspective.* Lanham, MD: Rowman & Littlefield.

Kidd, A. H. and R. M. Kidd. 1990. "Factors in children's attitudes toward pets." *Psychological Reports* 66:775–786.

Kidd, A. H. and R. M. Kidd. 1994. "Benefits and Liabilities of Pets for the Homeless." *Psychological Reports* 74(1):715–722.

Kilner, Dorothy. 1795. *The life and perambulation of a mouse.* London: J. Marshall.

Kim, Claire Jean. 2010. "Slaying the Beast: Reflections on Race, Culture and Species." *Kalfou*, Spring 2010.

Kipling, Rudyard and Peter Washington. 2007. *Poems.* New York: Alfred A. Knopf.

Kirsch, Sharon. 2008. *What Species of Creatures: Animals Relations from the New World.* Vancouver: New Star.

Kistler, John. 2002. *People Promoting and People Opposing Animal Rights.* Westport, CT: Greenwood.

Knight, John, ed. 2000. *Natural Enemies: People-Wildlife Conflicts in Anthropological Perspective.* London: Routledge.

Knight, John. 2002. *Wildlife in Asia: Cultural Perspectives.* London: RoutledgeCurzon.

Knight, John. 2002. *Wildlife in Asia: Cultural Perspectives.* London: RoutledgeCurzon.

Knight, John, ed. 2005. *Animals in Person: Cultural Perspectives on Human–Animal Intimacies.* Oxford: Berg.

Knight, Sarah and Louise Barnett. 2008. "Justifying Attitudes toward Animal Use: A Qualitative Study of People's Views and Beliefs." *Anthrozoös* 21(1):31–42.

Koebner, L. 1994. *Zoo Book: The Evolution of Wildlife Conservation Centres.* New York: T Doherty.

Kogan, Lori R., Sherry McConnell, Regina Schoenfeld-Tacher, and Pia Jansen-Lock. 2004. "Crosstrails: A Unique Foster Program to Provide Safety for Pets of Women in Safehouses." *Violence Against Women* 10(4):418–434.

Kowalski, Gary. 1999. *The Souls of Animals.* Walpole, NH: Stillpoint Publishing

Kunkel, H. O. 2000. *Human Issues in Animal Agriculture.* College Station: Texas A&M University Press.

Landry, Donna. 2009. *Noble Brutes: How Eastern Horses Transformed English Culture.* Baltimore: The Johns Hopkins University Press.

Langley, Gill, ed. 1989. *Animal Experimentation: The Consensus Changes.* London: MacMillan.

Lansbury, Coral. 1985. *The Old Brown Dog: Women, Workers, and Vivisection in Edwardian England.* Madison: University of Wisconsin Press.

Lawrence, Elizabeth A. 1982. *Rodeo: An Anthropologist Looks at the Wild and the Tame.* Chicago, IL: University of Chicago Press.

Lawrence, Elizabeth A. 1985. *Hoofbeats and Society: Studies of Human-Horse Interactions.* Bloomington: Indiana University Press.

Lawrence, Elizabeth A. 1994. "Rodeo Horses: The Wild and the Tame," in *Signifying Animals: Human Meaning in the Natural World*, ed. Roy Willis. London: Routledge, pp. 222–235.

Lawrence, Elizabeth A. 1997. *Hunting the Wren: Transformation of Bird to Symbol: A Study in Human-Animal Relationships.* Knoxville: University of Tennessee Press.

Lawrence, John, H. D. Symonds, and Charles Whittingham. 1802. *A Philosophical and Practical Treatise on Horses and on the Moral Duties of Man towards the Brute Creation.* London: Printed by C. Whittingham, Dean Stree, Fetter Lane for H.D. Symonds, Paternoster-Row.

Leach, Edmund. 1964. "Anthropological Aspects of Language: Animal Categories and Verbal Abuse," in *New Directions in the Study of Language*, ed. E Lenneberg. Cambridge, MA: MIT Press, pp. 23–63.

Leach, Edmund. 1970. *Claude Lévi-Strauss.* Chicago: University of Chicago Press.

Leahy, M. 1994. *Against Liberation: Putting Animals in Perspective.* London: Routledge.

LeDuff, Charlie. 2000. "At a Slaughterhouse, Some Things Never Die." *New York Times*, June 16, 2000, A1, A24–A25.

Lee, Keekok. 2006. *Zoos: A Philosophical Tour.* New York: Palgrave MacMillan.

Leist, Anton and Peter Singer. 2010. *J. M. Coetzee: Philosophical Perspectives on Literature.* New York: Columbia University Press.

Lemm, Vanessa. 2009. *Nietzsche's Animal Philosphy: Culture, Politics, and the Animality of the Human Being.* Bronx, NY: Fordham University Press.

Levinson, Boris and Gerald P. Mallon. 1997. *Pet-Oriented Child Psychotherapy.* Springfield, IL: Charles C. Thomas Publisher Ltd.

Lévi-Strauss, C. 1963. *Totemism.* Translated by R. Needham. Boston: Beacon.

Lightman, Bernard and Ann Shteir, eds. 2006. *Figuring it Out: Science, Gender and Visual Culture.* Hanover: University of Wisconsin Press.

Lilequist, J. 1992. "Peasants Against Nature: Crossing the Boundaries Between Man and Animals in 17th and 18th Century Sweden," in *Forbidden History, the State, Society and the Regulation of Sexuality in Modern Europe*, ed. J. Fout. Chicago: Chicago University Press, pp. 57–87.

Linden, E. 1976. *Apes, Men and Language.* Harmondsworth, UK: Penguin.

Linné, Carl von, M. S. J. Engel-Ledeboer, and Hendrik Engel. 1964. *Systema Naturae, 1735*. Nieuwkoop, the Netherlands: B. de Graaf.

Linzey, Andrew. 1987. *Christianity and the Rights of Animals*. New York: Crossroad.

Linzey, Andrew. 1995. *Animal Theology*. Urbana: University of Illinois Press.

Linzey, Andrew. 2007. *Creatures of the Same God: Explorations in Animal Theology*. New York: Lantern.

Linzey, Andrew and Dan Cohn-Sherbok. 1997. *After Noah: Animals and the Liberation of Theology*. New York: Cassell.

Lippit, Akira Mazuta. *Electric Animal: Toward a Rhetoric of Wildlife*. Minneapolis: University of Minnesota Press, 2000.

Loar, Lynn and Libby Coleman. 2004. *Teaching Empathy: Animal-Assisted Therapy Programs for Children and Families Exposed to Violence*. Alameda, CA: Latham Foundation.

Locke, John, Ruth Weissbourd Grant, Nathan Tarcov, and John Locke. 2007. *Some Thoughts Concerning Education; and, of the Conduct of the Understanding*. Indianapolis: Hackett Publishing Company.

Lockwood, Randy and Frank Ascione. 1998. *Cruelty to Animals and Interpersonal Violence: Readings in Research and Application*. West Lafayette, IN: Purdue University Press.

Lodrick, D. 1981. *Sacred Cows, Sacred Places: Origins and Survivals of Animal Homes in India*. Berkeley: University of California Press.

Lorenz, Konrad. 1952. *King Solomon's Ring; New Light on Animal Ways*. New York: Crowell.

Lorenz, Konrad. 1970. *Studies in Animal and Human Behavior*. London: Methuen.

Love, Rosaleen. 2001. *Reefscape: Reflections on the Great Barrier Reef*. Washington, DC: Joseph Henry.

Lowe, B. M. and C. F. Ginsberg. 2002. "Animal Rights as a Post-Citizenship Movement." *Society & Animals* 10(2):203–215.

Lucie-Smith, E. 1998. *Zoo: Animals in Art*. London: Aurum.

Luke, Brian. 2007. *Brutal: Manhood and the Exploitation of Animals*. Urbana: University of Illinois Press.

Lundin, S. 1999. "The Boundless Body: Cultural Perspectives on Xenotransplantation." *Ethnos* 64(1):5–31.

Macauley, D. 1987. "Political Animals: A Study of the Emerging Animal Rights Movement in the United States." *Between the Species* 3:66–74.

MacDonald, Helen. 2005. *Falcon*. London: Reaktion.

Mack, Arien, ed. 1999. *Humans and Other Animals*. Columbus: Ohio State University Press.

MacKenzie, John M. 1988. *The Empire of Nature: Hunting, Conservation and British Imperialism*. Manchester, UK: Manchester University Press.

Malamud, Randy. 1998. *Reading Zoos: Representations of Animals and Captivity*. New York: New York University Press.

Malamud, Randy. 2003. *Poetic Animals and Animal Souls*. New York: Palgrave MacMillan.

Malamud, Randy. 2010. "Animals on Film." *Spring: A Journal of Archetype and Culture*, 83(Spring):135–160.

Manfredo, Michael J. 2008. *Who Cares About Wildlife? Social Science Concepts for Exploring Human-Wildlife Relationships and Conservation Issues*. New York: Springer.

Manley, Frank. 1998. *The Cockfighter*. Minneapolis, MN: Coffee House.

Manning, Aubrey and James Serpell, eds. 1994. *Animals and Human Society: Changing Perspectives*. London: Routledge.

Manzo, Bettina. 1994. *The Animal Rights Movement in the U.S. 1975–1990: An Annotated Bibliography*. Metuchen, NJ: Scarecrow.

Marino, Lori, Scott Lilienfeld, Randy Malamud, Nathan Nobis, and Ron Broglio. 2010. "Do Zoos and Aquariums Promote Attitude Change in Visitors? A Critical Evaluation of the American Zoo and Aquarium Study." *Society & Animals* 18(2):126–138.

Marks, S. A. 1991. *Southern Hunting in Black and White: Nature, History, and Ritual in a Carolina Community*. Lawrenceville, NJ: Princeton University Press.

Martin, Stephen. 2009. *Penguin*. London: Reaktion.

Marvin, Garry. 1994. *Bullfight*. Urbana: University of Illinois Press.

Mason, Georgia and Jeff Rushen. 2007. *Stereotypic Animal Behavior: Fundamentals and Applications to Welfare*, 2nd ed. Wallingford, UK: CAB International.

Mason, Jennifer. 2005. *Civilized Creatures: Urban Animals, Sentimental Culture, and American Literature, 1850–1900*. Baltimore: The Johns Hopkins University Press.

Mason, Jim and Peter Singer. 1990. *Animal Factories: What Agribusiness Is Doing to the Family Farm, the Environment, and Your Health*. Rev. ed. New York: Harmony.

Masson, Jeffrey Moussaieff. 1997. *Dogs Never Lie About Love*. New York: Crown Publishers.

Masson, Jeffrey Moussaieff. 2003. *The Pig Who Sang to the Moon: The Emotional World of Farm Animals*. New York: Random House.

Masson, Jeffrey Moussaieff and Susan McCarthy. 1995. *When Elephants Weep: The Emotional Lives of Animals*. New York: Delta Trade Paperbacks.

Matheson, Megan, Lori Sheeran, Jin-Hua Li, and R. Steven Wagner. 2006. "Tourist Impact on Tibetan Macaques." *Anthrozoös* 19(2):158–168.

Mauer, Donna. 2002. *Vegetarianism: Movement or Moment?* Philadelphia: Temple University Press.

McConnell, Patricia. 2002. *The Other End of the Leash*. New York: Ballantine.

McCormick, Adele and Marlena McCormick. 1997. *Horse Sense and the Human Heart: What Horses Can Teach Us about Trust, Bonding, Creativity and Spirituality*. Deerfield Beach, FL: Health Communications.

McElroy, Susan Chernak. 1998. *Animals as Teachers and Healers*. New York: Ballantine.

McFarland, Sarah and Ryan Hediger, eds. 2009. *Animals and Agency*. Leiden, the Netherlands: Brill Academic Publishers.

McHugh, Susan. 2004. *Dog*. London: Reaktion.

McIntosh, Peggy. 1988. *White Privilege and Male Privilege: A Personal Account of Coming to See Correspondences through Work in Women's Studies*. Wellesley, MA: Wellesley College Center for Women.

McKenna, Erin and Andrew Light. 2004. *Animal Pragmatism: Rethinking Human-Nonhuman Relationships*. Bloomington: Indiana University Press.

McNicholas, J. and G. Collis. 2000. "Dogs as Catalysts for Social Interactions: Robustness of the Effect." *British Journal of Psychology* 91:61–70.

Melson, Gail F. 2001. *Why the Wild Things Are: Animals in the Lives of Children*. Cambridge, MA: Harvard University Press.

Melville, E. 1994. *A Plague of Sheep: Environmental Consequences of the Conquest of Mexico*. Cambridge: Cambridge University Press.

Melvin, V. A., W. McCormick, and A. Gibbs. 2004. "A Preliminary Assessment of How Zoo Visitors Evaluate Animal Welfare According to Enclosure Style and the Expression of Behavior." *Anthrozoös* 17(2):98–108.

Merchant, Carolyn. 1980. *The Death of Nature: Women, Ecology and the Scientific Revolution*. San Francisco: Harper and Row.

Merz-Perez, Linda and Kathleen Heide. 2004. *Animal Cruelty: Pathway to Violence against People*. Lanham, MD: Altamira Press/Rowman and Littlefield.

Midgley, Mary. 1978. *Beast and Man: The Roots of Human Nature*. Ithaca, NY: Cornell University Press.

Midgley, Mary. 1983. *Animals and Why They Matter*. Athens: University of Georgia Press.

Miller, H. and W. Miller, eds. 1983. *Ethics and Animals*. Clifton, NJ: Humana.

Miller, K. and J. Knutson. 1997. "Reports of Severe Physical Punishment and Exposure to Animal Cruelty by Inmates Convicted of Felonies and University Students." *Child Abuse and Neglect* 21:59–82.

Miller, W. I. 1997. *The Anatomy of Disgust*. Cambridge, MA: Harvard University Press.

Mills, C. Wright. 1959. *The Sociological Imagination*. Oxford: Oxford University Press.

Mitchell, Robert W., Nicholas S. Thompson, and H. Lyn Miles, eds. 1997. *Anthropomorphism, Anecdotes, and Animals*. Albany: State University of New York Press.

Mithen, S. 1999. "The Hunter-Gatherer Prehistory of Human Animal Interactions." *Anthrozoös* 12(4):195–204.

Mitman, Gregg. 1992. *The State of Nature: Ecology, Community, and American Social Thought, 1900–1950*. Chicago: University of Chicago.

Mitman, Gregg. 1999. *Reel Nature: America's Romance with Wildlife on Film*. Cambridge, MA: Harvard University Press.

Montgomery. Georgina M. and Linda Kalof. "History from Below: Animals as Historical Subjects" in *Teaching the Animal: Human Animal Studies across the Disciplines*, ed. Margo DeMello. New York: Lantern, pp. 35–47.

Morales, E. 1995. *The Guinea Pig: Healing, Food, and Ritual in the Andes*. Tucson: University of Arizona Press.

Morley, Christine and Jan Fook. 2005. "The Importance of Pet Loss and Some Implications for Services." *Mortality* 10(2):127–131.

Morris, B. 2000. *Animals and Ancestors: An Ethnography*. Oxford: Berg.

Morris, Desmond. 1967. *The Naked Ape: A Zoologist's Study of the Human Animal*. London: Jonathan Cape.

Morris, Desmond. 2009. *Owl*. London: Reaktion.

Morse, Deborah Denenholz and Martin A. Danahay, eds. 2007. *Victorian Animal Dreams*. Burlington, VT: Ashgate.

Morton, J. 1990. "Rednecks, Roos, and Racism: Kangaroo Shooting and the Australian Way." *Social Analysis* 7:30–49.

Moss, Cynthia. 2000. *Elephant Memories: Thirteen Years in the Life of an Elephant Family*. Chicago: University of Chicago Press.

Mullan, B. and Marvin Garry. 1999. *Zoo Culture*. Urbana: University of Illinois Press.

Mullin, Molly. 2010. "Anthropology's Animals" in *Teaching the Animal: Human Animal Studies across the Disciplines*, ed. Margo DeMello. New York: Lantern, pp. 145–201.

Mullin, Molly H. 1999. "Mirrors and Windows: Sociocultural Studies of Human-Animal Relationships." *Annual Review of Anthropology* 28:201–224.

Mullin, Molly H. 2002. "Animals and Anthropology." *Society & Animals* 10(4):387–393.

Munro, Lyle. 2001. "Caring about Blood, Flesh and Pain: Women's Standing in the Animal Rights Movement." *Society & Animals* 9(1):43–61.

Munro, Lyle. 2001. *Compassionate Beasts: The Quest for Animal Rights*. Westport, CT: Praeger.

Munro, Lyle. 2005. *Confronting Cruelty: Moral Orthodoxy and the Challenge of the Animal Rights Movement (Human-Animal Studies)*. Leiden, the Netherlands: Brill Academic Publishers.

Myers, Gene. 2006. *Children and Animals: Social Development and Our Connections to Other Species*. West Lafayette, IN: Purdue University Press.

Nadeau, Chantel. 2001. *Fur Nation: From Beaver to Brigitte Bardot*. New York: Routledge.

National Institutes of Health. 1987. "The Health Benefits of Pets." Workshop, September 10–11, 1987. http://www.ncbi.nlm.nih.gov/books/NBK15172. Accessed March 23, 2011.

National Survey of Fishing, Hunting, and Wildlife-Associated Recreation, U.S. Fish and Wildlife Service, May 2007. http://library.fws.gov/nat_survey2006.pdf.

Neate, Patrick. 2003. *The London Pigeon Wars*. London: Penguin.

Nebbe, Linda Lloyd. 1995. *Nature as a Guide*. Minneapolis, MN: Educational Media Corporation.

Nelson, Barney. 2000. *The Wild and the Domestic: Animal Representation, Ecocriticism, and Western American Literature*. Las Vegas: University of Nevada Press.

Nelson, Richard K. 1986. *Make Prayers to the Raven: A Koyukon View of the Northern Forest*. Chicago: University of Chicago Press.

Nelson, Richard K. 1997. *Heart and Blood: Living with Deer in America*. New York: Knopf.

Netting, E., J. New, and C. Wilson. 1987. "The Human-Animal Bond: Implications for Practice." *Social Work* 32(1):60–64.

Neumann, R. P. 1998. *Imposing Wilderness: Struggles over Livelihood and Nature Preservation in Africa*. Chicago: University of Chicago Press.

Nibert, David. 2002. *Animal Rights; Human Rights: Entanglements of Oppression and Domination*. Lanham, MD: Rowman & Littlefield.

Nibert, David. 2007. "The Promotion of 'Meat' and its Consequences," in *The Animals Reader: The Essential Classic and Contemporary Writings*, eds. Linda Kalof and Amy Fitzgerald. Oxford: Berg, pp. 182–189.

Nibert, David A. 1994. "Animal Rights and Human Social Issues." *Society & Animals* 2(2):115–124.

Nibert, David A. 2003. "Humans and Other Animals: Sociology's Moral and Intellectual Challenge." *International Journal of Sociology and Social Policy* 23(3):4–25.

Nimmo, Richie. 2010. *Milk, Modernity and the Making of the Human: Purifying the Social*. London: Routledge.

Niven, C. D. 1967. *History of the Humane Movement*. New York: Transatlantic Arts.

Noble, B. E. 2000. "Politics, Gender, and Worldly Primatology: The Goodall-Fossey Nexus" in *Primate Encounters: Models of Science, Gender and Society*, eds. Shirley C. Strum and Linda M. Fedigan. Chicago: University of Chicago Press, pp. 436–462.

Noddings, Nell. 1984. *Caring: A Feminist Approach to Ethics and Moral Education*. Berkeley: University of California Press.

Noelker, Frank. 2004. *Captive Beauty*. Urbana: University of Illinois Press.

Nordenfelt, L. 2006. *Animal and Human Health and Welfare: A Comparative Philosophical Analysis*. Wallingford, UK: CAB International.

Norris, Margot. 1985. *Beasts of the Modern Imagination: Darwin Nietzsche, Kafka, Ernst and Lawrence*. Baltimore: The Johns Hopkins University Press.

Noske, B. 1993. "The Animal Question in Anthropology." *Society & Animals* 1(2):185–190.

Noske, B. 1997. *Beyond Boundaries: Humans and Animals*. Montreal: Black Rose

Noske, Barbara. 1989. *Humans and Other Animals: Beyond the Boundaries of Anthropology*. London: Pluto.

Nussbaum, Martha. 2006. *Frontiers of Justice: Disability, Nationality, Species Membership*. Cambridge, MA: Harvard University Press.

Ohnuki-Tierney, E. 1987. *The Monkey as Mirror: Symbolic Transformations in Japanese History and Ritual*. Princeton, NJ: Princeton University Press.

Ohnuki-Tierney, E. 1990. "The Monkey as Self in Japanese Culture," in *Culture Through Time: Anthropological Approaches*, ed. E Ohnuki-Tierney. Stanford, CA: Stanford University Press, pp. 128–153.

Ojeda, Auriana. 2004. *The Rights of Animals*. San Diego: Greenhaven.

Ojoade, J. Olowo. 1994. "Nigerian Cultural Attitudes to the Dog," in *Signifying Animals: Human Meaning in the Natural World*, ed. Roy Willis. New York: Routledge, pp. 215–221.

Oliver, Kelly. 2009. *Animal Lessons: How They Teach Us to Be Human*. New York: Columbia University Press.

Orlan, Barbara. 1998. *The Human Use of Animals: Case Studies in Ethical Choice*. Oxford: Oxford University Press.

Ortner, Sherry B. 1974. "Is Female to Male as Nature Is to Culture?", in *Woman, Culture, and Society*, eds. M. Z. Rosaldo and L. Lamphere. Stanford, CA: Stanford University Press, pp. 68–87.

Orwell, George. 1946. *Animal Farm*. New York: Harcourt, Brace.

Oswald, John. 1791. *The Cry of Nature, or An Appeal to Mercy and to Justice, on Behalf of the Persecuted Animals*. London: J. Johnson.

Page, George. 1999. *Inside the Animal Mind*. New York: Broadway.

Palmer, Clare. 2010. *Animal Ethics in Context*. New York: Columbia University Press.

Palmeri, Frank. 2006. *Humans and Other Animals in Eighteenth-Century British Culture: Representation, Hybridity, Ethics*. Burlington, VT: Ashgate.

Pálsson, G. 1991. *Coastal Economies, Cultural Accounts: Human Ecology and Icelandic Discourse*. Manchester, UK: Manchester University Press.

Pálsson, G. 1994. "The Idea of Fish: Land and Sea in the Icelandic World-View," in *Signifying Animals: Human Meaning in the Natural World*. ed. R.G. Willis. London: Unwin Hyman.

Parker, S. T., R. W. Mitchell, and M. L. Boccia, eds. 1994. *Self-Awareness in Animals and Humans, Developmental Perspectives*. New York: Cambridge University Press.

Paterson, David, and Richard D. Ryder, eds. 1979. *Animals' Rights: A Symposium*. London: Centaur.

Patterson, Charles. 2002. *Eternal Treblinka: Our Treatment of Animals and the Holocaust*. New York: Lantern.

Patterson, Francine and Eugene Linden. 1981. *The Education of Koko*. New York: Holt, Rinehart and Winston.

Paul, Elizabeth and James Serpell. 1993. "Childhood Petkeeping and Attitudes in Young Adulthood." *Animal Welfare* 2:321–337.

Pavlov, I. P. 1927. *Conditioned Reflexes: An Investigation of the Physiological Activity of the Cerebral Cortex.* Trans. and ed. G. V. Anrep. London: Oxford University Press.

Payton, Brian. 2006. *Shadow of the Bear: Travels in Vanishing Wilderness.* New York: Bloomsbury.

Pedersen, Helena. 2010. *Animals in Schools: Processes and Strategies in Human-Animal Education.* Lafayette, IN: Purdue University Press.

Peek, Charles W., Charlotte Chorn Dunham, and Bernadette E. Dietz. 1997. "Gender, Relational Role Orientation, and Affinity for Animal Rights." *Sex Roles: A Journal of Research* 37(11–12):905–921.

Peek, Charles W., Mark A. Konty, and Terri E. Frazier. 1997. "Religion and Ideological Support for Social Movements: The Case of Animal Rights." *Journal for the Scientific Study of Religion* 36(3):429–439.

Pepperberg, Irene. 1999. *The Alex Studies: Cognitive and Communicative Abilities of Grey Parrots.* Cambridge, MA: Harvard University Press.

Perlo, Katherine. 2003. "Would You Let Your Child Die Rather Than Experiment on Nonhuman Animals? A Comparative Questions Approach." *Society & Animals* 11(1):51–67.

Perlo, Katherine Wills. 2009. *Kinship and Killing: The Animal in World Religions.* New York: Columbia University Press.

Phillips, Mary T. 2008. "Savages, Drunks and Lab Animals: The Researcher's Perception of Pain" in *Social Creatures*, ed. Clifton P. Flynn. New York: Lantern, pp. 317–333.

Philo, Chris and Chris Wilbert, eds. 2000. *Animal Spaces, Beastly Places: New Geographies of Human–Animal Relations.* New York: Routledge.

Pink, S. 1997. *Women and Bullfighting: Gender, Sex and the Consumption of Tradition.* Oxford: Berg.

Plooij, Frans. 2000. "A Slap in the Face," in *The Smile of a Dolphin: Remarkable Accounts of Animal Emotions*, ed. Marc Bekoff. New York: Discovery, pp. 88–89.

Plous, Scott. 1991. "An Attitude Survey of Animal Rights Activists." *Psychological Science* 2(3):194–196.

Plous, Scott. 1998. "Signs of Change Within the Animal Rights Movement: Results from a Follow-Up Survey of Activists." *Journal of Comparative Psychology* 112(1):48–54.

Pluhar, Evelyn. 1991. "The Joy of Killing." *Between the Species* 7(3):121–128.

Pluhar, Evelyn. 1996. *Beyond Prejudice: The Moral Significance of Human and Nonhuman Animals.* Durham, NC: Duke University Press.

Plumwood, Val. 1993. *Feminism and the Mastery of Nature.* London: Routledge.

Plumwood, Val. 2000. "Integrating Ethical Frameworks for Animals, Humans, and Nature: A Critical Feminist Eco-Socialist Analysis." *Ethics and the Environment*, 5(2):285–322.

Pluskowski, Aleksander, ed. 2007. *Breaking and Shaping Beastly Bodies: Animals as Material Culture in the Middle Ages.* Oxford: Oxbow.

Plutarch. 1874. "Of Eating of Flesh." *Plutarch's Morals*, translated from the Greek by R. Brown, corrected and revised by William W. Goodwin. Boston: Little, Brown and Co.

Plutarch. 1874. "On Affection for Offspring." *Plutarch's Morals*, translated from the Greek by R. Brown, corrected and revised by William W. Goodwin. Boston: Little, Brown and Co.

Podberscek, Anthony L., Elizabeth S. Paul, and James A. Serpell. 2000. *Companion Animals and Us*. Cambridge: Cambridge University Press.

Pollock, Mary S. and Catherine Rainwater, eds. 2005. *Figuring Animals: Essays on Animal Images in Art, Literature, Philosophy and Popular Culture*. New York: Palgrave.

Porter, Pete. 2010. "Teaching Animal Movies" in *Teaching the Animal: Human Animal Studies across the Disciplines*, ed. Margo DeMello. New York: Lantern, pp. 18–34.

Potts, Annie and Philip Armstrong. 2010. "Hybrid Vigor: Interbreeding Cultural Studies and Human-Animal Studies," in *Teaching the Animal: Human Animal Studies across the Disciplines*, ed. Margo DeMello. New York: Lantern, pp. 3–17.

Preece, Rod and David Fraser. 2000. "The Status of Animals in Biblical and Christian Thought: A Study in Colliding Values." *Society & Animals* 8(3):245–263.

Preylo, Brooke Dixon and Hiroko Arikawa. 2008. "Comparison of Vegetarians and Non-Vegetarians on Pet Attitude and Empathy." *Anthrozoös* 21(4):387–395.

Pryor, Karen. 1995. *On Behavior*. North Bend, WA: Sunshine.

Pryor, Karen. 1999. *Don't Shoot the Dog*. New York: Bantam.

Quammen, David. 2000. *The Boilerplate Rhino: Nature in the Eye of the Beholder*. New York: Scribner.

Quiatt, D. 1993. *Primate Behavior: Information, Social Knowledge and the Evolution of Culture*. Cambridge: Cambridge University Press.

Rachels, James. 1990. *Created from Animals: The Moral Implications of Darwinism*. Oxford: Oxford University Press.

Radford, Mike. 2001. *Animal Welfare Law in Britain: Regulation and Responsibility*. Oxford: Oxford University Press.

Ramirez, Michael. 2006. "'My Dog's Just Like Me': Dog Ownership as a Gender Display." *Symbolic Interaction* 29(3):373–391.

Randour, Mary Lou. 2000. *Animal Grace*. Novato, CA: New World Library.

Randour, Mary Lou and Howard Davidson. 2008. *A Common Bond: Maltreated Children and Animals in the Home: Guidelines for Practice and Policy*. Englewood, CO: American Humane Association.

Randour, Mary Lou, Susan Krinsk, and Joanne L. Wolf. 2002. *Anicare Child: An Assessment and Treatment Approach for Childhood Animal Abuse*. Ann Arbor, MI: Animals and Society Institute.

Raphael, Pamela, Libby Coleman, and Lynn Loar. 1999. *Teaching Compassion*. Alameda, CA: Latham Foundation.

Regan, Tom, ed. 1986. *Animal Sacrifices: Religious Perspectives on the Use of Animals in Science*. Philadelphia: Temple University Press.

Regan, Tom. 2001. *Defending Animal Rights*. Urbana: University of Illinois Press.

Regan, Tom. 2003. *Animal Rights, Human Wrongs: An Introduction to Moral Philosophy*. Lanham, MD: Rowman & Littlefield.

Regan, Tom. 2004. *The Case for Animal Rights*. Berkeley: University of California Press.

Regan, Tom. 2004. *Empty Cages: Facing the Challenge of Animal Rights*. Lanham, MD: Rowman & Littlefield.

Regan, Tom and Peter Singer, eds. 1989. *Animal Rights and Human Obligations*, 2nd ed. Englewood Cliffs, NJ: Prentice-Hall.

Ressler, Robert K., et al. 1988. *Sexual Homicide*. Lexington: DC Heath & Company.

Rew, Lynn. 2000. "Friends and Pets as Companions: Strategies for Coping with Loneliness Among Homeless Youth." *Journal of Child and Adolescent Psychiatric Nursing* 13(3):125–132.

Reynolds, Rebecca. 1995. *Bring Me the Ocean*. St. Louis, MO: VanderWyk & Burnham.

Richards, P. 1993. "Natural Symbols and Natural History: Chimpanzees, Elephants and Experiments in Mende Thought," in *Environmentalism: The View from Anthropology*, ed. Kay Milton. London: Routledge, pp. 144–159.

Riedweg, Christoph. 2005. *Pythagoras: His Life, Teaching, and Influence*. Translated by Steven Rendall in collaboration with Christoph Riedweg and Andreas Schatzmann. Ithaca, NY: Cornell University Press.

Risley-Curtiss, Christina. 2010. "Social Work and Other Animals: Living Up to Ecological Practice," in *Teaching the Animal: Human Animal Studies across the Disciplines*, ed. Margo DeMello. New York: Lantern, pp. 281–298.

Risley-Curtiss, Christina, Lynn C. Holley, and Shapard Wolf. 2006. "The Animal–Human Bond and Ethnic Diversity." *Social Work* 51(3):57–268.

Ritter, Erika, 2009. *The Dog by the Cradle, the Serpent Beneath: Some Paradoxes of Human-Animal Relationships*. Toronto: Key Porter.

Ritvo, Harriet. 1987. *The Animal Estate: The English and Other Creatures in the Victorian Age*. Cambridge, MA: Harvard University Press.

Ritvo, Harriet. 1991. "The Animal Connection," in *The Boundaries of Humanity*, ed. J. J. Sheehan and M. Sosna. Berkeley: University of California Press, pp. 68–84.

Ritvo, Harriet. 1997. *The Platypus and the Mermaid, and Other Figments of the Classifying Imagination*. Cambridge, MA: Harvard University Press.

Robbins, Louise E. 2002. *Elephant Slaves and Pampered Parrots: Exotic Animals in Eighteenth-Century Paris*. Baltimore: The Johns Hopkins University Press.

Robbins, Mary E. 1996. "The Truculent Toad in the Middle Ages," in *Animals in the Middle Ages: A Book of Essays*, ed. Nora C. Flores. New York: Garland, pp. 25–47.

Robbins, P. 1998. "Shrines and Butchers: Animals as Deities, Capital, and Meat in Contemporary North India," in *Animal Geographies*, eds. J. Wolch and J. Emel. London: Verso, pp. 218–240.

Roberts, A. 1995. *Animals in African Art*. New York: Museum for African Art.

Roberts, William A. 1998. *Principles of Animal Cognition*. Boston: McGraw Hill.

Robins, D. et al. 1991. "Dogs and Their People: Pet-Facilitated Interaction in a Public Setting." *Journal of Contemporary Ethnography* 20(1):3–25.

Rodd, R. 1990. *Biology, Ethics and Animals*. Oxford: Clarendon.

Rodgers, Diane M. 2008. *Debugging the Link between Social Theory and Social Insects*. Baton Rouge: Louisiana State University Press.

Rodrique, Christine M. 1992. "Can Religion Account for Early Animal Domestication? A Critical Assessment of the Cultural Geographic Argument, Based on Near Eastern Archaeological Data." *The Professional Geographer* 44(4):417–430.

Rogers, Katharine. 2001. *The Cat and the Human Imagination*. Ann Arbor: University of Michigan Press.

Rogers, Katharine M. 2006. *Cat*. London: Reaktion.

Rohlf, Vanessa and Pauleen Bennett. 2005. "Perpetration-Induced Traumatic Stress in Persons Who Euthanize Nonhuman Animals in Surgeries, Animal Shelters, and Laboratories." *Society & Animals* 13(3):201–219.

Rohman, Carrie. 2009. *Stalking the Subject: Modernism and the Animal*. New York: Columbia University Press.

Rohman, Carrie. 2010. "Animal Writes: Literature and the Discourse of Species," in *Teaching the Animal: Human Animal Studies across the Disciplines*, ed. Margo DeMello. New York: Lantern, pp. 48–59.

Rollin, Bernard. 1989. *The Unheeded Cry: Animal Consciousness, Animal Pain, and Science*. Oxford: Oxford University Press.

Rollin, Bernard E. 1992. *Animals Rights and Human Morality*, 3rd ed. Amherst, NY: Prometheus.

Rollin, Bernard E. 1995. *Farm Animal Welfare: Social, Bioethical, and Research Issues*. Ames: Iowa State University Press.

Roman, Joseph. 2005. *Whale*. London: Reaktion.

Rothfels, Nigel, ed. 2002. *Representing Animals*. Bloomington: Indiana University Press.

Rothfels, Nigel. 2002. *Savages and Beasts: The Birth of the Modern Zoo*. Baltimore: The Johns Hopkins University Press.

Rousseau, Jean-Jacques and Maurice Cranston. 1984. *A Discourse on Inequality*. Harmondsworth, UK: Penguin.

Rowland, Beryl. 1973. *Animals with Human Faces: A Guide to Animal Symbolism*. Knoxville: University of Tennessee Press.

Rowlands, Mark. 2002. *Animals Like Us*. London: Verso.

Rubin, James Henry. 2003. *Impressionist Cats and Dogs: Pets in the Painting of Modern Life*. New Haven, CT: Yale University Press.

Ruckert, Janet. 1987. *The Four-Footed Therapist*. Berkeley, CA: Ten Speed.

Rudacille, Deborah. 2000. *The Scalpel and the Butterfly: The Conflict between Animal Research and Animal Protection*. Berkeley: University of California Press.

Rush, Benjamin. 1812. *Medical Inquiries and Observations upon Diseases of the Mind*. Philadelphia: Kimber and Richardson.

Russ, A., ed. 1996. *Reaching Into Thought: The Minds of the Great Apes*. Cambridge: Cambridge University Press.

Ryder, Richard D. 1975. *Victims of Science: The Use of Animals in Research*. London: Davis-Poynter.

Ryder, Richard D. 1989. *Animal Revolution: Changing Attitudes Towards Speciesism*. Oxford: Blackwell.

Rynearson, E. 1980. "Pets as Family Members: An Illustrative Case History." *International Journal of Family Psychiatry* 1(2):263–68.

Sabloff, Annabell. 2001. *Reordering the Natural World: Humans and Animals in the City*. Toronto: University of Toronto Press.

Salisbury, Joyce. 1994. *The Beast Within: Animals in the Middle Ages*. New York: Routledge.

Salisbury, Joyce. 1997. "Human Beasts and Bestial Humans in the Middle Ages," in *Animal Acts: Configuring the Human in Western History*, eds. Jennifer Ham and Matthew Senior. London: Routledge, 9–22.

Salt, Henry. 1892. *Animals' Rights: Considered in Relation to Social Progress*. London: George Bell & Sons.

Sanders, Clinton. 1999. *Understanding Dogs: Living and Working with Canine Companions*. Philadelphia: Temple University Press.

Sanders, Clinton. 2000. "The Impact of Guide Dogs on the Identity of People with Visual Impairments." *Anthrozoös* 13(3):131–139.

Sanders, Clinton and Arnold Arluke. 1993. "If Lions Could Speak: Investigating the Animal-Human Relationship and the Perspectives of Nonhuman Others." *The Sociological Quarterly* 34(3):377–390.

Sanders, Clinton R. and Elizabeth C. Hirschman. 1996. "Involvement with Animals as Consumer Experience." *Society & Animals* 4(2):111–119.

Sanderson, Ivan Terence. 1946. *Animal Tales: An Anthology of Animal Literature of all Countries.* New York: A. A. Knopf.

Sandlos, John. 1998. "Savage Fields: Ideology and the War on the North American Coyote." *Nature* 9, 2(34):41–51.

Sandøe, Peter and Stine B. Christiansen. 2008. *Ethics of Animal Use.* Oxford: Blackwell Publishing.

Sapontzis, Steve F. 1987. *Morals, Reason, and Animals.* Philadelphia: Temple University Press.

Sapontzis, Steve F., ed. 2004. *Food for Thought: The Debate over Eating Meat.* New York: Prometheus.

Savage-Rumbaugh, S. K., K. Wamba, P. Wamba, and N. Wamba. 2007. "Welfare of Apes in Captive Environments: Comments on, and by, a Specific Group of Apes." *Journal of Applied Animal Welfare Science* 10(1):7–19.

Savage-Rumbaugh, Sue and R. Lewin 1994. *Kanzi: The Ape at the Brink of the Human Mind.* New York: Wiley.

Savage-Rumbaugh, Sue, Stuart G. Shanker, and Talbot J. Taylor. 1998. *Apes, Language and the Human Mind.* New York: Oxford University Press.

Sax, Boria. 1990. *The Frog King: On Legends, Fables, Fairy Tales and Anecdotes of Animals.* New York: Pace University Press.

Sax, Boria. 2000. *Animals in the Third Reich: Pets, Scapegoats and the Holocaust.* London: Continuum.

Sax, Boria. 2001. *The Mythical Zoo: An Encyclopedia of Animals in World Myth, Legend and Literature.* Santa Barbara, CA: ABC-Clio.

Sax, Boria. 2003. *Crow.* London: Reaktion.

Sax, Boria. 2009. "The Magic of Animals: English Witch Trials in the Perspective of Folklore." *Anthrozoös* 22(4):317–332.

Sax, Boria. 2010. *The Raven and the Sun: Poems and Stories.* Providence, RI: The Poet's Press.

Scarce, Rik. 2000. *Fishy Business: Salmon, Biology, and the Social Construction of Nature.* Philadelphia: Temple University Press.

Schenk, S. A., D. I. Templer, N. B. Peters, and M. Schmidt. 1994. "The Genesis and Correlates of Attitudes Toward Pets." *Anthrozoös* 7(1):60–68.

Schiebinger, Londa. 1993. *Nature's Body: Gender in the Making of Modern Science.* Boston: Beacon.

Schlosser, Eric. 2002. *Fast Food Nation.* New York: Perennial-HarperCollins.

Schmitt, J-C. 1983. *The Holy Greyhound: Guinefort, Healer of Children since the Thirteenth Century.* Cambridge: Cambridge University Press.

Schochet, Elijah Judah. 1984. *Animal Life in Jewish Tradition: Attitudes and Relationships.* New York: Ktav.

Schvaneveldt, Paul L., Margaret H. Young, Jay D. Schvaneveldt, and Vira R. Kivett. 2001. "Interactions of People and Pets in the Family Setting: A Life Course Perspective." *Journal of Teaching in Marriage and Family* 1(2):34–50.

Schwartz, M. 1997. *A History of Dogs in the Early Americas.* New Haven, CT: Yale University Press.

Scruton, R. 1996. *Animal Rights and Wrongs*. London: Demos.

Scully, Matthew. 2002. *Dominion: the Power of Man, the Suffering of Animals, and the Call to Mercy*. New York: St. Martin's.

Serpell, James. 1996. *In the Company of Animals: A Study of Human–Animal Relationships*. Cambridge, MA: Cambridge University Press.

Serpell, James A. 1981. "Childhood Pets and Their Influence on Adults' Attitudes." *Psychological Reports* 49:651–654.

Serpell, James A. 1993. "Childhood Pet Keeping and Humane Attitudes in Young Adulthood." *Animal Welfare* 2:321–337.

Serpell, James A. 2002. "Anthropomorphism and Anthropomorphic Selection: Beyond the 'Cute Response.'" *Society & Animals* 10(4):437–454.

Sewell, Anna. 1877. *Black Beauty*. New York: Scholastic Paperbacks.

Shanklin, E. 1985. "Sustenance and Symbol: Anthropological Studies of Domesticated Animals." *Annual Review of Anthropology* 14:375–403.

Shanks, Niall. 2002. *Animals and Science: A Guide to the Debates*. Santa Barbara, CA: ABC-CLIO.

Shapiro, Kenneth. 1998. *Animal Models of Human Psychology*. Cambridge, MA: Hogrefe & Huber.

Shapiro, Kenneth. 2002. "A Rodent for Your Thoughts: The Social Construction of Animal Models," in *Animals in Human Histories*, ed. Mary Henninger-Voss. Rochester: University of Rochester Press, pp. 439–469.

Shapiro, Kenneth. 2008. *Human-Animal-Studies: Growing the Field, Applying the Field*. Ann Arbor, MI: Animals and Society Institute.

Shapiro, Kenneth. 2010. "Psychology and Human-Animal Studies: Roads Not (yet) Taken," in *Teaching the Animal: Human Animal Studies across the Disciplines*, ed. Margo DeMello. New York: Lantern, pp. 254–280.

Shapiro, Kenneth J. 1983. "Psychology and Its Animal Subjects." *International Journal for the Study of Animal Problems* 4(3):188–191.

Shapiro, Kenneth J. 1990a. "Animal Rights Versus Humanism: The Charge of Speciesism." *Journal of Humanistic Psychology* 30(2):9–37.

Shapiro, Kenneth J. 1990b. "Understanding Dogs Through Kinesthetic Empathy, Social Construction, and History." *Anthrozoös* 3(3):184–195.

Sheehan, James and Morton Sosna, eds. 1990. *The Boundaries of Humanity: Humans, Animals, Machines*. Berkeley: University of California Press.

Shepard, Paul. 1996. *The Others: How Animals Made Us Human*. Washington, DC: Island.

Shepard, Paul. 1998. *Thinking Animals: Animals and the Development of Human Intelligence*. Athens: University of Georgia Press.

Shore, Elsie R., Connie L. Petersen, and Deanna K. Douglas. 2003. "Moving as a Reason for Pet Relinquishment: A Closer Look." *Journal of Applied Animal Welfare* 6(1):39–52.

Shukin, Nicole. 2009. *Animal Capital*. Minneapolis: University of Minnesota Press.

Silverstein, Helena. 1996. *Unleashing Rights: Law, Meaning, and the Animal Rights Movement*. Ann Arbor: University of Michigan Press.

Simmons, Laurence and Philip Armstrong, eds. 2007. *Knowing Animals*. Boston: Brill.

Simon, Clea. 2002. *The Feline Mystique: On the Mysterious Connection Between Women and Cats*. New York: St. Martin's.

Simons, John. 2002. *Animal Rights and the Politics of Literary Representation*. Basingstoke, UK: Palgrave.

Simoons, F. J. 1994. *Eat Not This Flesh: Food Avoidances from Prehistory to the Present*. 2nd ed. Madison: University of Wisconsin Press.

Singer, Isaac Bashevis. 1968. "The Letter Writer." *The New Yorker*, January 13, 1968, p. 26.

Singer, Peter, ed. 1985. *In Defense of Animals*. New York: Harper and Row.

Singer, Peter. 2002. *Animal Liberation*, rev. ed. New York: Harper Perennial.

Singer, Peter. 2006. *In Defense of Animals: The Second Wave*. Malden, MA: Blackwell.

Singer, Randall S., Lynette A. Hart, and R. Lee Zasloff. 1995. "Dilemmas Associated with Rehousing Homeless People Who Have Companion Animals." *Psychological Reports* 77:851–857.

Skaggs, Jimmy M. 1986. *Prime Cut: Livestock Raising and Meatpacking in the United States, 1607–1983*. College Station: Texas A&M University Press.

Skinner, B. F. 1938. *The Behavior of Organisms*. New York: Appleton-Century-Crofts.

Sleigh, Charlotte. 2003. *Ant*. London: Reaktion.

Slobodchikoff, C. N., B. Perla, and J. L. Verdolin. 2009. *Prairie Dogs: Communication and Community in an Animal Society*. Cambridge, MA: Harvard University Press.

Slobodichikoff, Con N., ed. 1988. *The Ecology of Social Behavior*. New York: Academic.

Smith, Bradley Philip and Carla Anita Litchfield. "How Well Do Dingoes, *Canis dingo*, Perform on the Detour Task?" *Animal Behavior* 80(1):155–162.

Smith, Julie Ann. 2003. "Beyond Dominance and Affection: Living with Rabbits in Post-Humanist Households." *Society & Animals* 11(2):181–197.

Smith, N. Kemp. 1952. *New Studies in the Philosophy of Descartes*. London: MacMillan.

Sorenson, John. 2009. *Ape*. London: Reaktion.

Spencer, Colin. 1990. *The Heretics Feast: A History of Vegetarianism*. London: Fourth Estate.

Sperling, Susan. 1988. *Animal Liberators: Research and Morality*. Berkeley: University of California Press.

Spiegel, Marjorie. 1996. *The Dreaded Comparison: Human and Animal Slavery*. New York: Mirror.

Steeves, H. Peter, ed. 1999. *Animal Others: On Ethics, Ontology, and Animal Life*. Albany: State University of New York Press.

Steiner, Gary. 2005. *Anthropomorphism and Its Discontents: The Moral Status of Animals in the History of Western Philosophy*. Pittsburgh: University of Pittsburgh Press.

Sterckx, Roel. 2002. *The Animal and the Daemon in Early China*, Albany: State University of New York Press.

Stewart, Desmond. 1972. "The Limits of Trooghaft." *Encounter* 38(2):3–7.

Stott, Rebecca. 2004. *Oyster*. London: Reaktion.

Strimple, Earl O. 2003. "A History of Prison Inmate-Animal Interaction Programs." *American Behavioral Scientist* 47(1):70–78.

Strum, Shirley C. and Linda M. Fedigan, eds. 2000. *Primate Encounters: Models of Science, Gender and Society*. Chicago: University of Chicago Press.

Stull, D. B, M. J. Broadway, and D. Griffith, eds. 1995. *Any Way You Cut It: Meat Processing and Small-Town America*. Lawrence: University Press Kansas.

Stutesman, Drake. 2005. *Snake*. London: Reaktion.

Sullivan, Robert. 2004. *Rats: Observations on the History and Habitat of the City's Most Unwanted Inhabitants*. New York: Bloomsbury.

Sunstein, Cass R. and Martha Nussbaum, eds. 2004. *Animal Rights: Current Debates and New Directions*. New York: Oxford University Press.

Sutherland, Amy. 2009. *Kicked, Bitten, and Scratched: What Shamu Taught Me about Life, Love, Marriage*. New York: Random House.

Swabe, Joanna. 1998. *Animals, Disease, and Human Society: Human-Animal Relations and the Rise of Veterinary Medicine*. London: Routledge.

Tambiah, Stanley J. 1969. "Animals Are Good to Think and Good to Prohibit." *Ethnology* 8:423–459.

Tanner, A. 1979. *Bringing Home Animals: Religious Ideology and Mode of Production of the Mistassini Cree Hunters*. New York: St. Martin's.

Tattersall, Ian. 1998. *Becoming Human: Evolution and Human Uniqueness*. Oxford: Oxford University Press.

Taylor, Angus. 2003. *Animals and Ethics: An Overview of the Philosophical Debate*. Peterborough, ON: Broadview.

Taylor, N. and T. D. Signal. 2005. "Empathy and Attitudes to Animals." *Anthrozoös* 18(1):18–27.

Tester, Keith. 1991. *Animals and Society: The Humanity of Animal Rights*. London: Routledge.

Thomas, Elizabeth Marshall. 1993. *Hidden Life of Dogs*. New York: Pocket.

Thomas, Elizabeth Marshall. 1994. *The Tribe of Tiger*. New York: Pocket.

Thomas, Keith. 1983. *Man and the Natural World: A History of the Modern Sensibility*. New York: Pantheon.

Thomas, Keith. 1991. *Man and the Natural World: Changing Attitudes in England 1500–1800*. Penguin.

Thompson, Paul B. 1997. *Food Biotechnology in Ethical Perspective*. London: Chapman and Hall.

Thompson, Paul B. 1998. *Agricultural Ethics: Research, Teaching, and Public Policy*. Ames: Iowa State University Press.

Thorpe, W. 1974. *Animal and Human Nature*. London: Methuen.

Thu, K. M. and P. E. Durrenberger. 1998. *Pigs, Profits, and Rural Communities*. Albany: State University New York Press.

Thurston, Mary Elizabeth. 1996. *The Lost History of the Canine Race*. Kansas City, MO: Andrews and McMeel.

Tinbergen, Niko. 1951. *The Study of Instinct*. Oxford: Clarendon

Toperoff, Shlomo Pesach 1995. *Animal Kingdom in Jewish Thought*. Northvale, NJ: Jason Aronson.

Trimmer, Sarah and Legh Richmond. 1977. *Fabulous Histories*. New York: Garland Pub.

Tuan, Yi-Fu. 1984. *Dominance and Affection: The Making of Pets*. New Haven, CT: Yale University Press.

Turner, James C. 2000. *Reckoning with the Beast: Animals, Pain, and Humanity in the Victorian Mind*. Baltimore: The Johns Hopkins University Press.

Twine, Richard. 2010. *Animals as Biotechnology—Ethics, Sustainability and Critical Animal Studies*. London: Earthscan.

Twine, Richard T. 2001. "Ma(r)king Essence—Ecofeminism and Embodiment." *Ethics and the Environment* 6(2):31–58.

Tyler, Tom, ed. 2006. *Animal Beings*. London: Routledge.

Tyler, Tom, ed. 2009. *Animal Encounters*. Leiden, the Netherlands: Brill Academic Publishers.

Tyson, Edward and Ashley Montagu. 1966. *Orang-Outang: Sive, Homo Sylvestris; or, The Anatomy of a Pygmie*. London: Dawsons.

U.S. Department of Agriculture. 2008. "Livestock slaughter 2008 summary." http://usda01.library.cornell.edu/usda/nass/LiveSlauSu/2000s/2009/LiveSlauSu-03-06-2009.pdf. Accessed April 2, 2012.

U.S. Department of Labor, Bureau of Labor Statistics, *Occupational Outlook Handbook, 2006–2007*, http://www.bls.gov/oco/ocos219.htm#earnings.

Van Dyke, Carolyn. 2006. *Chaucer's Agents*. Madison, NJ: Fairleigh Dickinson University Press.

Varner, Gary E. 1998. *In Nature's Interests? Interests, Animal Rights, and Environmental Ethics*. Oxford: Oxford University Press.

Vauclair, J. 1996. *Animal Cognition: An Introduction to Modern Comparative Psychology*. Cambridge, MA: Harvard University Press.

Vialles, Noelie. 1994. *Animal to Edible*. Cambridge: Cambridge University Press.

Vitebsky, Piers. 2005. *The Reindeer People: Living with Animals and Spirits in Siberia*. New York: Houghton Mifflin.

Voltaire. 1796. "Beasts." *The Philosophical Dictionary, for the Pocket*. Catskill, NY: J. Fellows and E. Duyckinsk.

Waisman, Sonia, Pamela D. Frasch, and Bruce A. Wagman, eds. 2006. *Animal Law: Cases and Materials*, 3rd ed. Durham, NC: Carolina Academic.

Waldau, Paul 2002. *Specter of Speciesism: Buddhist and Christian Views of Animals*. New York: Oxford University Press.

Waldau, Paul. 2010. "Law and Other Animals" in *Teaching the Animal: Human Animal Studies across the Disciplines*, ed. Margo DeMello. New York: Lantern, pp. 218–253.

Waldau, Paul. 2010. "Religion and Other Animals" in *Teaching the Animal: Human Animal Studies across the Disciplines*, ed. Margo DeMello. New York: Lantern, pp. 103–126.

Waldau, Paul and Kimberley C. Patton, eds. 2006. *A Communion of Subjects: Animals in Religion, Science, and Ethics*. New York: Columbia University Press.

Walters, Kerry S. and Lisa Portmess, eds. 1999. *Ethical Vegetarianism from Pythagoras to Peter Singer*. Albany: State University of New York Press.

Wand, Kelly. 2003. *The Animal Rights Movement*. San Diego: Greenhaven.

Warren, A and F. Goldsmith. 1983. *Conservation in Perspective*. Chichester: Wiley.

Warren, Karen J. 1996. *Ecological Feminist Philosophies*. Bloomington: Indiana University Press

Warren, Mary Anne. 1997. *Moral Status: Obligations to Persons and Other Living Things*. Oxford: Oxford University Press.

Warkentin, T., and L. Fawcett. 2010. "Whale and Human Agency in World-Making: Decolonizing Whale-Human Encounters" in *Metamorphoses of the Zoo: Animal Encounter after Noah*, ed. Ralph Acampora. Lanham, MD: Lexington, pp. 103–122.

Watt, Yvette. 2010. "Animals, Art, and Activism: An Investigation into Art as a Tool for Engaging an Ethical Consideration of Human-Animal Relationships." PhD exegesis, University of Tasmania.

Webb, Stephen H. 1997. *On God and Dogs: A Christian Theology of Compassion for Animals*. New York: Oxford University Press.

Webster, J. 2005. *Animal Welfare: Limping towards Eden*. Oxford: Blackwell Publishing.

Weil, Zoe. 2004. *The Power and the Promise*. Gabriola Island, BC: New Society Publishers.

Wells, D. L. 2004. "The Facilitation of Social Interactions by Domestic Dogs." *Anthrozoös* 17(4):340–352.

Wenzel, G. 1991. *Animal Rights, Human Rights: Ecology, Economy and Ideology in the Canadian Arctic*. Toronto: University of Toronto Press.

Wesley, Martin, Neresa Minatrea, and Joshua Watson. 2009. "Animal-Assisted Therapy in the Treatment of Substance Dependence." *Anthrozoös* 22(2):137–148.

West, T. 1972. *Heroes on Horseback: The Story of the Pony Express*. Glasgow: Blackie.

Wheatley, B. 1999. *The Sacred Monkeys of Bali*. Prospect Heights, IL: Waveland.

Whitman, Charles Otis and Oscar Riddle. 1919. *Postumous works of Charles Otis Whitman*. Washington, DC: Carnegie Institution of Washington.

Williams, Erin and Margo DeMello. 2007. *Why Animals Matter*. Amherst, NY: Prometheus.

Williams, Peter. 2009. *Snail*. London: Reaktion.

Willis, Roy, ed. 1990. *Signifying Animals: Human Meaning in the Natural World*. London: Routledge.

Wilson, Cindy and Dennis Turner, eds. 1998. *Companion Animals in Human Health*. Thousand Oaks, CA: Sage.

Wilson, Edward O. 1984. *Biophilia: The Human Bond with Other Species*. Cambridge, MA: Harvard University Press.

Wise, Steven M. 2000. *Rattling the Cage: Toward Legal Rights for Animals*. Cambridge, MA: Perseus.

Wise, Steven M. 2002. *Drawing the Line: Science and the Case for Animal Rights*. Cambridge, MA: Perseus.

Wittgenstein, Ludwig. 1994. *The Wittgenstein Reader*, ed. Anthony Kenny. Oxford: Blackwell.

Wolch, Jennifer and J. Emel, eds. 1998. *Animal Geographies: Place, Politics and Identity in the Nature–Culture Borderlands*. New York: Verso.

Wolf, David B. 2000. "Social Work and Speciesism." *Social Work* 45(1):88–93.

Wolfe, Cary. 2003. *Animal Rites: American Culture, the Discourse of Species, and Posthumanist Theory*. Chicago: University of Chicago Press.

Wolfe, Cary, ed. 2003. *Zoontologies: The Question of the Animal*. Minneapolis: University of Minnesota Press.

Wolfe, Cary. 2009. "Human, All too Human: 'Animal Studies' and the Humanities." *PMLA* 124(2):564–575.

Wollstonecraft, Mary. 1990. *Original Stories from Real Life*. Oxford: Woodstock.

Wood, Lisa, Billie Giles-Corti, Max K. Bulsara, and Darcy A. Bosch. 2007. "More Than a Furry Companion: The Ripple Effect of Companion Animals on Neighborhood Interactions and Sense of Community." *Society & Animals* 15(1):43–56.

Woodroffe, Rose, Simon Thirgood, and Alan Rabinowitz, eds. 2005. *People and Wildlife: Conflict or Co-existence?* Cambridge: Cambridge University Press.

Woodward, Wendy. 2008. *The Animal Gaze: Animal Subjectivities in Southern African Narratives*. Johannesburg: Wits University Press.

Wuensch, K. L., K. W. Jenkins, and G. M. Poteat. 2002. "Misanthropy, Idealism, and Attitudes towards animals." *Anthrozoös* 15(2):139–149.

Young, Peter. 2003. *Tortoise*. London: Reaktion.

Young, Thomas 1798/2001. *An essay on humanity to animals*, ed. Rod Preece. Lewiston, NY: Edwin Mellen.

Zeuner, Frederick E. 1963. *A History of Domesticated Animals*. London: Hutchinson.

Zilney, Lisa Anne and Mary Zilney. 2005. "Reunification of Child and Animal Welfare Agencies: Cross-Reporting of Abuse in Wellington County, Ontario." *Child Welfare* 84(1):47–66.

INDEX